T0235669

Lecture Notes in Computer Science 12446

More information about this series at http://www.springer.com/series/7412

Jaime Cardoso · Hien Van Nguyen ·
Nicholas Heller et al. (Eds.)

Interpretable and Annotation-Efficient Learning for Medical Image Computing

Third International Workshop, iMIMIC 2020
Second International Workshop, MIL3ID 2020
and 5th International Workshop, LABELS 2020
Held in Conjunction with MICCAI 2020
Lima, Peru, October 4–8, 2020
Proceedings

 Springer

Editors
Jaime Cardoso ⓘ
University of Porto
Porto, Portugal

Hien Van Nguyen ⓘ
University of Houston
Houston, TX, USA

Nicholas Heller ⓘ
University of Minnesota
Minneapolis, MN, USA

Additional Workshop Editors *see next page*

ISSN 0302-9743 ISSN 1611-3349 (electronic)
Lecture Notes in Computer Science
ISBN 978-3-030-61165-1 ISBN 978-3-030-61166-8 (eBook)
https://doi.org/10.1007/978-3-030-61166-8

LNCS Sublibrary: SL6 – Image Processing, Computer Vision, Pattern Recognition, and Graphics

This Springer imprint is published by the registered company Springer Nature Switzerland AG
The registered company address is: Gewerbestrasse 11, 6330 Cham, Switzerland

Additional Workshop Editors

iMIMIC 2020 Editors

Pedro Henriques Abreu ⓘ
University of Coimbra
Coimbra, Portugal

Wilson Silva ⓘ
University of Porto
Porto, Portugal

Jose Pereira Amorim ⓘ
University of Coimbra
Coimbra, Portugal

Ivana Isgum ⓘ
Amsterdam University Medical Center,
Amsterdam, The Netherlands

Ricardo Cruz ⓘ
University of Porto
Porto, Portugal

MIL3ID 2020 Editors

Vishal Patel
Johns Hopkins University
Baltimore, MD, USA

Kevin Zhou
Chinese Academy of Sciences
Beijing, China

Ngan Le
University of Arkansas
Fayetteville, AR, USA

Badri Roysam
University of Houston
Houston, TX, USA

Steve Jiang
UT Southwestern Medical Center
Dallas, TX, USA

Khoa Luu
University of Arkansas
Fayetteville, AR, USA

LABELS 2020

Raphael Sznitman ⓘ
University of Bern
Bern, Switzerland

Diana Mateus ⓘ
Technical University of Munich
Nantes, Germany

Samaneh Abbasi ⓘ
Eindhoven University of Technology
Eindhoven, The Netherlands

Veronika Cheplygina
Eindhoven University of Technology
Eindhoven, The Netherlands

Emanuele Trucco ⓘ
University of Dundee
Dundee, UK

iMIMIC 2020 Preface

It is our genuine honor and great pleasure to welcome you to the Third Workshop on Interpretability of Machine Intelligence in Medical Image Computing (iMIMIC 2020), a satellite event at the 23rd International Conference on Medical Image Computing and Computer Assisted Intervention (MICCAI 2020). Following in the footsteps of the two previous successful meetings in Granada, Spain (2018) and Shenzhen, China (2019), we gathered for this new edition.

iMIMIC is a single-track, half-day workshop consisting of high-quality, previously unpublished papers, presented either orally or as a poster, intended to act as a forum for research groups, engineers, and practitioners to present recent algorithmic developments, new results, and promising future directions in interpretability of machine intelligence in medical image computing. Machine learning systems are achieving remarkable performances at the cost of increased complexity. Hence, they become less interpretable, which may cause distrust, potentially limiting clinical acceptance. As these systems are pervasively being introduced to critical domains, such as medical image computing and computer assisted intervention, it becomes imperative to develop methodologies allowing insight into their decision making. Such methodologies would help physicians to decide whether they should follow and trust automatic decisions. Additionally, interpretable machine learning methods could facilitate defining the legal framework of their clinical deployment. Ultimately, interpretability is closely related to AI safety in healthcare.

This year's iMIMIC was held on October 4, 2020, virtually in Lima, Peru, and was hosted by INESC TEC and the University of Coimbra, with the support of University of Porto and CISUC, all located in Portgual. There was a very positive response to the call for papers for iMIMIC 2020. We received 18 full papers from 10 countries and 8 were accepted for presentation at the workshop, where each paper was reviewed by at least three reviewers. The accepted papers present fresh ideas of Interpretability in settings such as regression, multiple instance learning, weakly supervised learning, local annotations, classifier re-training, and model pruning.

The high quality of the scientific program of iMIMIC 2020 was due first to the authors who submitted excellent contributions and second to the dedicated collaboration of the International Program Committee and the other researchers who reviewed the papers. We would like to thank all the authors for submitting their contributions and for sharing their research activities.

We are particularly indebted to the Program Committee members and to all the reviewers for their precious evaluations, which permitted us to set up this publication.

We were also very pleased to benefit from the participation of the invited speakers Himabindu Lakkaraju, Harvard University, USA, and Wojciech Samek, Fraunhofer HHI, Germany. We would like to express our sincere gratitude to these world-renowned experts.

October 2020

Jaime Cardoso
Pedro Henriques Abreu
Ivana Isgum
Wilson Silva
Ricardo Cruz
Jose Pereira Amorim

The original version of the book was revised: the acronym was corrected to "MIL3ID" throughout the book. The correction to the book is available at
https://doi.org/10.1007/978-3-030-61166-8_30

iMIMIC 2020 Organization

General Chairs iMIMIC 2020

Jaime Cardoso   INESC TEC and University of Porto, Portugal
Pedro Henriques Abreu   CISUC and University of Coimbra, Portugal
Ivana Isgum   Amsterdam University Medical Center,
                             The Netherlands

Publicity Chair iMIMIC 2020

Jose Pereira Amorim   CISUC and University of Coimbra, Portugal

Program Chair iMIMIC 2020

Wilson Silva   INESC TEC and University of Porto, Portugal

Sponsor Chair iMIMIC 2020

Ricardo Cruz   INESC TEC and University of Porto, Portugal

Program Committee iMIMIC 2020

Ben Glocker   Imperial College, UK
Bettina Finzel   University of Bamberg, Germany
Carlos A. Silva   University of Minho, Portugal
Christoph Molnar   Ludwig Maximilian University of Munich, Germany
Claes Nøhr Ladefoged   Rigshospitalet, Denmark
Dwarikanath Mahapatra   Inception Institute of AI, UAE
George Panoutsos   The University of Sheffield, UK
Hrvoje Bogunovic   Medical University of Vienna, Austria
Isabel Rio-Torto   University of Porto, Portugal
Joana Cristo Santos   University of Coimbra, Portugal
Kelwin Fernandes   NILG.AI, Portugal
Luis Teixeira   University of Porto, Portugal
Miriam Santos   University of Coimbra, Portugal
Nick Pawlowski   Imperial College London, UK
Ricardo Cruz   University of Porto, Portugal
Sérgio Pereira   Lunit, South Korea

MIL3ID 2020 Preface

Welcome to the Second International Workshop on Medical Image Learning with Less Labels and Imperfect Data (MIL3ID 2020). The MIL3ID 2020 proceedings contain 11 high-quality papers of 8 pages that were selected through a rigorous peer-review process.

We hope this workshop will create a forum for discussing best practices in medical image learning with label scarcity and data imperfection. This forum is urgently needed because the issues of label noises and data scarcity are highly practical, but largely under-investigated in the medical image analysis community. Traditional approaches for dealing with these challenges include transfer learning, active learning, denoising, and sparse representation. The majority of these algorithms were developed prior to the recent advances of deep learning and might not benefit from the power of deep networks. The revision and improvement of these techniques in the new light of deep learning are long overdue.

This workshop potentially helps answer many important questions. For example, several recent studies found that deep networks are robust to massive random label noises but more sensitive to structured label noises. What implication do these findings have on dealing with noisy medical data? Recent work on Bayesian neural networks demonstrates the feasibility of estimating uncertainty due to the lack of training data. In other words, it enables our classifiers to be aware of what they do not know. Such a framework is important for medical applications where safety is critical. How can researchers of MICCAI community leverage this approach to improve their systems' robustness in the case of data scarcity? Our prior work shows that a variant of capsule networks generalizes better than convolutional neural networks with an order of magnitude fewer training data. This gives rise to an interesting question: Are there better classes of networks that intrinsically require less labeled data for learning? Humans always have an edge over deep networks when it comes to learning with small amounts of data. However, recent work on one-shot deep learning has surpassed human in an image recognition task using only a few training samples for each task. Do these results still hold for medical image analysis tasks?

The proceedings of the workshop are published as a joint LNCS volume alongside other satellite events organized in conjunction with MICCAI. In addition to the LNCS volume, to promote transparency, the papers' reviews and preprints are publicly available on the workshop website: https://www.hvnguyen.com/lesslabelsimperfect dataml2020. In addition to the papers, abstracts, slides, and posters presented during the workshop will be made publicly available on the MIL3ID website.

We would like to thank all the speakers and authors for joining our workshop, the Program Committee for their excellent work with the peer reviews, and the workshop chairs and editors for their help with the organization of the second MIL3ID workshop.

August 2020   Hien Van Nguyen
Vishal Patel
Badri Roysam
Kevin Zhou
Steve Jiang
Ngan Le
Khoa Luu

MIL3ID 2020 Organization

Chairs MIL3ID 2020

Hien Van Nguyen	University of Houston, USA
Vishal Patel	Johns Hopkins University, USA
Badri Roysam	University of Houston, USA
Kevin Zhou	Chinese Academy of Sciences, China
Steve Jiang	UT Southwestern Medical School, USA
Ngan Le	University of Arkansas, USA
Khoa Luu	University of Arkansas, USA

Program Committee MIL3ID 2020

Zhoubing Xu	Siemens Healthineers, USA
Pengyu Yuan	University of Houston, USA
Pengbo Liu	Chinese Academy of Sciences, China
Anjali Balagopal	UT Southwestern Medical School, USA
Jue Jiang	Memorial Sloan Kettering Cancer Center, USA
Jahandar Jahanipour	National Institutes of Health, USA
Pietro Antonio Cicalese	University of Houston, USA
Aryan Mobiny	University of Houston, USA
Pasawee Wirojwatanakul	New York University, USA
Samira Zare	University of Houston, USA
Pengfei Guo	Johns Hopkins University, USA
Xiaoyang Li	University of Houston, USA
Jeya Maria Jose Valanarasu	Johns Hopkins University, USA
Xiao Liang	UT Southwestern Medical School, USA
Li Xiao	Chinese Academy of Sciences, China
Siqi Liu	Siemens Healthineers, USA
Aditi Singh	University of Houston, USA

LABELS 2020 Preface

This volume contains the proceedings of the 5th International Workshop on Large-scale Annotation of Biomedical data and Expert Label Synthesis (LABELS 2020), which was held on October 8, 2020, in conjunction with the 23rd International Conference on Medical Image Computing and Computer Assisted Intervention (MICCAI 2020), originally planned for Lima, Peru, but ultimately held virtually, due to the COVID-19 pandemic. The first workshop in the LABELS series was held in 2016 in Athens, Greece. This was followed by workshops in Quebec City, Canada, in 2017, Granada, Spain, in 2018, and Shenzhen, China, in 2019.

As data-hungry methods continue to drive advancements in medical imaging, the need for high-quality annotated data to train and validate these methods continues to grow. Further, with the pressing need to address health disparities and to prevent learned systems from internalizing biases, there has never been a greater need for thorough study and discussion of best practices in data collection and annotation. For the past four years, LABELS has aimed to facilitate exactly this.

Following the success of the previous four LABELS workshops, the fifth workshop was planned for 2020. This year's edition of the workshop included invited talks by Anand Malpani (Johns Hopkins University, USA) and Amber Simpson (Queen's University, Canada), as well as several papers and abstracts. After peer review, a total of 10 papers and 3 abstracts were selected. The papers appear in this volume, and the abstracts are available on the workshop website: https://miccailabels.org. The research presented this year ranged from how to quantify and mitigate demographic biases, to probing the reproducibility of expert labels, to new tools for more efficient annotation of emerging image modalities. LABELS takes pride in the fact that theoretical novelty is not a prerequisite for work presented at the workshop, instead the event embraces the messy, tedious reality of medical image collection and annotation in an effort to expose and formalize its underlying principles.

We would like to thank all the speakers and authors for joining our workshop, the Program Committee for their excellent work with the peer reviews, our sponsors – Retinai and Auris Health – for their support, and the workshop chairs for their help with the organization of the fifth LABELS workshop.

August 2020

Nicholas Heller
Raphael Sznitman
Veronika Cheplygina
Diana Mateus
Emanuele Trucco
Samaneh Abbasi

LABELS 2020 Organization

Chairs LABELS 2020

Raphael Sznitman	University of Bern, Switzerland
Veronika Cheplygina	Eindhoven University of Technology (TU/e), The Netherlands
Diana Mateus	Technische Universität München (TUM), Germany
Emanuele Trucco	University of Dundee, UK
Samaneh Abbasi	Eindhoven University of Technology (TU/e), The Netherlands
Nicholas Heller	University of Minnesota, USA

Program Committee LABELS 2020

Florian Dubost	Erasmus University Medical Center, The Netherlands
Amelia Jimenez-Sanchez	Pompeu Fabra University, Spain
Obioma Pelka	University of Duisburg-Essen, Germany
Christoph Friedrich	University of Applied Sciences and Arts Dortmund, Germany
John Onofrey	Yale University, USA
Vinkle Srivastav	University of Strasbourg, France
Filipe Condessa	Carnegie Mellon University, USA
Roger Tam	University of British Columbia, Canada
Weidong Cai	The University of Sydney, Australia
Silas Ørting	University of Copenhagen, Denmark
Jack Rickman	University of Minnesota, USA
Alison O'Neil	Cannon Medical Research, UK
Bin Xie	Central South University, China
Fausto Milletari	NVIDIA, Germany
Holger Roth	NVIDIA, USA
Michael Goetz	DKFZ, Germany
Nishikant Deshmukh	Johns Hopkins University, USA
Tom Eelbode	KU Leuven, Belgium
Tuo Leng	Shanghai University, China
Wen Hui Lei	University of Electronic Science and Technology, China
Xiaoyu He	Central South University, China

Contents

MIL3ID 2020

iMIMIC 2020

Assessing Attribution Maps for Explaining CNN-Based Vertebral Fracture Classifiers

Eren Bora Yilmaz[1,4]([✉]), Alexander Oliver Mader[3], Tobias Fricke[2],
Jaime Peña[1], Claus-Christian Glüer[1], and Carsten Meyer[3,4,5]

[1] Section Biomedical Imaging, Department of Radiology and Neuroradiology,
University Hospital Schleswig-Holstein (UKSH), Campus Kiel, Kiel, Germany
eren.yilmaz@rad.uni-kiel.de
[2] Department of Radiology and Neuroradiology, UKSH, Campus Kiel, Kiel, Germany
[3] Institute of Computer Science, Kiel University of Applied Sciences, Kiel, Germany
[4] Department of Computer Science, Faculty of Engineering, Kiel University,
Kiel, Germany
[5] Department of Digital Imaging, Philips Research, Hamburg, Germany

Abstract. Automated evaluation of vertebral fracture status on computed tomography (CT) scans acquired for various purposes (opportunistic CT) may substantially enhance vertebral fracture detection rate. Convolutional neural networks (CNNs) have shown promising performance in numerous tasks but their black box nature may hinder acceptance by physicians. We aim (a) to evaluate CNN architectures for osteoporotic fracture discrimination as part of a pipeline localizing and classifying vertebrae in CT images and (b) to evaluate the benefit of using attribution maps to explain a network's decision. Training different model architectures on 3D patches containing vertebrae, we show that CNNs permit highly accurate discrimination of the fracture status of individual vertebrae. Explanations were computed using selected attribution methods: Gradient, Gradient * Input, Guided BackProp, and SmoothGrad algorithms. Quantitative and visual tests were conducted to evaluate the meaningfulness of the explanations (sanity checks). The explanations were found to depend on the model architecture, the realization of the parameters, and the precise position of the target object of interest.

Keywords: Explainable AI · Sanity checks · Osteoporosis

1 Introduction

Computed tomography (CT) images are taken for a variety of medical reasons, including diagnosis of bone fractures, internal bleedings, and tumors. Often CT scans show the spine or sections of the spine. Even for an experienced radiologist, manually searching for fractured vertebrae in such a CT image is a time-consuming task, and is often not conducted, unless it was the primary

© Springer Nature Switzerland AG 2020
J. Cardoso et al. (Eds.): iMIMIC 2020/MIL3ID 2020/LABELS 2020, LNCS 12446, pp. 3–12, 2020.
https://doi.org/10.1007/978-3-030-61166-8_1

purpose of the CT scan. Hence, an automated tool identifying fractures in CT images could indicate vertebral fractures to be checked by a radiologist and consequently uncover otherwise missed vertebral fractures, substantially enhancing fracture detection rate.

Tomita et al. [17] proposed a CNN/LSTM detection system automated on patient level, but made strong assumptions regarding the position and visible section of the spine in the CT image. Husseini et al. [8] combined an unsupervised auto-encoder with a supervised multi-layer perceptron to leverage larger quantities of unlabeled vertebrae, but require segmentation masks at training time for all vertebrae. Nicolaes et al. [12] trained a CNN to segment spinal CT images based on masks that are automatically derived from ground-truth coordinates and annotations of the vertebrae. However, these approaches [8,12,17] classified fractures based on Genant grading, and ignore whether a deformity is degenerative or constitutes an osteoporotic fracture. Another possible approach is an automated pipeline localizing and classifying vertebrae in CT images. For the localization task, successful methods were recently proposed, e.g. in [11]. In this work, we focus on the vertebral fracture classification task, for which we evaluate a pre-trained U-Net prefix and a custom CNN architecture[1]. In combination with the vertebra localization approach proposed in [11], this constitutes a fully automatic pipeline.

To improve the trust in the model's decision, particularly in a clinical context, it is desirable to shed light on the process how it came to its decision. Attribution methods—algorithms computing visual explanations (attribution maps) for interpreting models [3]—pose a step into this direction. A number of attribution methods has been suggested [3,14–16], but the interpretation of attribution maps is still under debate [2,3,18]. In order to evaluate the adequacy of attribution maps, [2] developed sanity checks that compare maps of trained models to (1) those of models with random weights and (2) those of models trained on random labels. Young et al. [18] compared attribution maps computed for melanoma classifiers of similar accuracy. We apply the analyses to the 3D vertebral osteoporotic fracture classification task, computing attribution maps using a set of selected attribution methods. Apart from assessing the sanity checks proposed in [2] we propose two additional checks. First, we compare explanations from independent training runs with the same architecture to test the dependence of the computed maps on the specific realization of the learned parameters. Additionally, we test the equivariance of attribution maps w.r.t. slight translations.

2   Methods

Given a 3D patch of a CT image containing a centered vertebra, the task of our proposed models is to decide if the vertebra has an osteoporotic fracture or not. Note that this means not only distinguishing between Genant score 0 and larger than 0, but also distinguishing whether a deformity is degenerative or

---

[1] We also conducted experiments with a 3D ResNet18 variant [7] that are out of scope for this publication but are in line with the presented results.

constitutes an osteoporotic fracture, since this is important for an early diagnosis of osteoporosis and decisions on therapeutic measures that may result from it.

2.1 CNN Architectures

Two architectures are evaluated: (1) A prefix of a 3D U-Net architecture that was pre-trained for brain tumor segmentation [9] with a classification head trained on task-related CT data, (2) a custom CNN architecture for 3D images based on the results of a hyperparameter search.

Each model has three parallel outputs (see Fig. 1 for the custom CNN as example) to calculate (i) a score between 0 and 1 to indicate the osteoporotic fracture status, (ii) the Genant score [5] ranging from 0 to 3 and indicating the severity of the deformity, and (iii) three "deformity percentages", continuous measures specifying the relative reduction in vertebral heights.

Output (i) uses the sigmoid activation function and is trained using the binary cross-entropy loss, outputs (ii) and (iii) use a linear activation function trained using the mean squared error (MSE) loss. At test time only output (i), in combination with a threshold value optimized on the validation data, is used to classify the vertebra. The additional outputs are used for regularization during training. As different models may require different regularization schemes, weighting the losses allows to control the influence of the individual outputs.

Although the architectures—as explained in the next sections—are quite different, they are trained in a similar setup. The Adam optimizer [10] is used in conjunction with Stochastic Weight Averaging (SWA) [13]. SWA averages the weights that occur after each 160 mini-batches, which we found to improve generalization with little computational overhead. The averaged models are evaluated on validation data for early stopping. To improve diversity in training data, random multiplicative and additive Gaussian noises, shifts by up to two voxels in each dimension, mirroring on the sagittal plane, rotation on the sagittal plane by up to 18 degrees and cropping (replacing borders of the input with zeros) are applied to the vertebral patches as data augmentation, motivated by prior work.

The hyperparameters listed in the following sub-sections were determined in manual and automatic hyperparameter searches.

2.2 Prefix of Pre-trained U-Net Encoder

As a well-known architecture for semantic segmentation we adopt the U-Net architecture to our task. Specifically, we use the encoder part of a 3D U-Net architecture applied to multi-modal MRI scans for the task of brain tumor segmentation [9]: We freeze the pre-trained weights of the first two residual blocks, omit the remaining layers of the 3D U-Net and instead append a new fully connected layer with 16 units, ReLU activation function and 25% dropout, followed by the three classification outputs described above. The model is trained on patches of size $32 \times 64 \times 96\,\text{mm}^3$ (longitudinal $\times$ anteroposterior $\times$ lateral). The large size in the lateral direction is required because this U-Net expects input shapes that are multiples of 32 and a resolution of $3\,\frac{\text{mm}}{\text{vx}}$ is used in the lateral

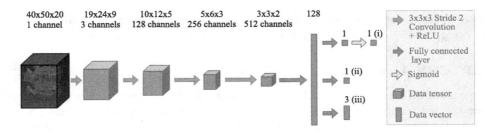

Fig. 1. Custom CNN; (i)–(iii): The outputs as named in Sect. 2.1

direction. A learning rate of 10^{-4} and a loss weight of 1 is used for all outputs, i.e., the total loss is the unweighted sum of the losses at all three output nodes.

2.3 Custom CNN

We propose a custom CNN consisting of 4 convolutional layers followed by one fully connected layer with 128 units and the output layers. The convolutional layers have a filter size of $3 \times 3 \times 3$ and a stride of $2 \times 2 \times 2$ (more details in Fig. 1). All convolutional layers use zero-padding, except the first which was added to reduce the input dimensionality for regularization and memory reasons. Building a task-specific architecture, the patch size can be changed to $40 \times 50 \times 60 \,\mathrm{mm}^3 = 40 \times 50 \times 20 \,\mathrm{vx}$ (longitudinal $\times$ anteroposterior $\times$ lateral) to better match the shape of the vertebrae. A learning rate of $5 \cdot 10^{-4}$ and loss weights 1, 1, and 100 for outputs (i)–(iii), respectively, were found by the hyperparameter search.

2.4 Attribution Maps

We select the following attribution methods, based on the realization of implementation invariance (see below) and on simplicity:

A simple method of highlighting important parts in the image is to compute the **Gradient** of the models activation (before sigmoid) w.r.t. the CNN input [3, 4]. Large values indicate regions in the input image where a small change in the input would have a comparably large effect on the model's output.

Element-wise multiplication of the gradient with the input (**Gradient * Input**) aims to show the contribution of the individual pixels and reduce noise in the explanations [14].

To highlight only areas that contribute positively to a model's decision, **Guided BackProp** was suggested, setting negative gradients at activation functions to zero during backpropagation [16].

SmoothGrad [15] addresses the issue of noisy gradients by averaging gradients on images that are augmented with random noise. In our experiment we use a noise level ($\sigma/(x_{\max} - x_{\min})$) of 15% and average across 50 noisy images.

When using fixed noise values for SmoothGrad and with the exception of Guided BackProp, the above methods satisfy implementation invariance: Two models that compute the same function should, given any fixed input image, have the same attribution maps.

2.5 Sanity Checks

To assess the quality of the explanations in form of attribution maps, we perform a number of sanity checks. Each sanity check compares attribution maps computed in two different settings (see below). As in [2,18], the similarity is evaluated using the structural similarity index (SSIM) computed between the whole 3D attribution maps. The reader is informed that the SSIM is a value between -1 and 1 where 0 corresponds to no correlation.

As suggested by [2], we compare attribution maps computed for a trained model to those of a model with re-initialized **"Random Weights"**. The frozen, pre-trained weights of the U-Net are not changed. The output layers of the model are also not re-initialized, since they were originally initialized with zeros leading to all-zero attribution maps. Strong similarity of the maps would indicate that learned parameters are independent from the explanations, while—as a result of learning—they clearly have strong influence on the model's output.

In the second setting proposed by [2], the attribution maps are compared for a model trained on **"Random Labels"** and a correctly trained model. More precisely, all labels in the training dataset are re-assigned to random vertebrae before training, preserving class ratio. A strong similarity in this setting indicates that attribution maps do not reflect task-specific supervision.

As a further check to assess the meaning of the explanations, **"Re-Training"**, we compare the explanations computed for two trained models with the same architecture on the same data, where only the random components during training vary. This includes the weight initialization, the choice of images in each mini-batch, dropout, and data augmentation. The total number of mini-batches applied during training may vary due to early stopping. If weak (or no) similarity is observed, either completely different features were learned—which seems unlikely—or the attribution maps are not consistent in the visualization of the learned features.

Finally, we address equivariance of attribution maps w.r.t. slight translations, **"2-Voxel-Shift"**: We compute attribution maps on patches of vertebrae that were translated by 2 voxels and compare them to translated attribution maps computed on original patches. As translation introduces zeros at the border of the patches, the respective borders are cropped from the patches and the explanations, in order to not influence the SSIM values. Since the object of interest is fully contained in the image patch in both scenarios, we expect both the model's output and attribution maps to be similar in both settings.

3   Experiments

The dataset used in this study contains 159 low-dose CT images of distinct patients (136 female), typically showing vertebrae T5–L4. The images were acquired in seven centers participating in the in the Diagnostik Bilanz study of the BioAsset project [6]. SpineAnalyzer^{TM} [1] was used by a radiologist to annotate—for each visible vertebra—the Genant score [5], "deformity percentages" indicating height reduction, vertebra centers on a 2D sagittal slice, and a differential diagnosis indicating if the vertebra shows either a "deformity" (1019 cases), an "osteoporotic fracture" (128 cases), is "unevaluable" (due to noise, 5 cases) or "normal" (802 cases). For the binary classification task, vertebrae with an "osteoporotic fracture" are labeled 1, "deformity" and "normal" correspond to 0, and "unevaluable" vertebrae were excluded. The lateral coordinate was computed using a state-of-the-art vertebra localization tool [11] and manually checked for correctness. Centered on these coordinates, for each vertebra, a 3D patch is extracted of fixed size from the CT image, serving as input for the respective CNN. As the only preprocessing steps, images were scaled to a resolution of $1 \times 1 \times 3 \frac{mm^3}{vx}$ (longitudinal × anteroposterior × lateral) and Hounsfield-values were divided by 2048. The images were split on patient level into four subsets defining a 4-fold cross-validation setup. In each run, two data subsets were used as training data, one as validation data (for early stopping and choosing the classification threshold) and one subset as test data. To address the class imbalance, at training time each vertebra in each mini-batch was chosen randomly, such that with 50% probability a vertebra labeled 1 was chosen. The data augmentation methods listed in Sect. 2.1 were applied to each vertebra in each mini-batch.

Table 1. Vertebral fracture discrimination results for the U-Net prefix and the Custom CNN. Mean and standard deviation are computed across 4 folds. **Bold**: Best per column. ROC-AUC: area under the ROC curve. AP: average precision.

Method	ROC-AUC ($\pm\sigma$)	AP ($\pm\sigma$)	Specificity ($\pm\sigma$)	Sensitivity ($\pm\sigma$)
U-Net prefix	0.939 ($\pm$0.011)	0.703 ($\pm$0.042)	0.912 ($\pm$0.030)	0.824 ($\pm$0.049)
Custom CNN	**0.989** ($\pm$0.0088)	**0.907** ($\pm$0.015)	**0.958** ($\pm$0.024)	**0.906** ($\pm$0.081)

3.1   Vertebral Fracture Discrimination

Both investigated architectures accurately discriminate osteoporotic fracture status of individual vertebrae, with ROC-AUCs of above 0.9 (results in Table 1). The custom CNN achieved on average a higher ROC-AUC (0.989) and average precision (0.907) than the U-Net prefix. The Pearson correlation coefficient of the architectures' outputs—the scores between 0 and 1—on vertebrae in the respective test sets was 0.72.

3.2   Attribution Maps

Some examples for attribution maps created in the different settings are displayed in Fig. 2 (top). The lower part of Fig. 2 plots SSIM scores for the sanity checks, together with the Pearson correlation of the respective models' output. The following observations are based on the results displayed in the figure and manual inspection of explanations for other vertebrae.

Similarly Accurate Models Can Have Different Explanations. Inspecting the Pearson correlation, we found the compared models' outputs to show stronger correlation in the settings "Re-Training", and "2-Voxel-Shift", than for

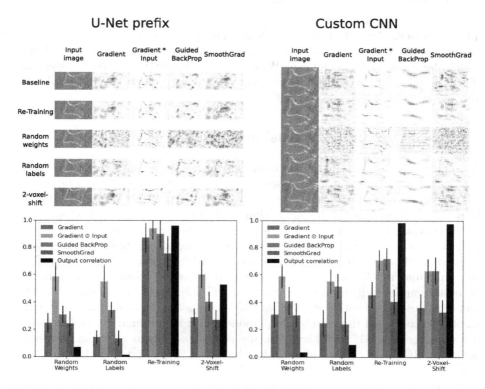

Fig. 2. Attribution maps and sanity checks for the U-Net prefix (left) and the Custom CNN (right). **Top:** Comparison of the central sagittal slices of attribution maps using a true positive L3 vertebra as example, where blue, white and red pixels correspond to negative, near-zero and positive values. The "Baseline" row shows the attribution maps for a trained instance of the corresponding CNN; the following rows show resulting maps when performing the described sanity checks. Images were normalized individually by division with the maximum absolute value. **Bottom:** A quantitative comparison, where colored bars display the SSIM values averaged across the vertebrae in the respective test sets and black bars show the Pearson correlation of the models' outputs. Error bars indicate the standard deviation across vertebrae.

"Random Weights" and "Random Labels". In contrast, the SSIM of the attribution maps was only slightly larger, if at all (see Fig. 2 and Sect. 3.1).

Gradient * Input Seems to Highlight Input Structures. In our experiments, Gradient * Input visually highlights the vertebra's outline in most settings, leading to comparably high SSIM values for the performed sanity checks.

Guided BackProp Shows Fracture-Relevant Features When Trained on Random Labels. While the Guided BackProp explanations for the Custom CNN seems to highlight fracture-relevant features (vertebral end plates), this is also the case in the "Random Labels" setting.

The Quantitative Results Do Not Always Match the Qualitative Impressions. One example is the comparison of "Random Weights" and "Random Labels" for the Custom CNN: While visually (inspecting the central sagittal slice) the similarity seems to be much higher in "Random Labels", the SSIM is almost the same in the two settings.

4   Discussion and Conclusions

Automatic vertebral fracture discrimination may help to improve the fracture detection rate by automatically assessing fracture status in CT scans taken for various other medical reasons. The investigated CNN architectures were found to permit highly accurate discrimination of the fracture status of vertebrae.

In contrast to previous publications [8,12], we focus on identifying osteoporotic fractures (as opposed to deformities in general). It must be noted that different authors worked on different datasets with varying size and difficulty, so results may not be comparable. Still, the quantitative results on our in-house dataset compare well to previous work: [8] and [12] report specificities of 0.905 and 0.669 and sensitivities of 0.938 and 0.854, respectively. In additional experiments, we combined the vertebra level models with an automatic vertebra localization algorithm [11] to achieve fracture status assessment on a patient level, achieving ROC-AUC values of 0.876 for the U-Net prefix and 0.940 for the Custom CNN.

Furthermore, to shed light on the complex decision process of the CNNs, we have computed attribution maps for the fracture status classification task on vertebra level. In addition to previously proposed sanity checks [2,18], we introduced two further tests for explanation methods, namely retraining of a given model configuration and testing equivariance of attribution maps with regard to slight translations. In both cases, the computed explanation maps were quantitatively (SSIM[2]) less similar than expected. We found that models

---

[2] Other correlation measures (Pearson and Spearman coefficients) lead to similar conclusions.

of similar accuracy can produce different attribution maps, which matches the results of [18]. Gradient * Input, designed to reduce noise in the explanations, highlights mostly the outline of the input structure, which was also observed by [2]. SmoothGrad, also designed to reduce noise, visually and quantitatively behaved similar to the original gradient.

In conclusion, the explanations exhibit a strong dependence on the model architecture, the realization of the parameters, and the precise position of the target object of interest. Since explanations of a model's decision would be most helpful to convince physicians that automated approaches perform trustworthy evaluations, future work should address the implications of these findings for the clinical practice.

References

1. SpineAnalyzer. Optasia Medical Ltd., Cheadle Hulme, United Kingdom (2013)
2. Adebayo, J., Gilmer, J., Muelly, M., Goodfellow, I., Hardt, M., Kim, B.: Sanity checks for saliency maps. In: Advances in Neural Information Processing Systems, Montréal, Canada, pp. 9525–9536. Curran Associates Inc. (2018)
3. Ancona, M., Ceolini, E., Öztireli, C., Gross, M.: Gradient-based attribution methods. In: Samek, W., Montavon, G., Vedaldi, A., Hansen, L.K., Müller, K.-R. (eds.) Explainable AI: Interpreting, Explaining and Visualizing Deep Learning. LNCS (LNAI), vol. 11700, pp. 169–191. Springer, Cham (2019). https://doi.org/10.1007/978-3-030-28954-6_9
4. Erhan, D., Bengio, Y., Courville, A., Vincent, P.: Visualizing higher - layer features of a deep network. Technical report, Univeristé de Montréal (2009)
5. Genant, H.K., Wu, C.Y., van Kuijk, C., Nevitt, M.C.: Vertebral fracture assessment using a semiquantitative technique. J. Bone Miner. Res. 8(9), 1137–1148 (1993)
6. Glüer, C.C., et al.: New horizons for the in vivo assessment of major aspects of bone quality microstructure and material properties assessed by Quantitative Computed Tomography and Quantitative Ultrasound methods developed by the BioAsset consortium. Osteologie 22, 223–233 (2013)
7. Haarburger, C., et al.: Multi scale curriculum CNN for context-aware breast MRI malignancy classification. In: Shen, D., et al. (eds.) MICCAI 2019. LNCS, vol. 11767, pp. 495–503. Springer, Cham (2019). https://doi.org/10.1007/978-3-030-32251-9_54
8. Husseini, M., Sekuboyina, A., Bayat, A., Menze, B.H., Loeffler, M., Kirschke, J.S.: Conditioned variational auto-encoder for detecting osteoporotic vertebral fractures. In: Cai, Y., Wang, L., Audette, M., Zheng, G., Li, S. (eds.) CSI 2019. LNCS, vol. 11963, pp. 29–38. Springer, Cham (2020). https://doi.org/10.1007/978-3-030-39752-4_3
9. Isensee, F., Kickingereder, P., Wick, W., Bendszus, M., Maier-Hein, K.H.: Brain tumor segmentation and radiomics survival prediction: contribution to the BRATS 2017 challenge. In: Crimi, A., Bakas, S., Kuijf, H., Menze, B., Reyes, M. (eds.) BrainLes 2017. LNCS, vol. 10670, pp. 287–297. Springer, Cham (2018). https://doi.org/10.1007/978-3-319-75238-9_25
10. Kingma, D.P., Ba, J.L.: Adam: a method for stochastic optimization. In: 3rd International Conference on Learning Representations, San Diego, May 2015

11. Mader, A.O., Lorenz, C., von Berg, J., Meyer, C.: Automatically localizing a large set of spatially correlated key points: a case study in spine imaging. In: Shen, D., et al. (eds.) MICCAI 2019. LNCS, vol. 11769, pp. 384–392. Springer, Cham (2019). https://doi.org/10.1007/978-3-030-32226-7_43

12. Nicolaes, J., et al.: Detection of vertebral fractures in CT using 3D convolutional neural networks. In: Cai, Y., Wang, L., Audette, M., Zheng, G., Li, S. (eds.) CSI 2019. LNCS, vol. 11963, pp. 3–14. Springer, Cham (2020). https://doi.org/10.1007/978-3-030-39752-4_1

13. Izmailov, P., Podoprikhin, D., Garipov, T., Vetrov, D., Wilson, A.G.: Averaging weights leads to wider optima and better generalization. In: Uncertain Artificial Intelligence, Monterey, California, pp. 876–885. AUAI Press, Corvallis, March 2018

14. Shrikumar, A., Greenside, P., Shcherbina, A., Kundaje, A.: Not Just a Black Box: Learning Important Features Through Propagating Activation Differences (2016)

15. Smilkov, D., Thorat, N., Kim, B., Viégas, F., Wattenberg, M.: SmoothGrad: removing noise by adding noise (2017). arXiv:Learning

16. Springenberg, J., Dosovitskiy, A., Brox, T., Riedmiller, M.: Striving for simplicity: the all convolutional net. In: International Conference on Learning Representations (2015)

17. Tomita, N., Cheung, Y.Y., Hassanpour, S.: Deep neural networks for automatic detection of osteoporotic vertebral fractures on CT scans. Comput. Biol. Med. **98**, 8–15 (2018)

18. Young, K., Booth, G., Simpson, B., Dutton, R., Shrapnel, S.: Deep neural network or dermatologist? In: Suzuki, K., et al. (eds.) ML-CDS/IMIMIC -2019. LNCS, vol. 11797, pp. 48–55. Springer, Cham (2019). https://doi.org/10.1007/978-3-030-33850-3_6

Projective Latent Interventions for Understanding and Fine-Tuning Classifiers

Andreas Hinterreiter[1,2]([✉]), Marc Streit[2], and Bernhard Kainz[1]

[1] Biomedical Image Analysis Group, Imperial College, London, UK
{a.hinterreiter,b.kainz}@imperial.ac.uk
[2] Institute of Computer Graphics, Johannes Kepler University Linz, Linz, Austria
{andreas.hinterreiter,marc.streit}@jku.at

Abstract. High-dimensional latent representations learned by neural network classifiers are notoriously hard to interpret. Especially in medical applications, model developers and domain experts desire a better understanding of how these latent representations relate to the resulting classification performance. We present Projective Latent Interventions (PLIs), a technique for retraining classifiers by back-propagating manual changes made to low-dimensional embeddings of the latent space. The back-propagation is based on parametric approximations of t-distributed stochastic neighbourhood embeddings. PLIs allow domain experts to control the latent decision space in an intuitive way in order to better match their expectations. For instance, the performance for specific pairs of classes can be enhanced by manually separating the class clusters in the embedding. We evaluate our technique on a real-world scenario in fetal ultrasound imaging.

Keywords: Latent space · Non-linear embedding · Image classification

1 Introduction

The interpretation of classification models is often difficult due to a high number of parameters and high-dimensional latent spaces. Dimensionality reduction techniques are commonly used to visualise and explain latent representations via low-dimensional embeddings. These embeddings are useful to identify problematic classes, to visualise the impact of architectural changes, and to compare new approaches to previous work. However, there is a lot of debate about how well such mappings represent the actual decision boundaries and the resulting model performance.

In this work, we aim to change the paradigm of passive observation of mappings to active interventions during the training process. We argue that such interventions can be useful to mentally connect the embedded latent space with the classification properties of a classifier. We show that in some situations, such as class-imbalanced problems, the manual interventions can also be used

© Springer Nature Switzerland AG 2020
J. Cardoso et al. (Eds.): iMIMIC 2020/MIL3ID 2020/LABELS 2020, LNCS 12446, pp. 13–22, 2020.
https://doi.org/10.1007/978-3-030-61166-8_2

for fine-tuning and targeted performance gains. This means that practitioners can prioritise the decision boundary for certain classes over the others simply by manipulating the embedded latent space. The overall idea of our work is outlined in Fig. 1. We use a neural-network-based parametric implementation of t-distributed stochastic neighbourhood embeddings (t-SNE) [11,12,15] to inform the training process by back-propagating the manual manipulations of the embedded latent space through the classification network.

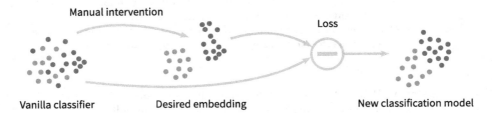

Fig. 1. PLIs define a desired embedding, which is subsequently used to inform the training or fine-tuning process of a classification model in an end-to-end way.

Related Work: Low dimensional representations of high dimensional latent spaces have been subject to scientific research for many decades [12–14,20,22]. Commonly these methods are treated as independent modules and applied to a selected part of the representation, e.g., the penultimate layer of a discriminator network. However, these embeddings are often spatially inconsistent during training from epoch to epoch and cannot inform the training process through back-propagation. Van der Maaten et al. [11,15] proposed to learn mappings through a neural network. This approach has the advantage that it can be directly integrated into an existing network architecture enabling end-to-end forward and backward updates. While unsupervised dimensionality reduction techniques have been used as part of deep learning workflows [4,10,19,21] and for visualising latent spaces [6,18], we are not aware of any previous work that exploited parametric embeddings for a direct manipulation of learned representations. This shaping of the latent space relates our approach to metric learning [2,8]. Metric learning makes use of specific loss functions to automatically constrain the latent space, but does not allow manual interventions. PLIs are general enough to be combined with concepts of metric learning.

Contribution: We introduce Projective Latent Interventions (PLIs), a technique for (a) understanding the relationship between a classifier and its learned latent representation, and (b) facilitating targeted performance gains by improving latent space clustering. We discuss an application of PLIs in the context of anatomical standard plane classification during fetal ultrasound imaging.

2   Method

Projective Latent Interventions (PLIs) can be applied to any neural network classifier. Consider a dataset $X = \{x_1, \ldots, x_N\}$ with N instances belonging to K classes. A neural network C was trained to predict the ground truth labels g_i of x_i, where $g_i \in \{\gamma_1, \ldots, \gamma_K\}$. Let $C_l(x_i)$ be the activations of the network's lth layer, and let the network have L layers in total.

Given C, PLIs consist of three steps: (1) training of a secondary network $\tilde{E}$ that approximates a given non-linear embedding $E = \{y_1, \ldots, y_N\}$ for the outputs $C_l(x_i)$ of layer l; (2) modifying the positions y_i of embedded points, yielding new positions y_i'; and (3) retraining C, such that $\tilde{E}(C_l(x_i)) \approx y_i'$. In the following sections, we will discuss these three steps in detail.

2.1   Parametric Embeddings

The embeddings used for PLIs are parametric approximations of t-SNE. For t-SNE, distances between high-dimensional points z_i and z_j are converted to probabilities of neighbourhood p_{ij} via Gaussian kernels. The variance of each kernel is adjusted such that the perplexity of each distribution matches a given value. This perplexity value is a smooth measure for how many nearest neighbours are covered by the high-dimensional distributions. Then, a set of low-dimensional points is initialised and likewise converted to probabilities q_{ij}, this time via heavy-tailed t-distributions. The low-dimensional positions are then adjusted by minimising the Kullback–Leibler divergence $\mathrm{KL}(p_{ij}||q_{ij})$ between the two probability distributions.

Given a set of d-dimensional points $z_i \in \mathbb{R}^d$, t-SNE yields a set of d'-dimensional points $z' \in \mathbb{R}^{d'}$. However, it does not yield a general function $E : \mathbb{R}^d \to \mathbb{R}^{d'}$ defined for all $z \in \mathbb{R}^d$. It is thus impossible to add new points to existing t-SNEs or to back-propagate gradients through the embeddings.

In order to allow out-of-sample extension, van der Maaten introduced the idea of approximating t-SNE with neural networks [11]. We adapt van der Maaten's approach and introduce two important extensions, based on recent advancements related to t-SNE [17]: (1) *PCA initialisation* to improve reproducibility across multiple runs and preserve global structure; and (2) *approximate nearest neighbours* [5] for a more efficient calculation of the distance matrix without noticeable effects on the embedding quality.

Our approach is an unsupervised learning workflow resulting in a neural network that approximates t-SNE for a set of input vectors $\{z_1, \ldots, z_N\}$ given a perplexity value Perp. We only take into account the k approximate nearest neighbours, where $k = \min(3 \times \text{Perp}, N - 1)$. In contrast to the simple binary search used by van der Maaten [11], we use Brent's method [3] for finding correct variances of the kernels. Optionally, we pre-train the network such that its 2D output matches the first two principal components of z_i. In the actual training phase, we calculate low-dimensional pairwise probabilities q_{ij} for each input batch, and use the KL-divergence $\mathrm{KL}(p_{ij}||q_{ij})$ as a loss function.

While van der Maaten used a network architecture with three hidden layers of sizes 500, 500, and 2000 [11], we found that much smaller networks (e.g., two hidden layers of sizes 300 and 100) are more efficient and yield more reliable results. The t-SNE-approximating network can be connected to any complex neural network, such as CNNs for medical image classification.

2.2   Projective Latent Constraints

Once the network $\tilde{E}$ has been trained to approximate the t-SNE, new constraints on the embedded latent space can be defined. This is most easily done by visualising the embedded points, $y_i = E(C_l(x_i))$, in a scatter plot with points coloured categorically by their ground truth labels g_i. For our applications, we chose only simple modifications of the embedding space: shifting of entire class clusters[1], and contraction of class clusters towards their centres of mass. The modified embedding positions y_i' are used as target values for the subsequent regression learning task.

In this work, we focus on class-level interventions because their effect can be directly measured via class-level performance metrics and they do not require domain-specific interactive tools that would lead to additional cognitive load. In principle, arbitrary alterations of the embedded latent space are possible within our technique.

2.3   Retraining the Classifier

In the final step, the original classifier is retrained with an adapted loss function $\mathcal{L}_{\text{PLIs}}$ based on the modified embedding:

$$\mathcal{L}_{\text{PLIs}}(x_i, g_i, y_i') = (1 - \lambda)\,\mathcal{L}_{\text{class}}(C_L(x_i), g_i) + \lambda\,\mathcal{L}_{\text{emb}}(\tilde{E}(C_l(x_i)), y_i'). \tag{1}$$

The new loss function combines the original classification loss function $\mathcal{L}_{\text{class}}$, typically a cross-entropy term, with an additional term $\mathcal{L}_{\text{emb}}$. Minimisation of $\mathcal{L}_{\text{emb}}$ causes the classifier to learn new activations that yield embedded points similar to y' (using the given embedding function $\tilde{E}$). As $\tilde{E}$ is simply a neural network, back-propagation of the loss is straightforward. In our experiments, we use the squared euclidean distance for $\mathcal{L}_{\text{emb}}$ and test different values for the weighting coefficient λ. We also experiment with only counting the embedding loss for instances of classes that were altered in the embedding.

3   Experiments

3.1   MNIST and CIFAR

As a proof of concept, we applied PLIs to simple image classifiers: a small MLP for MNIST [9] images and a simple CNN for CIFAR-10 [7] images. For MNIST,

---

[1] The class cluster for class γ_j is simply the set of points $\{y_i = E(C_l(x_i)) \mid g_i = \gamma_j\}$.

the embedded latent space after retraining generally preserved the manipulations well, when class clusters were contracted and/or translated. The classification accuracy only changed insignificantly (within a few percent over wide ranges of λ). Typical results for the CIFAR-10 classifier are shown in Fig. 2, where the goal of the Projective Latent Interventions was to reduce the model's confusion between the classes *Truck* and *Auto*, by separating the respective class clusters. When comparing a classifier trained for $5 + 4$ epochs with $\mathcal{L}_{class}$ to one trained for 5 epochs $\mathcal{L}_{class} + 4$ epochs $\mathcal{L}_{PLIs}$, the latter showed a relative increase of target-class-specific F_1-scores by around 5 %, with the overall accuracy improving or staying the same. The embeddings after retraining, as seen in Fig. 2, reflected the manual interventions well, but not as closely as in the case of MNIST. We also found that, in the case of CNNs, using the activations of the final dense layer ($l = L$) yielded the best results.

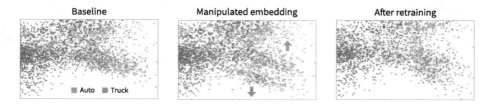

Fig. 2. Detail views of the embedded latent space before (left), during (centre) and after (right) Projective Latent Interventions for classification of CIFAR-10 images, focusing on the classes *Truck* and *Auto*.

3.2   Standard Plane Detection in Ultrasound Images

We tested our approach on a challenging diagnostic view plane classification task in fetal ultrasound screening. The dataset consists of about 12,000 2D fetal ultrasound images sampled from 2,694 patient examinations with gestational ages between 18 and 22 weeks. Eight different ultrasound systems of identical make and model (GE Voluson E8) were used for the acquisitions to eliminate as many unknown image acquisition parameters as possible. Anatomical standard plane image frames were labelled by expert sonographers as defined in the UK FASP handbook [16]. We selected a subset of images that tend to be confused by established models [1]: Four Chamber View (4CH), Abdominal, Femur, Spine, Left Ventricular Outflow Tract (LVOT) and Right Ventricular Outflow Tract (RVOT)/Three Vessel View (3VV). RVOT and 3VV were combined into a single class after clinical radiologists confirmed that they are identical. We split the resulting dataset into 4,777 training and 1,024 test images.

The architecture of our baseline classifier is SonoNet-64 [1]. The network was trained for 5 epochs with pure classification loss, i.e., $\mathcal{L} = \mathcal{L}_{class}$. We used Kaiming initialization, a batch size of 100, a learning rate of 0.1, and 0.9 Nesterov

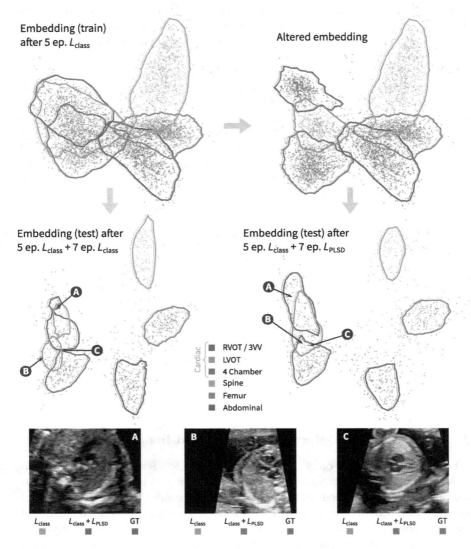

Fig. 3. Projective Latent Interventions for standard plane classification in fetal ultra-sound images. Top left: embedding of the baseline network's output (train) after 5 epochs of classification training ($\mathcal{L} = \mathcal{L}_{\mathrm{class}}$). Top right: altered output embedding (train) with manually separated cardiac classes. Centre left: Output embedding (test) after resuming standard classification training for 7 epochs ($\mathcal{L} = \mathcal{L}_{\mathrm{class}}$), starting from the baseline classifier (top left). Centre right: embedding (test) after resuming training with an updated loss function ($\mathcal{L} = \mathcal{L}_{\mathrm{PLIs}} = 0.9\,\mathcal{L}_{\mathrm{class}} + 0.1\,\mathcal{L}_{\mathrm{emb}}$), starting again from the baseline classifier (top left). For easier comparability, class-specific contour lines at a density threshold of $1/N$ are shown, where N is the total number of train or test images, respectively. Performance measures for the classifiers are given in Table 1. Bottom: Three example images that were successfully classified after applying PLSD. For each image, the positions in both embeddings are indicated.

momentum. During these first five training epochs, we used random affine transformations for data augmentation ($\pm 15°$ rotation, ± 0.1 shift, 0.7 to 1.3 zoom).

The 6-dimensional final-layer logits for the non-transformed training images were used as inputs for the training of the parametric t-SNE network. We used a fully connected network with two hidden layers of sizes 300 and 100. The t-SNE network was trained for 10 epochs with a learning rate of 0.01, a batch size of 500 and a perplexity of 50. We pre-trained the network for 5 epochs to approximate a PCA initialisation.

The ultrasound dataset is imbalanced, with 1,866 images in the three cardiac classes, and 2,911 images in the three non-cardiac classes. There are about twice as many 4CH images as RVOT/3VV, and three times as many 4CH images as LVOT. As a result, after five epochs of classification learning, our vanilla classifier could not properly distinguish between the three cardiac classes. This is apparent in the baseline embedding shown in Fig. 3 (top left).

We experimented with PLIs to improve the performance for the cardiac classes, in particular for RVOT/3VV and LVOT. Figure 3 (top right) shows the case of contracting and shifting the class clusters of RVOT/3VV and LVOT.

Table 1. Global and class-specific performance measures for standard plane classification in fetal ultrasound images with and without PLIs, evaluated on the test set. The last two columns are weighted averages of the values for the three cardiac and the three non-cardiac classes, respectively. (*The class labelled as RVOT also includes 3VV.)

		RVOT*	4CH	LVOT	Abd.	Femur	Spine	Cardiac	Other
Precision	Class. only	**0.82**	0.82	0.42	0.93	0.98	0.97	0.77	0.96
	PLIs	0.78	**0.85**	**0.61**	0.91	0.97	0.96	**0.80**	0.95
Recall	Class. only	0.38	0.94	**0.46**	0.96	0.97	0.94	0.76	0.96
	PLIs	**0.73**	0.94	0.28	0.96	0.97	0.94	**0.81**	0.96
F_1-score	Class. only	0.56	0.88	**0.44**	0.95	0.97	0.95	0.75	0.96
	PLIs	**0.76**	**0.89**	0.41	0.94	0.97	0.95	**0.80**	0.95

After the latent interventions, training was resumed for 7 epochs with the mixed loss function defined in Eq. 1. We experimented with different values for λ; all results given in this section are for $\lambda = 0.1$, which was found to be a suitable value in this application scenario. For a fair comparison, training of the baseline network was also resumed for 7 epochs with pure classification loss. In both cases, the remaining training epochs were performed without data augmentation, but with all other hyperparameters kept the same as for the vanilla classifier.

The outputs were then embedded with the parametric t-SNE learned on the baseline outputs (see Fig. 3, centre). By resuming the training with included embedding loss, the clusters for the three cardiac classes assume relative positions that are closer to those in the altered embedding. The contraction constraint also led to more convex clusters for the test outputs. Figure 3 (bottom) shows three exemplary images that were misclassified in case of the pure classification loss model, but correctly classified after applying PLIs. Further inspection showed that most of the images that were correctly classified after PLIs (but not before) had originally been embedded close to decision boundaries.

Table 1 lists the class-specific precision, recall, and F_1-scores for the two different networks. By applying PLIs, the average quality for the cardiac classes could be improved without negatively affecting the performance for the remaining classes. In some experiments, we observed much larger quality improvements for individual classes. For example, in one case the F_1-score for LVOT improved by a factor of two. In these extreme cases, however, local improvements were often accompanied by significant performance drops for other classes.

4   Discussion

The insights gained from PLIs about the relationship between a classifier and its latent space are based on an assessment of the model's response to the interventions. This response can be evaluated on two axes: the *embedding response* and the *performance response*.

Simple classifications tasks, for which the baseline classifier already works well (e.g., MNIST) often show a considerable embedding response with only a minor performance response. This means that the desired alterations of the latent space are well reflected after retraining without strong effects on the classification performance. Such classifiers are flexible enough to accommodate the latent manipulations, likely because they are overparameterised. In more complex cases, such as CIFAR, the embedding response is weaker, but often accompanied by a more pronounced class-specific performance increase. For these cases, the learned representation seems to be more rigidly connected with the classification performance. Finally, the standard plane detection experiments showed that sometimes a minimal change in the embedding is accompanied by a considerable performance increase for the targeted classes. Here, the overall structure of the embedding seems to be fixed, but the classification accuracy can be redistributed between classes by injecting additional domain knowledge while allowing non-targeted classes to move freely.

In general, we found that too severe alterations of the latent space cannot be preserved well since the embeddings are based on local information. Furthermore, seemingly obvious changes made in the embedding may contradict the original classification task due to the non-linearity of the embedding. The strength of PLIs is that a co-evaluation of the two components of the loss function can reveal these discrepancies. As a result, even when PLIs cannot be used for improving a classifier's performance, it can still lead to a better understanding of the flexibility of the model and/or the trustworthiness of the embedding.

In future work, we would like to experiment with parametric versions of different dimensionality reduction techniques and explore the potential of instance-level manipulations controlled via an interactive visualisation.

5 Conclusion

We introduced Projective Latent Interventions, a promising technique to inject additional information into neural network classifiers by means of constraints derived from manual interventions in the embedded latent space. PLIs can help to get a better understanding of the relationship between the latent space and a classifier's performance. We applied PLIs successfully to obtain a targeted improvement in standard plane classification for ultrasound images without negatively affecting the overall performance.

Acknowledgments. This work was supported by the State of Upper Austria (Human-Interpretable Machine Learning) and the Austrian Federal Ministry of Education, Science and Research via the Linz Institute of Technology (LIT-2019-7-SEE-117), and by the Wellcome Trust (IEH 102431 and EPSRC EP/S013687/1.).

References

1. Baumgartner, C.F., et al.: SonoNet: real-time detection and localisation of fetal standard scan planes in freehand ultrasound. IEEE Trans. Med. Imaging **36**(11), 2204–2215 (2017). https://doi.org/10.1109/TMI.2017.2712367
2. Bellet, A., Habrard, A., Sebban, M.: A survey on metric learning for feature vectors and structured data. arXiv preprint arXiv:1306.6709 (2013)
3. Brent, R.P.: Algorithms for minimization without derivatives. Courier Corporation (2013)
4. Chen, X., Weng, J., Lu, W., Xu, J., Weng, J.: Deep manifold learning combined with convolutional neural networks for action recognition. IEEE Trans. Neural Netw. Learn. Syst. **29**(9), 3938–3952 (2017). https://doi.org/10.1109/TNNLS.2017.2740318
5. Dong, W., Moses, C., Li, K.: Efficient k-nearest neighbor graph construction for generic similarity measures. In: Proceedings of the 20th International Conference on World Wide Web, pp. 577–586 (2011). https://www.cs.princeton.edu/cass/papers/www11.pdf
6. Erhan, D., Bengio, Y., Courville, A., Manzagol, P.A., Vincent, P., Bengio, S.: Why does unsupervised pre-training help deep learning? J. Mach. Learn. Res. **11**, 625–660 (2010). http://jmlr.org/papers/volume11/erhan10a/erhan10a.pdf
7. Krizhevsky, A., Nair, V., Hinton, G.: CIFAR-10 (Canadian Institute for Advanced Research). http://www.cs.toronto.edu/~kriz/cifar.html. Accessed 16 Mar 2020
8. Kulis, B., et al.: Metric learning: a survey. Found. Trends Mach. Learn. **5**(4), 287–364 (2012)
9. LeCun, Y., Cortes, C.: The MNIST database of handwritten digits (2005). http://yann.lecun.com/exdb/mnist/. Accessed 16 Mar 2020
10. Lee, C.Y., Xie, S., Gallagher, P., Zhang, Z., Tu, Z.: Deeply-supervised nets. In: Artificial intelligence and statistics, pp. 562–570 (2015). https://www.proceedings.mlr.press/v38/lee15a.pdf
11. van der Maaten, L.: Learning a parametric embedding by preserving local structure. In: Artificial Intelligence and Statistics, pp. 384–391 (2009). http://proceedings.mlr.press/v5/maaten09a.html

12. van der Maaten, L., Hinton, G.: Visualizing data using t-SNE. J. Mach. Learn. Res. **9**, 2579–2605 (2008). https://lvdmaaten.github.io/publications/papers/JMLR_2008.pdf
13. McInnes, L., Healy, J., Melville, J.: UMAP: uniform manifold approximation and projection for dimension reduction, December 2018. arXiv:1802.03426
14. Mead, A.: Review of the development of multidimensional scaling methods. J. Roy. Stat. Soc. Ser. D (Stat.) **41**(1), 27–39 (1992). https://doi.org/10.2307/2348634
15. Min, M.R., van der Maaten, L., Yuan, Z., Bonner, A.J., Zhang, Z.: Deep supervised t-distributed embedding. In: Proceedings of the 27th International Conference on Machine Learning (ICML 2010) (2010). https://www.cs.toronto.edu/~cuty/DSTEM.pdf
16. NHS: Fetal Anomaly Screening Programme: Programme Handbook June 2015. Public Health England (2015)
17. Poličar, P.G., Stražar, M., Zupan, B.: openTSNE: a modular Python library for t-SNE dimensionality reduction and embedding. bioRxiv, August 2019. https://doi.org/10.1101/731877. http://biorxiv.org/lookup/doi/10.1101/731877
18. Rauber, P.E., Fadel, S.G., Falcão, A.X., Telea, A.C.: Visualizing the hidden activity of artificial neural networks. IEEE Trans. Visual Comput. Graphics **23**(1), 101–110 (2017)
19. Rusu, A.A., et al.: Meta-learning with latent embedding optimization (2018). arXiv:1807.05960
20. Tenenbaum, J.B.: Science, pp. 2319–2323. (2000). https://doi.org/10.1126/science.290.5500.2319. http://www.sciencemag.org/cgi/doi/10.1126/science.290.5500.2319
21. Tomar, V.S., Rose, R.C.: Manifold regularized deep neural networks. In: Fifteenth Annual Conference of the International Speech Communication Association (2014)
22. Wold, S., Esbensen, K., Geladi, P.: Principal component analysis. Chemometr. Intell. Lab. Syst. **2**(1–3), 37–52 (1987). https://doi.org/10.1016/0169-7439(87)80084-9

Interpretable CNN Pruning for Preserving Scale-Covariant Features in Medical Imaging

Mara Graziani[1,3]([✉]), Thomas Lompech[2], Henning Müller[1,3],
Adrien Depeursinge[1,4], and Vincent Andrearczyk[1]

[1] University of Applied Sciences Western Switzerland (HES-SO), Sierre, Switzerland
MaraGrmara.graziani@hevs.ch
[2] INP-ENSEEIHT, Toulouse, France
[3] University of Geneva (UNIGE), Geneva, Switzerland
[4] Centre Hospitalier Universitaire Vaudois (CHUV), Lausanne, Switzerland

Abstract. Image scale carries crucial information in medical imaging, e.g. the size and spatial frequency of local structures, lesions, tumors and cell nuclei. With feature transfer being a common practice, scale-invariant features implicitly learned from pretraining on ImageNet tend to be preferred over scale-covariant features. The pruning strategy in this paper proposes a way to maintain scale covariance in the transferred features. Deep learning interpretability is used to analyze the layer-wise encoding of scale information for popular architectures such as InceptionV3 and ResNet50. Interestingly, the covariance of scale peaks at central layers and decreases close to softmax. Motivated by these results, our pruning strategy removes the layers where invariance to scale is learned. The pruning operation leads to marked improvements in the regression of both nuclei areas and magnification levels of histopathology images. These are relevant applications to enlarge the existing medical datasets with open-access images as those of PubMed Central. All experiments are performed on publicly available data and the code is shared on GitHub.

Keywords: Interpretability · Scale · Histopathology · Transfer

1 Introduction

Transfer learning has become a standard approach in tasks with a limited amount of training data [35]. In medical imaging, it has led to significant improvements in various applications in terms of accuracy and speed of convergence [17,21,25, 26]. Scale invariance is required and learned implicitly by Convolutional Neural Networks (CNNs) in the object recognition task on ImageNet, as they normally appear at different distances from the observation point. Despite the controlled viewpoint and the considerable domain shift (i.e. reduced number of classes, less color, texture and object variety [25]), medical imaging applications often

© Springer Nature Switzerland AG 2020
J. Cardoso et al. (Eds.): iMIMIC 2020/MIL3ID 2020/LABELS 2020, LNCS 12446, pp. 23–32, 2020.
https://doi.org/10.1007/978-3-030-61166-8_3

reuse basic features from pretraining on natural images, i.e. color, edges and textures [13, 17].

The implicitly learned invariances could have different impacts in medical imaging. Global and local rotation invariance, for instance, were shown to be relevant [1, 31]. The scale invariance, however, could be detrimental. The viewpoint in medical images is controlled and the pixel (or voxel) size has a corresponding physical dimension. The size of an object of interest within an image carries relevant information [6, 7]. Nuclei size histopathology applications is a clear example of a discriminant factor of tumor regions [8, 14]. Approaches introducing scale analysis in the sense of either scale covariant networks [33] or multi-scale learning [3, 16, 24, 32] showed that analyzing tissue at various magnifications benefits from the combination of fine-grained details and global tissue information. Histopathology is not the only application that benefits from information about scale. Nodule detection and classification in computed tomography is another example in the medical domain [16]. From a larger perspective, other applications can be remote sensing, defect detection, material recognition and biometrics (e.g. iris recognition) [28]. It is thus relevant to analyze the role of information about scale in state-of-the-art CNNs that are often used for transfer learning such as inception-based [27] and residual-learning networks [15].

A key question is how to quantify the degree of scale invariance at each layer in the network. Taking inspiration from previous research in concept-based interpretability of CNNs [12, 20], we define the layer-wise quantification of scale-covariance as an interpretability task. Image scale is seen as a concept that is learned during training. This is analyzed with Regression Concept Vectors (RCVs) [12]. RCVs extend previous research on binary-expressed concept interpretability (where the concept is either present or not present) [5, 20] and were already used to analyze the effects of transfer in [13]. They are particularly suited for our task since they allow us to measure scale with continuous values obtained from the bounding box annotations in the publicly available PASCAL-VOC dataset. The degree of invariance at each layer is evaluated as a regression task of the scale measures. Besides, the layer-wise quantification of scale covariance is used to implement a pruning strategy that preserves the scale-covariance of the features. Differently from the scale-covariant designs that explicitly model the requirements of specific applications [4, 10, 19, 23, 30, 33], this pruning can be applied to state-of-the-art CNNs. In this way, ImageNet pretrained weights[1] can be used without the need of retraining from scratch. Being based on the interpretability analysis, the pruning has an interpretation that promotes its algorithmic transparency. It removes, in fact, the layers that introduce scale-invariance to the features. The experimental results on Estrogen Receptor-positive Breast Cancer (ERBCA+) images show a marked benefit in the magnification regression of open-access histopathology images [24]. This can help predicting the magnification range of images where the physical dimension of voxels is unknown, e.g.

---

[1] Downloadable at https://keras.io/api/applications/.

the large open-access biomedical data repository PubMed Central[2], to extend existing medical datasets.

2   Methods

2.1   Notations

We consider an input image $X \in \mathbb{R}^{w \times h}$, where w is the image width and h is the height. The function $\phi(\cdot)$, defined as $\phi : \mathbb{R}^{h \times w} \to \mathbb{R}^d$ maps the input image to a vector of arbitrary dimension d. For instance, it transforms X into a collection of d scalars obtained from averaged feature maps at a given intermediate layer. At the final fully-connected layer, $\phi(\cdot)$ transforms X into a set of predictions. When analyzing scale information, we are interested in covariance[3], thus whether we can find a transformation $g' : \mathbb{R}^d \to \mathbb{R}^d$ that predicts the transformation $g : \mathbb{R}^{h \times w} \to \mathbb{R}^{h \times w}$ of the input image X in the feature space obtained by $\phi(g(X))$. The scaling transformations are expressed as $g_\sigma(\cdot)$, being parameterized by a scale factor σ.

2.2   Representation of Scale Information

Our interest is in finding a linear transformation $g'_\sigma(\cdot)$ that is a predictable transformation of the scaling operation $g_\sigma(\cdot)$. To this end, a regression vector $\mathbf{v}$ can be searched in the feature space to predict the scaling factor σ as[4]:

$$\sigma = \sum_i v_i \phi_i(g_\sigma(X)) = \mathbf{v} \cdot \phi(g_\sigma(X)). \tag{1}$$

Therefore, $g'_\sigma(\cdot)$ can be represented as a translation matrix (in $\mathbb{R}^d$) by σ along $\mathbf{v}$, so that $g'_\sigma(\phi(X)) = \phi(X) + \mathbf{v} \cdot \sigma$.

2.3   Bounding-Box Size vs. Image Size

This section clarifies our definition and measurement of the image scale. Indications of scale are commonly used to relate the dimensions of two objects. In design modeling and cartography, the scale is the ratio comparing the length of the represented segment to the one in the real world (i.e. 1 cm:1000 km). Computer vision and image processing mostly refer to the act of scaling, namely the transformation that generates a new image with a larger or smaller number of pixels. If the input size is changed with the scaling, however, the transformation

---

[2] https://www.ncbi.nlm.nih.gov/pmc/tools/openftlist/.

[3] Following the same terminology, the equivariance, as opposed to covariance, implies that the function $\phi(\cdot)$ maps an input image to a point in the same domain, i.e. $\phi : \mathbb{R}^{h \times w} \to \mathbb{R}^{h \times w}$.

[4] For simplicity, we omit the intercept. In Eq. (1), the intercept would be v_0 with $\phi_0(g_\sigma(X)) = 1$.

causes the "train-test" resolution discrepancy in [29] during network inference. For this reason, it is recommendable to fix the input size to the default model input size $S_i = 299 \times 299$ when measuring scale information as shown in [22]. By focusing only on ImageNet-like images that only contain a single object, image scale can be pragmatically defined as the solid angle of the object in the image, namely the proportion of the field of view occupied by an object [34]. More directly, we measure the bounding-box area S_b occupied by the object in the image. The image has area $S_o = h_o \times w_o$, where h_o and w_o are respectively the original image width and height. A small bounding box corresponds to a smaller space in the field of view of the camera, and thus a smaller solid angle. Scale measures are thus defined as the ratio $r = \frac{S_b}{S_o} = \frac{h_b \times w_b}{h_o \times w_o}$, where h_b and w_b are the bounding box height and width. Figure 1 shows an example of scale measures on input images from the same class appearing at different scales.

2.4   Network Architectures and Tools

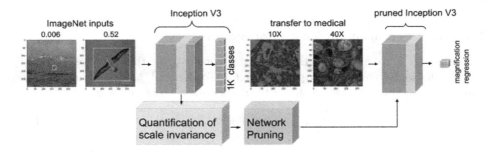

Fig. 1. Scale quantification and network pruning for better transfer in the medical domain. The bounding boxes for the ImageNet class *albatross* and the segmentation masks for the ERBCa+ inputs are overlaid in yellow on the images. The bounding box ratios r are reported on top of the inputs. Images are shown at magnifications 10X and 40X. The layer evidenced in yellow encodes the most of information about scale. The pruned network drops the layers after this for solving the medical task. Best seen on screen. (Color figure online)

ImageNet-Weights Initialization. InceptionV3 [27] and ResNet50 [15] are used for the analysis with pretrained ImageNet weights.

Regression of Scale. The regression of the scale of multiple objects of the same class that appear naturally at various scales is sought to approximate $g'_\sigma(\cdot)$ as in Eq. 1. This corresponds to computing the RCV representing "scale" [22]. The regression is sought at several layers in the network to compare different depths. Aggregation is performed on the feature maps in the form of Global Average Pooling (GAP) as in [11] to obtain the feature vector $\phi(X)$ (except for the prediction layer which is already pooled). The determination coefficient R^2 is used to evaluate the prediction of the scale ratio r on unseen test data of the same class[5]. This evaluation is informative about the scale-covariance of the features. The R^2 is a measure between zero and one when the regression is evaluated on the training data. The R^2, however, could take negative values when evaluated on the test data (test R^2). Differently from what one may think, this is not due to a bad choice in the evaluation technique but it rather shows that the prediction on the test samples is far worse than predicting their mean. To address this issue, a normalization of the test R^2 is performed by evaluating $\frac{e^{R^2}}{e}$. In this way, the performance of the RCV on test data is kept in a [0,1] range, with values below $\frac{1}{e}$ evidencing bad performance.

Pruning Strategy. Network pruning is performed by comparing the test R^2 to identify the layer where the scale covariance is the highest. This evaluation is averaged across different object categories to remove the dependence on the class of the inputs. The layer with the highest test R^2 (the yellow layer in Fig. 1) is where the scale covariance is the highest. Layers deeper than this one are pruned off the architecture and a Global Average Pooling operation (GAP) is added to obtain a vector of the aggregated features.

Transfer and Network Pruning. Transfer is performed from both the original and pruned architectures. To predict the average nuclei area, a single-unit dense layer is trained with the mean squared error loss between the true areas and the predicted ones. The nuclei area is expressed for each image as the average number of pixels within the segmentation of the nuclei. The regression is evaluated by the Mean Average Error (MAE). The magnification category (i.e. 5X, 8X, 10X, 15X, 20X, 30X, 40X) is also obtained from the average nuclei areas. The predicted areas are mapped to the magnification category that has the closest mean average value of the nuclei areas in the training set. This approach outperformed the direct classification of the magnification in [24]. Cohen's kappa coefficient is used to measure the inter-rater reliability of the magnification prediction. The networks are implemented in Keras and trained for five epochs with standard hyperparameters ($lr = 1e - 4$). The full pipeline is reported in Fig. 1 and the source code is available on github for reproducibility[6].

2.5   Datasets

The experiments in this paper involve two different datasets since the scale analysis is performed on inputs of natural images and the proposed final architecture is evaluated on a medical imaging task. For the scale quantification part, images with manual annotations of bounding boxes are selected from the publicly available PASCAL-VOC dataset [9]. We restrict our analysis to three object

---

[5] We compute $R^2 = \frac{\sum_{i=1}^{N}(\hat{r}_i - \bar{r})}{\sum_{i=1}^{N} r_i - \bar{r}}$, were N is the number of test data samples, $\hat{r}$ is the ratio predicted by the regression model, $\bar{r}$ is the mean of the true ratios $\{r_i\}_{i=1}^{N}$.

[6] https://bit.ly/2N6teMA.

categories and to images containing a single bounding box, chosen among the available annotated classes. These are *albatross* (ID: n02058221, 441 images), *kite* (ID: n01608432, 406 images) and *racing car* (ID: n04037443, 365 images).

The data for the histopathology application consist of 141 whole slide images of $2K \times 2K$ pixels taken at a maximum magnification of 40x ERBCa+ images. In these, 12,000 nuclei boundaries were manually annotated [18]. Patch sampling was performed at 5, 8, 10, 15, 20, 30, and 40x magnification. A total of 69,019 patches with nuclei segmentation masks were split into training, validation and test partitions (approximately 60%, 20%, 20% respectively) as shown in Table 1. The imbalance in the different magnification categories is due to the area covered by each magnification level, with the least number of patches being extracted at 5x and 8x. The average nuclei area is extracted for each input image by computing the average number of pixels in the relative nuclei segmentation mask. Example images with overlaid segmentation masks are displayed in Fig. 1.

Table 1. Number of ERBCa+ patches extracted per magnification and partition.

Split/# patches	5X	8X	10X	15X	20X	30X	40X	Total
Train	94	2,174	4,141	7,293	9,002	10,736	11,638	45,078
Validation	8	588	1,197	2,132	2,604	3,504	3,150	12,733
Test	36	428	900	1,728	2,198	2,802	3,166	11,208
Total	138	3,190	6,238	11,153	13,804	16,592	17,904	69,019

3   Experiments and Results

3.1   Layer-Wise Quantification of Scale Invariance

The layerwise analysis of scale representation in InceptionV3 and ResNet50 is shown in Fig. 2[7]. The object categories *racing-car*, *albatross* and *kite* are used for the analysis. For each class, 70% of the available images are used for learning the regression and the rest for the evaluation of the R^2. The evaluation was performed for ten splits of images. To remove the dependency of the evaluation on the image selection (by multiple split) and category (by analyzing multiple classes) we average the 10 repetitions for all classes (a total of 30 evaluations). The regression of scale in a randomly initialized network (orange line) is compared to a pretrained model (blue line) in Fig. 2. An additional baseline (green line) shows the performance of regressing random scale measures, i.e. the scale ratios were shuffled to break the true image-label correspondence. Values of R^2 close to one reflect the linear covariance of the intermediate layers to object scale as defined in Sect. 2.1. Individual results for each class were discussed in [22], while the generalization on different test classes is further analyzed in [2].

---

[7] Layer names refer to the Keras implementation names.

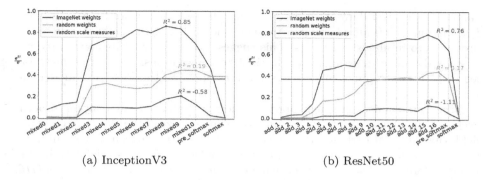

(a) InceptionV3   (b) ResNet50

Fig. 2. Regression of scale measures on test data (performance of the RCV) at different layers. (Color figure online)

3.2   Improvement of Transfer

The performance of the networks predicting average nuclei areas is compared between the original and pruned versions in Table 2. We report the MAE of ten repetitions[8] and the relative standard deviation. In the same table, we also report the kappa for the prediction of the magnification category.

Table 2. Mean Average Error (MAE) of the nuclei area regression (in pixels) and Cohen's kappa coefficient between the true and predicted magnification categories. Results are averaged over ten repetitions, the standard deviation is reported in brackets.

Model	Layer	MAE (std)	Kappa (std)
InceptionV3	mixed10	81.85 (11.08)	0.435 (0.02)
Pruned InceptionV3	mixed8	**54.93** (4.32)	**0.571** (0.05)
ResNet50	add16	70.08 (12.49)	0.610 (0.03)
Pruned ResNet50	add15	**54.76** (3.10)	**0.623** (0.04)

4   Discussion

The experiments were designed for analyzing the presence of scale-informative features in state-of-the-art CNNs pretrained on ImageNet. Our results in Sect. 3.1, particularly highlight the linear covariance of medium-deep layers, with invariance being learned before the classification layer. The scale of unseen objects (test data) is regressed with the highest determination coefficient $R^2 = 0.85$ in InceptionV3 (blue line in Fig. 2a), independently from the object class being tested. In comparison, the RCV learned from random scale ratios

---

[8] Different seeds were used to initialize the dense connections to the last dense layer.

cannot predict test data. This sanity check provides a lower bound $R^2 = -0.58$. For randomly initialized weights (orange line in Fig. 2a), the performance is around zero in almost all layers. This suggests that architectures with random weights do not contain linearly covariant representations, that hence must be learned during training. Similar observations apply to the ResNet50 models in Fig. 2b.

One important comment is about the low R^2 values at early layers of both architectures. We attribute this result to the limited size of the effective receptive field, that being at early layers only contains information from a very small fraction of the image. This affects the prediction of the scale ratios. The drop in the layers before the class prediction, namely in *mixed10* (for InceptionV3), *add16* (for ResNet50) and in *pre_softmax* (for both architectures), shows that deep network features learn scale invariance to classify image categories. Invariance to scale is thus achieved gradually in these layers preceding the last layer.

The quantification of scale invariance is applied to the image magnification regression as described in Sect. 3.2. The pruning strategy drops the layers with scale-invariant features. For InceptionV3, the pruned features are a result of a GAP on top of the *mixed8* features. As shown in Table 2, the MAE = 54.93 of the nuclei area regression in *mixed8* is markedly lower than the MAE = 81.85 in *mixed10*. This corresponds to a better prediction of the magnification range, hence to a higher kappa coefficient.

5   Conclusions and Future Work

This paper proposed the analysis of scale covariance in state-of-the-art CNNs pretrained on ImageNet and a pruning strategy to mantain such covariance for better transfer. Feature extraction and finetuning are very diffused techniques, and the pruned features can lead to improved performances on imaging tasks where scale carries crucial information, as for example the medical task of nuclei area regression and scale magnification prediction shown in our application. This work shows, in addition, that research in deep learning interpretability can be actively used to improve model development. Other transformations could also be analyzed, e.g. rotation, to improve the feature extraction process without the need for explicit equivariant designs. Such analysis could be relevant not only in other medical imaging tasks, but also in remote sensing, defect detection, material analysis and biometrics.

A limitation of this work is that the regression only captures linear correlations in the data, whereas nonlinear relationships could be necessary to model other transformations. In future work, we will investigate non-linear regression and manifold learning of the feature space to formally address this point.

Acknowledgements. This work was partially possible thanks to the project PRO-CESS, part of the European Union's Horizon 2020 research and innovation program (grant agreement No 777533). This work was also supported by the Swiss National Science Foundation (grant 205320_179069).

References

1. Andrearczyk, V., Fageot, J., Oreiller, V., Montet, X., Depeursinge, A.: Exploring local rotation invariance in 3D CNNs with steerable filters. In: International Conference on Medical Imaging with Deep Learning (2019)
2. Andrearczyk, V., Graziani, M., Müller, H., Depeursinge, A.: Consistency of scale equivariance in internal representations of CNNs. In: Irish Machine Vision and Image Processing (2020)
3. Bejnordi, B.E., et al.: Context-aware stacked convolutional neural networks for classification of breast carcinomas in whole-slide histopathology images. J. Med. Imaging **4**(4), 044504 (2017)
4. Bruna, J., Mallat, S.: Invariant scattering convolution networks. IEEE Trans. Pattern Anal. Mach. Intell. **35**(8), 1872–1886 (2013)
5. Cai, C.J., et al.: Human-centered tools for coping with imperfect algorithms during medical decision-making. In: Proceedings of the 2019 CHI Conference on Human Factors in Computing Systems, pp. 1–14 (2019)
6. Depeursinge, A., Foncubierta-Rodriguez, A., Van De Ville, D., Müller, H.: Three-dimensional solid texture analysis in biomedical imaging: review and opportunities. Med. Image Anal. **18**(1), 176–196 (2014)
7. Depeursinge, A.: Multi-scale and multi-directional biomedical texture analysis: finding the needle in the haystack. In: Biomedical Texture Analysis: Fundamentals. Applications and Tools, Elsevier-MICCAI Society Book Series, pp. 29–53. Elsevier (2017)
8. Elston, C.W., Ellis, I.O.: Pathological prognostic factors in breast cancer. I. The value of histological grade in breast cancer: experience from a large study with long-term follow-up. Histopathology **19**(5), 403–410 (1991)
9. Everingham, M., Van Gool, L., Williams, C.K., Winn, J., Zisserman, A.: The pascal visual object classes (VOC) challenge. Int. J. Comput. Vision **88**(2), 303–338 (2010)
10. Ghosh, R., Gupta, A.K.: Scale steerable filters for locally scale-invariant convolutional neural networks. In: Workshop on Theoretical Physics for Deep Learning at the International Conference on Machine Learning (2019)
11. Graziani, M., Andrearczyk, V., Marchand-Maillet, S., Müller, H.: Concept attribution: explaining CNN decisions to physicians. Comput. Biol. Med. **123**, 103865 (2020)
12. Graziani, M., Andrearczyk, V., Müller, H.: Regression concept vectors for bidirectional explanations in histopathology. In: Stoyanov, D., et al. (eds.) MLCN/DLF/IMIMIC-2018. LNCS, vol. 11038, pp. 124–132. Springer, Cham (2018). https://doi.org/10.1007/978-3-030-02628-8_14
13. Graziani, M., Andrearczyk, V., Müller, H.: Visualizing and interpreting feature reuse of pretrained CNNs for histopathology. In: Irish Machine Vision and Image Processing (IMVIP) (2019)
14. Graziani, M., Müller, H., Andrearczyk, V.: Interpreting intentionally flawed models with linear probes. In: SDL-CV Workshop at the IEEE International Conference on Computer Vision (2019)
15. He, K., Zhang, X., Ren, S., Sun, J.: Deep residual learning for image recognition. In: Proceedings of the IEEE Conference on Computer Vision and Pattern Recognition, pp. 770–778 (2016)
16. Hu, Z., Tang, J., Wang, Z., Zhang, K., Zhang, L., Sun, Q.: Deep learning for image-based cancer detection and diagnosis-a survey. Pattern Recogn. **83**, 134–149 (2018)

17. Huh, M., Agrawal, P., Efros, A.A.: What makes ImageNet good for transfer learning? In: Workshop on Large Scale Computer Vision Systems at NeurIPS 2016 (2016)
18. Janowczyk, A., Madabhushi, A.: Deep learning for digital pathology image analysis: a comprehensive tutorial with selected use cases. J. Pathol. Inform. **7** (2016)
19. Kanazawa, A., Sharma, A., Jacobs, D.W.: Locally scale-invariant convolutional neural networks. In: Advances in Neural Information Processing Systems (2014)
20. Kim, B., et al.: Interpretability beyond feature attribution: quantitative testing with concept activation vectors (TCAV). In: International Conference on Machine Learning, pp. 2673–2682 (2018)
21. Litjens, G., et al.: A survey on deep learning in medical image analysis. Med. Image Anal. **42**, 60–88 (2017)
22. Lompech, T., Graziani, M., Otálora, S., Depeursinge, A., Andrearczyk, V.: On the scale invariance in state of the art CNNs trained on ImageNet (2020, submitted)
23. Marcos, D., Kellenberger, B., Lobry, S., Tuia, D.: Scale equivariance in CNNs with vector fields. In: FAIM workshop at the International Conference on Machine Learning (2018)
24. Otálora, S., Atzori, M., Andrearczyk, V., Müller, H.: Image magnification regression using DenseNet for exploiting histopathology open access content. In: Stoyanov, D., et al. (eds.) OMIA/COMPAY-2018. LNCS, vol. 11039, pp. 148–155. Springer, Cham (2018). https://doi.org/10.1007/978-3-030-00949-6_18
25. Raghu, M., Zhang, C., Kleinberg, J., Bengio, S.: Transfusion: understanding transfer learning with applications to medical imaging. arXiv preprint arXiv:1902.07208 (2019)
26. Shin, H.C., et al.: Deep convolutional neural networks for computer-aided detection: CNN architectures, dataset characteristics and transfer learning. IEEE Trans. Med. Imaging **35**(5), 1285–1298 (2016)
27. Szegedy, C., Vanhoucke, V., Ioffe, S., Shlens, J., Wojna, Z.: Rethinking the inception architecture for computer vision. In: Proceedings of the IEEE Conference on Computer Vision and Pattern Recognition, pp. 2818–2826 (2016)
28. Szeliski, R.: Computer Vision: Algorithms and Applications. Springer, London (2010). https://doi.org/10.1007/978-1-84882-935-0
29. Touvron, H., Vedaldi, A., Douze, M., Jégou, H.: Fixing the train-test resolution discrepancy. In: Advances in Neural Information Processing Systems (2019)
30. Van Noord, N., Postma, E.: Learning scale-variant and scale-invariant features for deep image classification. Pattern Recogn. **61**, 583–592 (2017)
31. Veeling, B.S., Linmans, J., Winkens, J., Cohen, T., Welling, M.: Rotation equivariant CNNs for digital pathology. In: Frangi, A.F., Schnabel, J.A., Davatzikos, C., Alberola-López, C., Fichtinger, G. (eds.) MICCAI 2018. LNCS, vol. 11071, pp. 210–218. Springer, Cham (2018). https://doi.org/10.1007/978-3-030-00934-2_24
32. Wan, T., Cao, J., Chen, J., Qin, Z.: Automated grading of breast cancer histopathology using cascaded ensemble with combination of multi-level image features. Neurocomputing **229**, 34–44 (2017)
33. Worrall, D.E., Welling, M.: Deep scale-spaces: equivariance over scale. arXiv preprint arXiv:1905.11697 (2019)
34. Yan, E., Huan, Y.: Do CNNs encode data augmentations? arxiv.org/2003.08773 (2020)
35. Yosinski, J., Clune, J., Bengio, Y., Lipson, H.: How transferable are features in deep neural networks? In: Advances in Neural Information Processing Systems, pp. 3320–3328 (2014)

Improving the Performance and Explainability of Mammogram Classifiers with Local Annotations

Lior Ness$^{(\boxtimes)}$, Ella Barkan, and Michal Ozery-Flato

IBM Research, Haifa University, Mount Carmel, 31905 Haifa, Israel
`lior.ness@ibm.com`

Abstract. Cancer prediction models, which deeply impact human lives, must provide explanations for their predictions. We study a simple extension of a cancer mammogram classifier, trained with image-level annotations, to facilitate the built-in generation of prediction explanations. This extension also enables the classifier to learn from local annotations of malignant findings, if such are available. We tested this extended classifier for different percentages of local annotations in the training data. We evaluated the generated explanations by their level of agreement with (i) local annotations of malignant findings, and (ii) perturbation-based explanations, produced by the LIME method, which estimates the effect of each image segment on the classification score. Our results demonstrate an improvement in classification performance and explainability when local annotations are added to the training data. We observe that training with only 20–40% of the local annotations is sufficient to achieve improved performance and explainability comparable to a classifier trained with the entire set of local annotations.

Keywords: Image classification · Object detection · Explainable Artificial Intelligence (XAI) · Deep convolutional neural networks · Breast cancer · Mammography

1 Introduction

In recent years, artificial intelligence (AI) systems for medical imaging interpretation have shown tremendous potential to offer a solution to the shortage in the radiologist workforce. In the context of breast cancer screening, AI systems demonstrated accuracies comparable to those of human radiologists [1–5]. To gain trust, these AI systems must provide explanations for the predicted cancers. An expected explanation would be a visualization that highlights possibly malignant findings. This kind of visualization can be provided by object detection models, trained with *local annotations* that specify the location and extent of malignant findings within mammograms. However, local annotations are costly, since their generation requires the manual work of a human expert (e.g., breast imaging specialist), and they are often noisy and subjective. On the other hand, *global annotations* that indicate which mammograms have biopsy-proven

© Springer Nature Switzerland AG 2020
J. Cardoso et al. (Eds.): iMIMIC 2020/MIL3ID 2020/LABELS 2020, LNCS 12446, pp. 33–42, 2020.
https://doi.org/10.1007/978-3-030-61166-8_4

cancers, are reliable and can readily be extracted from electronic radiology and pathology reports.

In this work, we explore a simple yet effective extension to an image classifier architecture: adding a segmentation branch that shares the same weights as the classification branch. This extended architecture builds on previous studies of weakly-supervised localization, in which object detectors are trained using only global labels [6, 7]. Our extended classifier allows us to quickly generate heatmap explanations and train the AI model with local annotations of malignant findings, if such exist. When no local annotations are available, the modified classifier is identical to the original classifier in terms of output classification scores. Multi-task architectures, with classification and segmentation/ detection branches, have been applied in the past to jointly optimize cancer classification of mammograms and localization of malignant findings [8, 9]. However, these architectures do not share weights between the two task branches, and therefore cannot be trained when local annotations are unavailable. In a recent study, Bakalo et al. [10] presented an architecture for cancer classification and localization that can be trained in a weakly- or semi-supervised setting, similar to our proposed architecture. However, the architecture in [10], which is based on the weakly-supervised dual-stream network of Bilen and Vedaldi [11], is much more complex than the one presented here.

The extensive research on explainable AI has yielded numerous methods for highlighting regions in an image that are relevant to its predicted class [12]. In particular, these methods can be used to generate explanations for mammography classifiers. Basic sanity checks of such methods include sensitivity to data and model perturbations [13, 14]. Analyzing the effect of input image perturbations on the output score is a common model-agnostic approach for identifying influential regions in the image. This approach has been used by the popular LIME [15] and SHAP [16] methods to generate a heatmap explanation per image. Unfortunately, applying these methods can be very time consuming, since they involve running the model on thousands of image perturbations to create a single explanation.

We analyze the ability of the extended classifier to detect cancerous mammograms and provide explanations for the predicted cancers. We tested this classifier using different percentages of local annotations in the training data, ranging from no local annotations to fully annotated data. We evaluated the generated heatmap explanations in two different manners. First, we tested the accuracy of these maps with respect to ground-truth local annotations. Second, we measured their agreement with explanations generated by image perturbations using the LIME method. In our case study, using 20% to 40% of the local annotations was sufficient to boost both the classification performance and the explainability of the predictions.

2   Methods

2.1   The Extended Classifier

Our baseline image classifier is a convoluted neural network (CNN) that uses a global pooling operator such as global-max-pooling (GMP) or global-average-pooling (GAP). Most image classifiers fit the description of our baseline image classifier, and are based on common architectures such as ResNet [17], Inception [18], and VGG [19]. We refer to the network part before the global pooling as the *backbone network*; and to the network part following the global pooling as the *classification head*. We add a new branch to the backbone network in which the classification head is applied to each position of the feature map before the global pooling operator. As a result, the extended classifier produces two outputs: (i) the original (global) classification score, and (ii) a score-map whose cells localize to regions in the input image and has the dimensions of the final feature-map before the pooling operator (See Fig. 1).

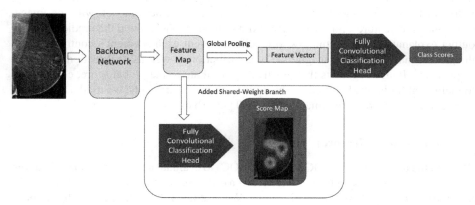

Fig. 1. Architecture of the extended image classifier. The detection branch shares the weights of the classification branch

In this study we used an InceptionResNet-V2 architecture [18] as the backbone network due to its suitable trade-off between accuracy and memory requirements. Nevertheless, in order to fit high-resolution mammograms on a GPU memory, we truncated the backbone network after 14 (out of 42) Inception blocks. The output feature map of our backbone network had 384 channels and spatial dimensions that are roughly 1/16 of the input image dimensions. The classification head consisted of a single hidden layer of size 256, as the number of channels in the feature-map before the pooling operator.

During training, whenever local annotations were available for an input image, we created a binary ground truth map, with the same dimensions as the output score-map, that depicts the annotated findings. We then compared each position in the score-map to its corresponding position in the ground truth map and computed the average cross-entropy loss. Expressed formally, the training loss of the extended model is:

$$L\left(y^{\text{global}}, \hat{y}^{\text{global}}, y^{\text{local - map}}, \hat{y}^{\text{local - map}}\right)$$
$$= L_{CE}\left(y^{\text{global}}, \hat{y}^{\text{global}}\right) + L_{CE}\left(y^{\text{local - map}}, \hat{y}^{\text{local - map}}\right)$$

where y^{global} is the global label; $\hat{y}^{\text{global}}$ is the predicted cancer score; $\hat{y}^{\text{local - map}}$ is the output score-map; $y^{\text{local - map}}$ is a binary ground truth map of the same dimension as $\hat{y}^{\text{local - map}}$, indicating positions localized within annotated findings; L_{CE} is the average cross-entropy loss. When $y^{\text{global}} = 0$ or when $y^{\text{local - map}}$ is unavailable, the second term in the loss, i.e., the one involving $y^{\text{local - map}}$, is ignored. Sharing the weights between the two branches allows the extended model to output score-maps even when no local annotations were used for training.

2.2 Perturbation-Based Explanations

We applied the LIME [15] method with its Python package implementation[1] to generate perturbation-based explanations for the predicted global scores. LIME operates by running an input image through a black-box model many times, each time randomly hiding different image segments. Then it learns a sparse linear model, with the image segments as its features, for locally approximating the model around the prediction. The output explanation is a heatmap whose cells represent each segment's contribution to the global image score: the magnitude of contribution as well as its direction (negative or positive). Image segments that positively contribute to the global score, correspond to malignant findings identified by the model. In our experiments, we used a grid of 10 × 5 segments, and ran each image with 1000 perturbations.

2.3 Performance Measures

We used the area under the ROC curve (AUROC) to evaluate and compare classification performance. We assessed the significance of a difference between two paired AUROCs using 10,000 paired bootstrap replications. We evaluated the agreement with the local annotations by applying the following procedure for K = 1, 2, 3, where K is the number of predicted regions per score-map. First, we considered only positions of the top 1% scores in each score-map and zeroed out the rest. Then, we identified the regions corresponding to non-zero connected components, ranked them by the maximal score within each region, and selected the K top-scored regions as our predicted regions (or fewer if the number of connected components was less than K). We scaled the predicted region scores for each image such that the maximal region score was equal to the global score from the classification branch. Similar to Ribli et al. [20], we considered a predicted region as a true positive if its center of mass was contained within a local annotation. Finally, we calculated the average precision (AP), which is the area under the precision-recall curve. To evaluate the agreement with LIME explanations, we first resized LIME maps to the same dimensions as the model's output score-maps. Then, we defined a *hotspot-match* when the maximal-score cells in the predicted score-map coincided with the maximal-score cell, or its adjacent cells, in the LIME map.

3 Experiments and Results

3.1 Datasets

For our experiments, we used two in-house mammogram datasets: "Data A" and "Data B". Data A was gathered from four medical centers and Data B was acquired from a separate single medical center. We randomly split Data A into patient-disjoint train and test datasets, Train A and Test A, in an 80:20 ratio. Data B was used only for testing.

---

[1] https://github.com/marcotcr/lime.

Statistics on the number of images per dataset are given in Table 1. Both Data A and Data B were globally annotated. Positive images had a confirmed malignant biopsy during the year following the imaging date. Negative images had a two-year follow-up with a subsequent normal screening mammogram. Most of the negative images were identified by radiologists as normal, while the remaining contained suspicious findings that were found to be benign on a subsequent biopsy. All positive images in Data A were locally annotated with ground truth contours marked around malignant lesions. Since our goal was to detect malignant findings, we ignored the local annotations of negative images and benign findings, if such existed, during the training process. Data B had no local annotations. All datasets contained a mixture of both craniocaudal (CC) and mediolateral-oblique (MLO) viewpoints.

Table 1. Number of images in train and test datasets

	Negative		Positive	Total
	Normal	Benign	Malignant	
Train A	3555 (45%)	1704 (21%)	2684 (34%)	7943 (100%)
Test A	921 (47%)	430 (22%)	621 (31%)	1972 (100%)
Data B	1560 (78%)	262 (13%)	171 (9%)	1993 (100%)

3.2 Implementation Details

We trained all of our models for 100 epochs using NVIDIA Tesla V100 GPUs. Each minibatch consisted of 2 mammogram images, 1 positive and 1 negative, with random oversampling of the smaller positive class. The images were cropped around the breast area and resized to a resolution of 2200×1200 while maintaining original aspect ratio. We added a zero-padding of 60 pixels at each dimension, for a final resolution of 2260×1260. We also applied random augmentations during training, both affine transformations (translation, rotation, flipping and zooming) and color perturbations (contrast, gamma and intensity value multiplication/addition). We used an Adam optimizer with an initial learning rate of 10^{-4}, which was divided by 10 whenever the training loss did not improve for 5 consecutive epochs.

3.3 Classification Performance

We compared the classification performance of our architecture to a baseline architecture that does not share weights, as in [8, 9]. We evaluated each of the two architectures with global-max-pooling (GMP) and global-average-pooling (GAP). Since the non-shared weights architecture requires local annotations in the training process, all models were trained with the entire set of local annotations. As shown in Table 2, the shared-weights model with GMP consistently demonstrated higher AUROC on the two test datasets.

Table 2. Classification performance (AUROC) for the proposed architecture, which shares weights in the classification and segmentation branches, and the baseline architecture, which does not share these weights. The two architectures are tested with global max pooling (GMP) and global average pooling (GAP). All models were trained with complete local annotations

Model	Non-shared, GAP	Shared, GAP	Non-shared, GMP	Shared, GMP
AUROC Test A	0.85	0.84	0.86	**0.87**
AUROC Data B	0.81	0.82	0.83	**0.84**

To examine the contribution of local annotations to classification performance, we trained five models with all global labels and varying percentages of local annotations. As shown in Table 3, the classification performance improved with the addition of local annotations. The model trained with a complete set of local annotations showed a significantly higher AUROC than the model trained without local annotations. Nevertheless, the largest increase was obtained with the addition of 20 to 40% of the local annotations. The models trained with only 40% to 60% of the local annotations were on a par with the model trained with the complete set of local annotations.

Table 3. Classification performance when training with varied percentages of local annotations. P-values for the decrease in AUROC when not training with complete local annotations: *($p < 0.01$); **($p < 10^{-4}$)

% Local annotations	0%	20%	40%	60%	100%
AUROC Test A	0.83**	0.86	0.86	0.86	**0.87**
AUROC Data B	0.79*	0.8*	0.83	0.83	**0.84**

3.4 Agreement with Local Annotations

We tested the agreement between the ground truth local annotations and the score-maps of the five models trained with various percentages of local annotations. The agreement was measured by the average precision (AP), using the K top-scored regions extracted from score-maps, with K = 1, 2, 3. (See Sect. 2.2 for details.) The results are shown in Table 4. For all values of K, the AP of the model we trained with 100% local annotations almost doubled itself compared to the model trained without local annotations. The biggest increase in the AP was achieved for the model trained with 40% of the local annotations. After adding 60% of the local annotations, the accuracy of detections seemed to be saturated.

3.5 Agreement with Perturbation-Based Explanations

Due to the long running time required for generating perturbation-based explanations, we analyzed all positive-labeled images but only a random subset of the negative-labeled

Table 4. Detection performance (average precision, AP) in Test A for different percentages of locally annotated images

% Local annotations	0%	20%	40%	60%	100%
AP, K = 1	22.5	24.0	35.1	**38.3**	36.9
AP, K = 2	24.2	26.7	38.7	**42.6**	40.7
AP, K = 3	24.4	26.9	39.6	**43.1**	41.8

images. Overall, we analyzed: 621 positive, 300 benign, and 400 normal images from Test A; and 171 positive, 83 benign, and 110 normal images from Data B. Figure 2 presents the rate of hotspot-matches between LIME explanations and predicted score-maps for different ranges of the global score. (See Sect. 2.3 for definition of a hotspot-match). It is noteworthy that the agreement rate between LIME explanations and the output local score-maps increases for larger predicted global-scores. Specifically, for the top 10% of the highest global-score images, the rate of hotspot matches was above 0.9 for all models. In both Test A and Data B, the models trained with 100% and 40% of the local annotations exhibited a similar rate of agreement with LIME.

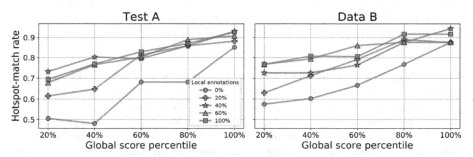

Fig. 2. Agreement with perturbation-based (LIME) explanations. Agreement increases with the global score.

Figure 3 presents the output score-maps of the five models for two images from Test A: a cancer image and a benign (negative) image. It appears that with the addition of more local annotations, the output score-maps become more "focused" on a single region, with less background noise.

4 Discussion

This work explored a simple yet effective extension of a mammogram classifier, enabling it to generate heatmap explanations, as well as learn from local annotations, if any exist. The problem of generating these explanations can be viewed as a variant of multiple instance learning (MIL): each image is a "bag" of regions ("instances"); every region is represented by a cell in the image explanation heatmap; a positive image indicates at least

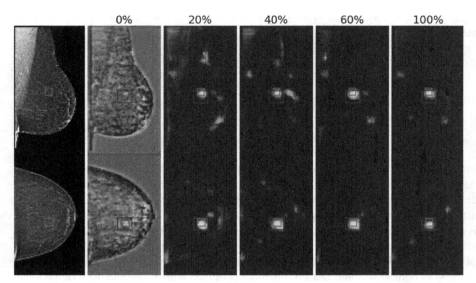

Fig. 3. An example of local-score maps for models trained with various percentage of local annotations. Local annotations are delineated with red boxes; blue boxes indicate the maximum-scored regions in the perturbation-based (LIME) explanations. The top row is a cancer positive image; the bottom row is a cancer negative image with a benign finding (a benign calcification). Benign findings were not used during the model's training. (Color figure online)

one positive region (i.e., a malignant finding); a negative image implies all its regions to be negative. Our extended classifier is motivated by the connection between MIL and the weakly supervised learning of object localization [6]. The pooling layer in the extended classifier is a complete analog of the pooling operator in MIL. The choice of global-max-pooling (GMP) rather than the more commonly used global-average-pooling (GAP) can also be justified by the observation that, unlike typical natural images where the object of interest occupies a significant part of the image, mammography lesions are much smaller and frequently occupy less than 5% of the breast area.

Previous weakly-supervised localization studies focused mostly on improving localization accuracy. Conversely, working on a case study of breast cancer screening, our primary evaluation measure is the accuracy of image classification. Furthermore, in our data, image-level labels are assumed to be accurate, while local annotations are likely to be noisy, inconsistent, and even biased. However, we demonstrated that our simple technique allows image classifiers to also learn from such local annotations and improve the accuracy of image classification. To evaluate the ability of the output score maps to explain the predicted global scores, we measured their agreement with the (noisy and subjective) local annotations, and additionally with perturbation-based, model-agnostic LIME explanations.

Finally, we showed a cost-effectiveness analysis that implied that annotating lesions in 20% to 40% of cancer-positive images may be sufficient for improving classification performance, and potentially the quality of generated explanations. Our model's inherent ability to produce score-map explanations for its global predictions eliminates the need for additional, time-consuming post-hoc analysis for generating such explanations. Compared to other complex architectures, occasionally customized for specific benchmarks, our method can be easily be adopted and tested to improve and explain medical imaging classifiers in other domains.

References

1. Wu, N., et al.: Deep neural networks improve radiologists' performance in breast cancer screening. IEEE Trans. Med. Imaging **39**, 1184–1194 (2019)
2. Rodriguez-Ruiz, A., et al.: Stand-alone artificial intelligence for breast cancer detection in mammography: comparison with 101 radiologists. JNCI J. Nat. Cancer Inst. **111**, 916–922 (2019)
3. Akselrod-Ballin, A., et al.: Predicting breast cancer by applying deep learning to linked health records and mammograms. Radiology **292**, 331–342 (2019)
4. Kim, H.-E., et al.: Changes in cancer detection and false-positive recall in mammography using artificial intelligence: a retrospective, multireader study. Lancet Digit. Health **2**, e138–e148 (2020)
5. McKinney, S.M., et al.: International evaluation of an AI system for breast cancer screening. Nature **577**, 89–94 (2020)
6. Oquab, M., Bottou, L., Laptev, I., Sivic, J.: Is object localization for free?-weakly-supervised learning with convolutional neural networks. Presented at the Proceedings of the IEEE Conference on Computer Vision and Pattern Recognition (2015)
7. Hwang, S., Kim, H.-E.: Self-transfer learning for weakly supervised lesion localization. In: Ourselin, S., Joskowicz, L., Sabuncu, M.R., Unal, G., Wells, W. (eds.) MICCAI 2016. LNCS, vol. 9901, pp. 239–246. Springer, Cham (2016). https://doi.org/10.1007/978-3-319-46723-8_28
8. Le, T.-L.-T., Thome, N., Bernard, S., Bismuth, V., Patoureaux, F.: Multitask classification and segmentation for cancer diagnosis in mammography. arXiv preprint arXiv:1909.05397 (2019)
9. Gao, F., Yoon, H., Wu, T., Chu, X.: A feature transfer enabled multi-task deep learning model on medical imaging. Expert Syst. Appl. **143**, 112957 (2020). https://doi.org/10.1016/j.eswa.2019.112957
10. Bakalo, R., Goldberger, J., Ben-Ari, R.: A dual branch deep neural network for classification and detection in mammograms. arXiv preprint arXiv:1904.12589 (2019)
11. Bilen, H., Vedaldi, A.: Weakly supervised deep detection networks. Presented at the Proceedings of the IEEE Conference on Computer Vision and Pattern Recognition (2016)
12. Guidotti, R., Monreale, A., Ruggieri, S., Turini, F., Giannotti, F., Pedreschi, D.: A survey of methods for explaining black box models. ACM Comput. Surv. **51**, 93:1–93:42 (2018). https://doi.org/10.1145/3236009
13. Samek, W., Binder, A., Montavon, G., Lapuschkin, S., Müller, K.-R.: Evaluating the visualization of what a deep neural network has learned. IEEE Trans. Neural Netw. Learn. Syst. **28**, 2660–2673 (2017). https://doi.org/10.1109/TNNLS.2016.2599820
14. Adebayo, J., Gilmer, J., Muelly, M., Goodfellow, I., Hardt, M., Kim, B.: Sanity checks for saliency maps. In: Bengio, S., Wallach, H., Larochelle, H., Grauman, K., Cesa-Bianchi, N., Garnett, R. (eds.) Advances in Neural Information Processing Systems, vol. 31, pp. 9505–9515. Curran Associates, Inc. (2018)

15. Ribeiro, M.T., Singh, S., Guestrin, C.: "Why should i trust you?": explaining the predictions of any classifier. In: Proceedings of the 22nd ACM SIGKDD International Conference on Knowledge Discovery and Data Mining, pp. 1135–1144. Association for Computing Machinery, New York (2016). https://doi.org/10.1145/2939672.2939778

16. Lundberg, S.M., Lee, S.-I.: A unified approach to interpreting model predictions. In: Guyon, I., et al. (eds.) Advances in Neural Information Processing Systems, vol. 30, pp. 4765–4774. Curran Associates, Inc. (2017)

17. He, K., Zhang, X., Ren, S., Sun, J.: Deep residual learning for image recognition. Presented at the Proceedings of the IEEE Conference on Computer Vision and Pattern Recognition (2016)

18. Szegedy, C., Ioffe, S., Vanhoucke, V., Alemi, A.A.: Inception-v4, Inception-ResNet and the impact of residual connections on learning. Presented at the Thirty-First AAAI Conference on Artificial Intelligence (2017)

19. Simonyan, K., Zisserman, A.: Very deep convolutional networks for large-scale image recognition. arXiv preprint arXiv:1409.1556 (2014)

20. Ribli, D., Horváth, A., Unger, Z., Pollner, P., Csabai, I.: Detecting and classifying lesions in mammograms with deep learning. Sci. Rep. **8**, 1–7 (2018)

Improving Interpretability for Computer-Aided Diagnosis Tools on Whole Slide Imaging with Multiple Instance Learning and Gradient-Based Explanations

Antoine Pirovano[1,2]([✉]), Hippolyte Heuberger[1], Sylvain Berlemont[1], Saïd Ladjal[2], and Isabelle Bloch[2]

[1] Keen Eye, Paris, France
{antoine.pirovano,hippolyte.heuberger,sylvain.berlemont}@keeneye.tech
[2] LTCI, Télécom Paris, Institut Polytechnique de Paris, Paris, France
said.ladjal@telecom-paristech.fr, isabelle.bloch@telecom-paris.fr

Abstract. Deep learning methods are widely used for medical applications to assist medical doctors in their daily routines. While performances reach expert's level, interpretability (highlight how and what a trained model learned and why it makes a specific decision) is the next important challenge that deep learning methods need to answer to be fully integrated in the medical field. In this paper, we address the question of interpretability in the context of whole slide images (WSI) classification. We formalize the design of WSI classification architectures and propose a piece-wise interpretability approach, relying on gradient-based methods, feature visualization and multiple instance learning context. We aim at explaining how the decision is made based on tile level scoring, how these tile scores are decided and which features are used and relevant for the task. After training two WSI classification architectures on Camelyon-16 WSI dataset, highlighting discriminative features learned, and validating our approach with pathologists, we propose a novel manner of computing interpretability slide-level heat-maps, based on the extracted features, that improves tile-level classification performances by more than 29% for tile level AUC.

Keywords: Histopathology · WSI classification · Explainability · Interpretability · Heat-maps

1 Introduction

Since their successful application for image classification [1] on ImageNet [2], deep learning methods (especially Convolutional Neural Network (CNN) deep architectures) have been extensively used and adapted to tackle efficiently a wide range of health issues [3,4].

© Springer Nature Switzerland AG 2020
J. Cardoso et al. (Eds.): iMIMIC 2020/MIL3ID 2020/LABELS 2020, LNCS 12446, pp. 43–53, 2020.
https://doi.org/10.1007/978-3-030-61166-8_5

Along with these new methods, the recent emergence of Whole Slide Imaging (WSI), microscopy slides digitized at a high resolution, represents a real opportunity for the development of efficient Computer-Aided Diagnosis (CAD) tools to assist pathologists in their work. Indeed, over the last three years, notably due to the WSI publicly available datasets, such as Camelyon-16 [5] and TCGA [6], and in spite of the very large size of these images (generally around 10 giga pixels per slide), deep learning architectures for WSI classification have been developed and proved to be really efficient.

In this work, we are interested in WSI classification architectures that use only the global label (e.g. diagnosis) to train and require no intermediate information such as cell labeling or tissue segmentation (which are time-consuming annotations). The training is regularized by introducing prior knowledge by design in the architectures which, in addition, makes the result interpretable. But the interpretability beyond the architectural design is still pretty shallow.

However, interpretability (capacity to provide explanations that are relevant and interpretable by experts in the field) for medical applications are critical in many ways. (i) For routine tools where useful features are well known and are subject to a consensus among experts, it is important to show that the same features are used by the trained model in order to gain confidence of practitioners. (ii) A good explainability would enable to get the most out of the architectural interpretability and thus assist more efficiently medical doctors in their slide reviews. (iii) The ability to train using only slide level supervision opens a new field we call discovery which consists in predicting, based on easier access (e.g. less intrusive) data, outputs that generally requires heavy processes or waiting such as surgery (e.g. prognosis, treatment response). In order to be able to guide experts towards new discoveries the need for reliable interpretability is obviously high.

In this work, after formalizing the architectural design of most WSI architectures, we propose a piece-wise interpretability approach, that provides cell-level features that prove to be highly relevant and interpretable by pathologists. We also propose a new way of computing explanation slide-level heat-maps based on cell-level identified features and measure their interpretability relevance.

2   Related Work and Motivations

All successful WSI classification architectures deal with these very large images by cutting them into tiles, which is close to the workflow of pathologists who generally analyze these slides at levels of magnification between 5X and 40X.

Recently, as explained in Sect. 1, architectures that are able to learn using only global slide-level labels have been proposed. They rely on a context of Multiple Instance Learning (MIL), i.e. slides are represented by bags of tiles with positive bags containing at least one positive tile and negative bags containing only negatives tiles. For example, CHOWDER [7] is an extension of WELDON [8] solution for WSI classification that uses min-max strategy to guide the training and make the decision. This approach reaches an AUC of 0.858 on Camelyon-16 and 0.915 on TCGA-lung (subset of TCGA dataset related to lung cancer).

In [9], an attention module [10] is used instead of a min-max layer. AUC of 0.775 for a breast cancer dataset and 0.968 for a colon cancer dataset were reported. Recently, more works on large datasets proposed architectures that follow the same design [11,12]. Heat-maps based on intermediate scores computed in these architectures are what we call architectural explainability that results from prior knowledge on WSI problems that is introduced by design in the architecture. They are of great interest and have proved to be really efficient to the point of being able to spot cancerous lesions that had been missed by experts (in [11]). However explanations are relying on a single "medical" score which might limit the interpretability regarding complex tissue structures that can be found on these slides.

While interpretability for deep learning CNN models is still at its beginning, some methods arise from the literature. "Feature Visualization" has been extensively developed in [13]. It consists of methods that aim at outputting visualizations to express in the most interpretable manner features associated with a single neuron or a group of neurons. It can be used to understand the general training of a model. For example, the question of transferring features learned from natural images (ImageNet) to medical images has only recently been investigated [14] while widely used and yet surprisingly good. It has also been used to measure how robust a learned feature is [15]. Other explainability methods are called attribution methods, i.e. methods that output values reflecting, for each input, its contribution to the prediction. They are performed either through perturbation [16] or gradient computation (i.e. measure of the gradient of the output with respect to the input). This second group of methods is gaining more and more interest. In [17], the authors show that gradient is a good approximation of the saliency of a model and even put forward a potential to perform weakly supervised localization. This work opened a new way of accessing explanations in deep neural networks and motivated a lot of interesting researches [18–20]. Mixed together these explanation methods can provide meaningful and complementary interpretability.

To the best of our knowledge, a lot of explainability is still to be introduced in WSI classification architectures. In the next section, we present our approach to improve interpretability of a model trained for WSI classification in histopathology. We rely on gradient-based methods to identify and attribute the importance of features in intermediate descriptors, and on patch visualization for cell-level feature explanations. We also extend feature explanation to a slide level, thus drastically improving tumor localization and medical insights.

3   Proposed Methods

As introduced in Sect. 2, WSI classification architectures have a common design that we formalize here. Let i be the slide index and j the tile index for each slide. There are four distinct blocks in a typical WSI classification architecture:

1. A feature extractor module f_e (typically a CNN architecture) that encodes each tile $x_{i,j}$ into a descriptor $d_{i,j} \in \mathbb{R}^N$ with N the descriptor size (depending on the feature extractor): $d_{i,j} = f_e(x_{i,j})$;

2. A tile scoring module f_s that, based on each tile descriptor $d_{i,j}$, assigns a single score per tile $s_{i,j} \in \mathbb{R}$: $s_{i,j} = f_s(d_{i,j})$;
3. An aggregation module f_a that, based on all tile scores $s_{i,j}$, and sometimes their tile descriptors $d_{i,j}$, computes a slide descriptor $D_i \in \mathbb{R}^M$ with M the slide descriptor size (depending on the aggregation module): $D_i = f_a(s_{i,j}, d_{i,j})$;
4. A decision module f_{cls} that, based on the slide descriptor D_i, makes a class prediction $P_i \in \mathbb{R}^C$ with C the number of classes: $P_i = f_{cls}(D_i)$.

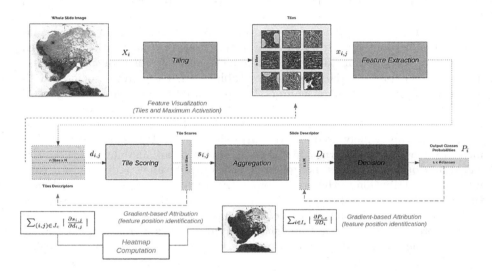

Fig. 1. Overview of the proposed method.

Our approach (illustrated in Fig. 1) consists in rewinding explanations from the decision module to tile information by applying interpretability methods and by answering successively the following three questions:

1. Which features of slide descriptors are relevant for a class prediction?
2. With regards to the aggregation module, which features of tile descriptors are responsible for previously identified relevant slide descriptor features?
3. Are these features of tile descriptors relevant medically and representative of histopathological information?

The first question is answered using attribution vectors $A_c \in \mathbb{R}^M$ (one for each class c) computed as the gradient of the component of index c of P_i (noted $P_{i,c}$) with respect to D_i. It enables us to identify a set of relevant positions $K_c = \{K_{c,1}, ..., K_{c,L}\}$ in slide descriptors, i.e. the L (empirically determined) positions in A_c with highest attributions over the slide predicted in class c:

$$A_c = \sum_{i \in I_c} | \frac{\partial P_{i,c}}{\partial D_i} | = \sum_{i \in I_c} | \frac{\partial f_{cls}(D_i)_c}{\partial D_i} |,$$

with I_c the set of slides predicted to be in class c.

Then, the second question is also answered using an attribution vector $a_c \in \mathbb{R}^N$ computed as the gradient of tile score $s_{i,j}$ with respect to tile descriptor $d_{i,j}$. This enables to identify features positions $k_c = \{k_{c,1}, ..., k_{c,l}\}$ in tile descriptors, i.e. the l (empirically determined) tile descriptors that are responsible for high activation at previously identified K_c positions in slide descriptor:

$$a_c = \sum_{(i,j) \in J_c} | \frac{\partial s_{i,j}}{\partial d_{i,j}} | = \sum_{(i,j) \in J_c} | \frac{\partial f_s(d_{i,j})}{\partial d_{i,j}} |$$

with J_c the set of tile positions (i,j) that most activate K_c positions in slide descriptors.

To answer the third question, we rely on feature activation to highlight features identified as being discriminative to the task by selecting tiles $x_{i,j}$ that have the highest activation per feature in k_c identified over the whole test set. Along with these tiles, we display a maximum activation $\mathcal{X}$ image obtained by iteratively tuning pixels values to activate the feature by gradient ascent as follows: for each k in k_c, $\mathcal{X}$ is initialized as a uniformly distributed image $\mathcal{X}_0$; then while $f_e(\mathcal{X}_{n-1})_k$ increases, iterate over $n > 0$:

$$\mathcal{X}_n(k) = \mathcal{X}_{n-1} + \frac{\partial f_e(\mathcal{X}_{n-1})_k}{\partial \mathcal{X}_{n-1}}.$$

Finally, we also propose a new way to compute heat-maps for each slide i. We note $H_{c,i}$ the map that highlights regions on slide i that explain what has been learned to describe class c based on the identified features. For each slide i and tile j, the heat-map value $H_{c,i,j}$ is computed as the average of activations $d_{i,j,k}$ (normalized per feature over all tiles of all slides) over identified features k in k_c for class c:

$$H_{c,i,j} = \frac{1}{|k_c|} \cdot \sum_{k \in k_c} \frac{d_{i,j,k} - \min_k}{\max_k - \min_k}$$

with $\max_k = \max_{i,j}(d_{i,j,k})$ and $\min_k = \min_{i,j}(d_{i,j,k})$.

This heat-map values (between 0 and 1) can be considered as a prediction scoring system, and thus we propose to compute the Area Under the ROC (Receiver Operating Characteristic) Curve to measure how relevant is the interpretability brought by our automatic feature extraction approach using ground truth lesion annotations when given. This localization AUC measures the separability between the class of interest (e.g. "tumor" and other classes using heat-maps, indeed for a good heat-map we expect tiles that are representative of the class of interest to have a high score and all other tiles to have a low score.

4   Experiments and Results

Architectures. We validate our approach on two WSI classification trained architectures: CHOWDER and Attention-based classification.

CHOWDER [7] uses a 1×1 convolution layer to turn each tile descriptor into a single tile score. These scores are then aggregated using a min-max layer, that keeps the top-R and bottom-R scores (e.g. empirically $R = 5$ gives the best results), to give a slide descriptor ($M = 2 \times R$).

Attention-based architecture [9] uses an attention module (two 1×1 convolution layers with respectively 128 and 1 channels and a softmax layer) to compute competitive and normalized (sum to 1) tile scores from tile descriptors. Then, the slide descriptor is computed as the weighted (by tile scores) sum of tile descriptors ($M = N$).

Note that in our experiments the feature extractor is a ResNet-50 [21] ($N = 2048$) trained on ImageNet and the decision module is a two layers fully connected network with 200 and 100 hidden neurons, respectively.

Datasets. We validate our approach using Camelyon-16 dataset that contains 345 WSI divided into 209 "normal" cases and 136 "tumor" cases. This dataset contains slides digitized at 40X magnification from which we extract, with regard to a non-overlapping grid, 224×224 pixels at 20X magnification and pre-compute 2048-tile descriptors for each tile (using the ResNet-50 model trained on ImageNet). 216 slides are used to train our models while 129 slides form the test set to evaluate performances of models.

Results on CHOWDER. Both architectures trained on Camelyon-16 show similar classification performances (AUC of 0.82 for the CHOWDER model and 0.83 for the Attention-based model). Let us now illustrate and detail the results of our approach on the CHOWDER model guided by the three questions raised in Sect. 3.

The first question is "Which slide descriptors features are relevant for a class prediction?" i.e. for CHOWDER given the $M = 10$ ($R = 5$) tile scores given as slide descriptor (the 5 minimum tile scores and the 5 maximum tile scores), what is the contribution of each of these values to the prediction?

The distribution of the (5-)min and (5-)max scores w.r.t. predictions over the whole 129 test slides shows that min scores are the ones that contribute to discriminate between the two classes (i.e. the lower min scores, the more the slide is predicted as being "tumor"). A Mann-Whitney U-Test between scores (min and max independently) distributions reveals that min scores distributions per predicted class are statistically different ($p < 10^{-3}$) while max scores are not ($p = 0.23$). The attribution of min and max distributions validates this assertion.

After explaining that min scores are the ones describing tumorous regions and thus that max scores are used for the "normal" class, we are interested in identifying which features of tile descriptors are mostly responsible for minimum and maximum scores, i.e. to describe each class. To address this second question, we use the same gradient-based explanation method.

Most minimal tile scores are under -5 and most maximal tile scores are above 11. For each of these groups of tiles, we compute the average attribution of each of the $N = 2048$ features in tile descriptors (extracted by a ResNet-50 trained on ImageNet). The distribution of features hence activated allows us to identify which features are mostly responsible for min and max tile scores, i.e. highest attribution for min and max scored tiles.

Thus we are able to claim that features (defined by their position in the descriptor) that are mostly useful for the trained model for each class are: 242, 420, 602, 1154, 1644, 1652 and 1866 for "tumor" class, and 565, 628, 647, 1158 and 1247 for "normal" class.

Interpretability. As exposed in the previous paragraph, based on explanations on decision blocks, we have been able to identify 7 and 5 features that are mostly used by the trained CHOWDER model to make decisions (and we did the same for the attention-based model). Now, we are interested in interpretable information to return to pathologists so that they can use their expertise to understand what these features put forward histopathologically speaking. We benefited from discussions with two experienced pathologists and report their overall feedback on the interpretable visualization we proposed.

Figure 2 shows the 7 tiles that activate the most (over all tiles) each feature and the max activation image, that we expect to reveal what the feature reveals with regards to the histopathological problem it has been trained on.

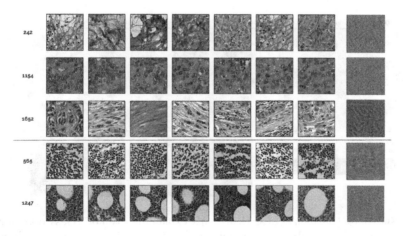

Fig. 2. Patch-based visualizations obtained for features 242, 1154 and 1652 (for min-scores features); 565 and 1247 (for max-scores features); tiles and max activation images (right).

Pathologists agreed that patch-based tiles visualizations are highly interpretable and reveal features that are indeed related to each class [22]. For example, feature 1652 tends to trigger spindle-shaped cells that indeed can be a metastasic tissue organization. For "normal" tissue features, feature 565 describes mainly clustered lymphocytes that are preponderant in normal tissues.

Coherence between patches exposed for a better interpretability led us to think about another way to present features to pathologists. Indeed, since tissues have coherent and somehow organized structure, a relevant feature for histological problems would be activated in a coherent and somehow organized way over slides. Thus, along with patch-based visualization, we propose to access features activation heat-maps $H_{c,i}$ over slides as presented in Sect. 3.

Figure 3 illustrates qualitative results. Quantitatively, we report a tile-level localization AUC of 0.884 for CHOWDER model and 0.739 for Attention-based model, using this average normalized activation as a "tumor" prediction score and using lesion annotation provided by Camelyon-16 dataset to get the ground-truth label per tile. Both AUCs are significantly high, which validates our approach of identifying features that are relevant and of computing heat-maps for interpretation and explanation. Note that the AUC computed using tile scores is 0.684 for CHOWDER model and 0.421 for Attention-based model (see Table 1). We can also note that there is a gap in interpretability between CHOWDER model and Attention-based model while classification performances are comparable. The gap can be explained by the fact that, in the context of Camelyon-16, identifying one tumorous tile is enough to label a slide as "tumor", so implicit tile classification does not need to be exhaustive to provide meaningful information to the slide level decision module, however if so interpretability will decrease.

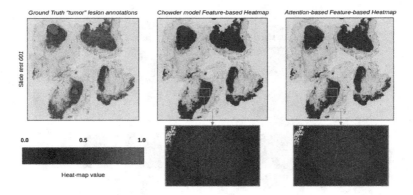

Fig. 3. Slide-based visualizations: Heat-maps explaining "tumor" class obtained by computing average normalized activation over identified features; ground-truth annotations for "tumor" tissue (left); CHOWDER model feature-based heat-maps (middle); attention-based model feature-based heat-maps (right).

Table 1. Results: classification and localization AUC using both methods (improvement of localization AUC by 29.2% for CHOWDER and 75.5% for Attention-based model).

Model	Classification AUC	Heat-map method	Localization AUC
CHOWDER	0.82	Tile scores	0.684
		Feature-based (ours)	**0.884**
Attention-based	**0.83**	Tile scores	0.421
		Feature-based (ours)	0.739

5 Conclusion

In this paper, we presented our interpretability approach and researches for WSI classification architectures. We proposed a unified design that gathers a large majority of WSI classification methods relying on MIL learning, and applied a gradient-based attribution method to identify features that have been learned to be relevant in intermediate (tile and slide) descriptors. Then we showed the relevance of these features by visualization, and validated it with the help of pathologists. We finally proposed explainability heat-maps over whole slides taking into account only identified features. This contribution considerably improved tile-level classification AUC. Allying patch-based and slide-based visualization took interpretability to a next level for pathologists to understand histological meanings of features used by trained models. More generally, this method could be used to extract meaningful and interpretable information for any medical application when using a MIL context.

Acknowledgment. We would like to gratefully thank Jacques Bosq and Yaelle Harrar for their important medical inputs and feedbacks during this work, along with our colleagues at Keen Eye for their work and support. This work was partially funded by ANRT, Grant N° 2017/1076.

References

1. Krizhevsky, A., Sutskever, I., Hinton, G.E.: ImageNet classification with deep convolutional neural networks. In: Advances in Neural Information Processing Systems, vol. 25, pp. 1097–1105 (2012)
2. Deng, J., Dong, W., Socher, R., Li, L.-J., Li, K., Fei-Fei, L.: ImageNet: a large-scale hierarchical image database. In: Conference on Computer Vision and Pattern Recognition (2009)
3. Pasa, F., Golkov, V., Cremers, D., Pfeiffer, F.: Efficient deep network architectures for fast chest x-ray tuberculosis screening and visualization. Nature Sci. Rep. **9**, 62–68 (2019)
4. Pratt, H., Coenen, F., Broadbent, D.M., Harding, S.P., Zheng, Y.: Convolutional neural networks for diabetic retinopathy. Procedia Comput. Sci. **90**, 200–205 (2016)

5. Ehteshami Bejnordi, B., et al.: Diagnostic assessment of deep learning algorithms for detection of lymph node metastases in women with breast cancer. J. Am. Med. Assoc. **312**(22), 2199–2210 (2017)
6. Tomczak, K., Czerwińska, P., Wiznerowicz, M.: The cancer genome atlas (TCGA): an immeasurable source of knowledge. Contemp. Oncol. (Pozn) **19**(1A), A68–A77 (2015). https://doi.org/10.5114/wo.2014.47136
7. Courtiol, P., Tramel, E.W., Sanselme, M., Wainrib, G.: Classification and disease localization in histopathology using only global labels: a weakly-supervised approach. Computing Research Repository (CoRR), Arxiv (2018). https://arxiv.org/abs/1802.02212
8. Durand, T., Thome, N., Cord, M.: WELDON: weakly supervised learning of deep convolutional neural networks. In: 29th IEEE Conference on Computer Vision and Pattern Recognition, June 2016
9. Ilse, M., Tomczak, J.M., Welling, M.: Attention-based deep multiple instance learning. In: Proceedings of the International Conference on Machine Learning (ICML) (2018)
10. Raffel, C., Ellis, D.P.W.: Feed-forward networks with attention can solve some long-term memory problems. Computing Research Repository (CoRR), Arxiv (2015). https://arxiv.org/abs/1512.08756
11. Campanella, G., Hanna, M.G., Geneslaw, L., Miraflor, A., Silva, V.W.K., Busam, K.J.: Clinical-grade computational pathology using weakly supervised deep learning on whole slide images. Nature Med. **25**, 1 (2019). https://doi.org/10.1038/s41591-019-0508-1
12. Li, J., Li, W., Gertych, A., Knudsen, B.S., Speier, W., Arnold, C.W.: An attention-based multi-resolution model for prostate whole slide image classification and localization. Computing Research Repository (CoRR), Arxiv (2019). https://arxiv.org/abs/1905.13208
13. Olah, C., et al.: The building blocks of interpretability. Distill **2**, e10 (2018)
14. Raghu, M., Zhang, C., Kleinberg, J., Bengio, S.: Transfusion: understanding transfer learning for medical imaging. In: Proceedings of the Neural Information Processing Systems, pp. 3342–3352 (2019)
15. Couteaux, V., Nempont, O., Pizaine, G., Bloch, I.: Towards interpretability of segmentation networks by analyzing DeepDreams. In: Suzuki, K., et al. (eds.) ML-CDS/IMIMIC -2019. LNCS, vol. 11797, pp. 56–63. Springer, Cham (2019). https://doi.org/10.1007/978-3-030-33850-3_7
16. Fong, R.C., Vedaldi, A.: Interpretable explanations of black boxes by meaningful perturbation. In: IEEE International Conference on Computer Vision (ICCV), pp. 3449–3457, October 2017. https://doi.org/10.1109/ICCV.2017.371
17. Simoyan, K., Vedaldi, A., Zissermn, A.: Deep inside convolutional networks: visualising image classification models and saliency maps. Computing Research Repository (CoRR), Arxiv, December 2013. https://arxiv.org/abs/1312.6034
18. Sundararajan, M., Taly, A., Yan, Q.: Axiomatic attribution for deep networks. In: 34th International Conference on Machine Learning, vol. 70, pp. 3319–3328, March 2017
19. Smilkov, D., Thorat, N., Kim, B., Viégas, F., Wattenberg, M.: SmoothGrad: removing noise by adding noise. Computing Research Repository (CoRR), Arxiv, June 2017. https://arxiv.org/abs/1706.03825
20. Goh, G.S.W., Lapuschkin, S., Weber, L., Samek, W., Binder, A.: Understanding integrated gradients with SmoothTaylor for deep neural network attribution. Computing Research Repository (CoRR), Arxiv, April 2020. https://arxiv.org/abs/2004.10484

21. He, K., Zhang, X., Ren, S., Sun, J.: Deep residual learning for image recognition. In: IEEE Conference on Computer Vision and Pattern Recognition (CVPR), pp. 770–778, June 2016. https://doi.org/10.1109/CVPR.2016.90
22. Hoon Tan, P., et al., The WHO Classification of Tumours Editorial Board: Breast Tumours, WHO Classification of tumours. International Collaboration on Cancer Reporting (ICCR), 5th edition, December 2019

Explainable Disease Classification via Weakly-Supervised Segmentation

Aniket Joshi[✉], Gaurav Mishra, and Jayanthi Sivaswamy

IIIT Hyderabad, Hyderabad 500032, India
aniket.joshi@research.iiit.ac.in

Abstract. Deep learning based approaches to Computer Aided Diagnosis (CAD) typically pose the problem as an image classification (Normal or Abnormal) problem. These systems achieve high to very high accuracy in specific disease detection for which they are trained but lack in terms of an explanation for the provided decision/classification result. The activation maps which correspond to decisions do not correlate well with regions of interest for specific diseases. This paper examines this problem and proposes an approach which mimics the clinical practice of looking for an evidence prior to diagnosis. A CAD model is learnt using a mixed set of information: class labels for the entire training set of images plus a rough localisation of suspect regions as an extra input for a smaller subset of training images for guiding the learning. The proposed approach is illustrated with detection of diabetic macular edema (DME) from OCT slices. Results of testing on a large public dataset show that with just a third of images with roughly segmented fluid filled regions, the classification accuracy is on par with state of the art methods while providing a good explanation in the form of anatomically accurate heatmap /region of interest. The proposed solution is then adapted to Breast Cancer detection from mammographic images. Good evaluation results on public datasets underscores the generalisability of the proposed solution.

Keywords: CAD · OCT · DME · Breast cancer

1 Introduction

Deep Learning (DL) based prediction systems which are black-boxes lack an explicit and declarative knowledge representation. Hence, despite their wide use for classification [1–3], such systems have difficulty in generating the underlying explanatory structures [4] which consequently impedes clinical adoption. Providing an evidence which might have led to the decision can mitigate this situation. In the case of diseases characterised by presence of lesions/abnormalities, assuming only image data is available, a natural option for this evidence is in the form of predicted regions of interest (ROI) which should be well aligned with locations of actual lesions/abnormalities (as annotated by experts).

© Springer Nature Switzerland AG 2020
J. Cardoso et al. (Eds.): iMIMIC 2020/MIL3ID 2020/LABELS 2020, LNCS 12446, pp. 54–62, 2020.
https://doi.org/10.1007/978-3-030-61166-8_6

A common attempt towards explanation has been to use the activations of the last layer of the network used for the classification (using well known architectures such as Inception-V3, Resnet, AlexNet etc.) to get a heatmap which mark those regions in the image which might have led the model to give a particular prediction [5,6]. Since the focus of such work is on classification, training is done with only labeled data and the reported evaluation is also of the classification accuracy and not of the heatmaps and their explanatory accuracy. Since these models were trained only on label information, there is no guarantee that the model will output a clinically accurate heatmap. An approach to get both class labels and accurate explanation would be to train the model using images which are annotated at both the image level and pixel/region/local level. This however poses a logistical challenge in the medical domain as labeled data are easier to extract from medical records and are available in abundance whereas region-level annotated data is not readily available to carry out a fully supervised segmentation.

In this paper, we propose a novel approach to address the above problem. We propose a neural network architecture which can do both: learn to classify an image and leverage limited annotations to give an accurate heatmap. A novel training regime is designed to enable flexibility in the model building to accommodate and use varying levels of information that may be available.

2   Method

We illustrate our proposed method using Diabetic macular edema image classification in OCT image.

2.1   Dataset(s)

OCT images (containing speckle noise) of retinal layers are used for DME detection. The fluid filled regions (FFR) in the retina can vary in size affecting the layer morphology. A publicly available set [7] was chosen for our experiments. It has 84,495 OCT slices assigned one of 4 classes (NORMAL, CNV, DME and DRUSEN). We used the DME and NORMAL classes for our experiments. These images had no localizations for the fluid filled region. For generating the localizations, we trained a UNET model [8] on the other 2 datasets - 2015_BOE_Chiu [9] and RETOUCH [10] having 71 and 935 OCT slices with labelled segmentation maps for fluid filled regions respectively. This trained UNET model was used to predict the rough segmentation maps for dataset from [7] which was used later in all our experiments. A total of 16440 OCT slices (B-scans) were used in our experiments out of which 9332 were NORMAL cases and 7118 were DME affected. This was divided into Train, Validation and Test sets in the ratio 60:25:15. The normal images had empty segmentation masks while it represented the fluid filled regions in case of DME images.

Each of the OCT slice was preprocessed to identify the retinal layers which occupy only a small part of the OCT image. This was done using simple steps: row sum to vectorise the image and thresholding to extract the layered part.

2.2 Model

The problem at hand is to build a model that is fully supervised for classification while also generating anatomically accurate heatmaps as an explanation for the class outputs using *limited* region-level annotations. The assumption of limited availability of annotations constrains the approach to heatmap generation to be weakly supervised. Our solution is to design a network, that produces class label as a main output and an auxiliary output in the form of a rough segmentation map for a given input. Figure 1 shows the proposed network architecture for achieving this task. It consists of encoder and decoder segments, where the former is used to produce the classification output while the latter is used to generate the rough segmentation maps. A novel training methodology described in the next section ensures that the derived heatmaps using CAM [11] are not only guided by labels but also by the rough localisations. Fewer filter layers (relative to a normal encoder) are employed at the end of the encoder for better localisation of the heatmap. After the last encoder layer ($6 \times 6 \times 16$ output), the output is flattened and a dense layer is added to obtain a class label as the output and heatmaps are derived using the method described in Sect. 2.4.

2.3 Training

Training of the model was done with images from the training set using P images with only classification labels and $Q < P$ with both classification label and segmentation map. Thus, $Q : P$ indicates the proportion of different type of images used in training. 1:3 indicates that one third of the total number of training images had both classification label and segmentation map (See Table 1).

Table 1. Public OCT data specifications [7]

Data type	Train		Validation		Test	
	Normal	DME	Normal	DME	Normal	DME
Images with segmentation	1866	1425	773	583	468	364
Images without segmentation	3732	2850	1547	1168	936	728

A special training regime is designed for the given task. The model has two branches (See Fig. 1 with two outputs, namely, the class label and a rough segmentation map). The left branch (encoder before the classification output) together with the dense layers is a unit (referred to as ED) whose output is the class label.

In the first phase, the ED unit is trained for a few (5) epochs following which the whole model is trained for some (10) epochs. At the end of this phase, the ED part of the network will learn the features aiding both classification and segmentation tasks. In the second phase, only the dense layers of ED unit are trained keeping all the other weights constant (for 20 epochs). This second phase forces the model to predict the class label using the features that were learned during the whole model training for segmentation and classification. This also serves to boost the classification accuracy of the model. In the third and final phase of training, the whole model is trained again so that the encoder part learns more of the features that will be used for segmentation (for 50 epochs). This phase of training is made to be the longest so that the model is forced to learn more of the segmentation features so as to get the best heatmaps using the regime described in Sect. 2.4.

In terms of loss functions, binary cross entropy loss is used while training only the ED unit of the model whereas while training the entire model, a weighted sum of dice coefficient loss and binary cross entropy loss is used. The weight is the hyper-parameter and is taken as 1 during the training phase. Higher weight for dice coefficient loss will give us better region of interests(heatmaps) but at the cost of lower classification accuracy.

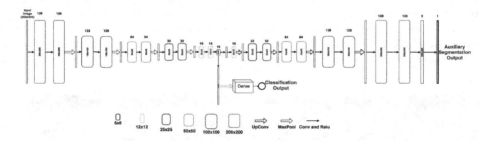

Fig. 1. Proposed network architecture with 2 outputs - Classification and Auxiliary Segmentation. Left Branch before Classification output is called ED branch.

2.4 Deriving the Heatmaps

The desired heatmaps (as explanation for the predicted class label) were derived using the following steps. The dense layers from the ED unit were removed and a global average pooling layer was added to the final encoder layer ($6 \times 6 \times 16$ output) to give a 1×16 feature map. The dense layer was attached finally to give the classification output. Only the 16 neurons of the dense layer were trained in this phase and all the other weights were kept constant from the last phase of the training. Heatmaps were obtained using the Class Activation Mapping (CAM) [11]. As the convolution layer weights were taken from the last phase of the training, it had the features which were learned for both segmentation and classification. In our case, the adapted CAM approach was weakly supervised, as opposed to the normal CAM where heatmaps are simply an extracted by-product of the main classification task. Using the adapted CAM, model design and training methodology, it was ensured that the heatmaps are more accurately localised then the normal CAM approach whose results can be seen in the last column (M_b) and last row of Fig. 2 and Table 2 respectively.

3   Experiments

Several experiments were done to assess the proposed idea for generating explainable classification. Here, ED unit upon training with only the classification labelled data forms the base classification model, i.e. M_b with $Q = 0$. In the first experiment, training was done with different values for 1:R, R = 1, 2, 3, 4 to assess the effect of lowering the value of Q on the performance, namely accuracy of classification and the generated heatmaps. In the second experiment, we wished to understand if the degree of accuracy of local annotation affects the model's performance. Training was done with 4 types of annotations: roughly accurate segmentation boundary of each FFR; a bounding box for each FFR; randomly generated image patches and finally the whole image (i.e. no localisation at all). Quantitative assessment of the classification task is done using Accuracy, Sensitivity, Specificity and AUC. Correctness of ROI prediction for a particular image is accessed using a method described later and the accuracy of detection which is the number of images in which ROIs are predicted correctly to the total number of images is reported in %.

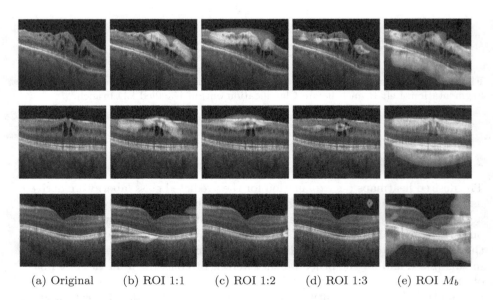

(a) Original   (b) ROI 1:1   (c) ROI 1:2   (d) ROI 1:3   (e) ROI M_b

Fig. 2. DME detection. Sample ROI outputs for training with different ratios of labeled images and local annotation for DME (rows 1, 2) and normal (row 3) cases.

Table 2. DME detection results with different ratios of local annotation

Training	Classification				ROI prediction		
Data ratio 1:R	Accuracy	AUC	Specificity	Sensitivity	Total images	Images with correct ROIs	Accuracy of detection
1:1	91.38	96.56	94.65	87.17	993	951	**95.7**
1:2	92.06	96.39	95.44	87.72	1004	947	94.3
1:3	97.71	98.95	98.57	96.61	1056	**978**	92.6
1:4	**97.86**	98.95	**98.6**	96.93	1061	610	57.49
M_b	97.83	**99.27**	97.50	**98.26**	1061	395	37.2

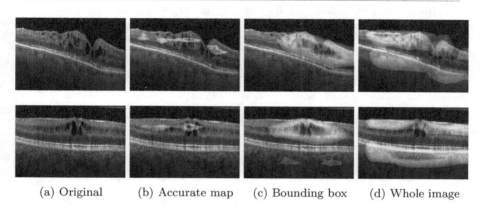

(a) Original   (b) Accurate map   (c) Bounding box   (d) Whole image

Fig. 3. ROI obtained for training with different types of local annotation.

4 Results

The ROIs derived with models trained using different ratios (1:R) of images with labels and with labels + localisation annotation are shown for 3 sample test images in Fig. 2. The results for M_b (last column) are diffuse ROI covering almost the entire image. With the addition of more and more images with localisation information during the training phase, the ROIs improve progressively and we get the best overlap with the FFRs when R reaches 1. The intersection over union (IOU) metric was used to help quantitatively assess the derived ROIs against the ground truth for FFR. An IOU threshold of 0.3 is taken to declare correct detection of ROI. Table 2 lists the number of correctly detected ROIs and the accuracy of detection(Correct ROIs/Total Images). These results are consistent with the qualitative results showing an increasing trend in accuracy of detection as R value approaches 1. Lowest ROI detection accuracy is obtained by M_b due to lack of information about suspect regions during training.

Next, we present the results of experimenting with different types of localization of suspect regions during training. Figure 3 shows the derived ROIs for 2 sample images. It can be seen that ROI is less and less localised as the precision with which local annotations used in training data is compromised, which is to be expected. Quantitatively, accuracy of classification remains above 90 % for all types of local annotations used (see Table 3). However, there is a fall in accuracy of correct ROI predictions for a bounding box type of annotation and a steep degradation when the whole image or random patches are used as annotations. More results are shown in Fig. 4 for 1:3 training regime. A comparison of the classification accuracy of the proposed method with 2 state of the art (SOTA) methods are given in Table 4. Our method is seen to be almost on par with [12] for 4 metrics, when tested on the same large dataset [7].

Table 3. DME detection. Results with different types of local annotations.

Annotation	Classification				ROI Prediction		
Type	Accuracy	AUC	Specificity	Sensitivity	Total images	Correct ROIs	Accuracy of detection
Accurate map	**97.71**	98.95	**98.57**	**96.61**	1056	**978**	**92.6**
Bounding box	94.91	**99.01**	96.86	92.39	1035	803	77.5
Whole image	93.38	98.32	95.01	91.30	1019	345	33.8
Random	92.42	97.49	94.65	90.56	1010	432	42.77

Table 4. DME detection performance comparison with SOTA

Method	Dataset	Accuracy	AUC	Specificity	Sensitivity
Kermany et al. [12]	[7]	98.2	99.87	99.6	96.8
Our method (1:3)	[7]	97.71	98.95	98.57	96.61
Srinvasan et al. [13]	Duke	93.335	–	93.8	68.8

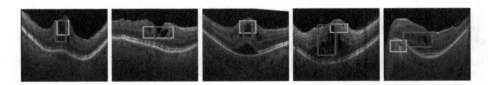

Fig. 4. Predicted ROI (in green) for DME detection. Ground Truth is in red. (Color figure online)

The proposed approach was also applied to breast cancer detection (for screening) from mammograms. Unlike the DME problem, evidence for breast cancer is not easily discernible to the naked, untrained eye and hence is particularly challenging. The 2 classes of interest were defined to be normal and abnormal. The latter includes benign and malignant cases as discrimination between these cases is difficult and best done by a specialised model. The ROI prediction aimed at are suspect regions regardless of whether they are benign or malignant.

The architecture used for the DME problem was used and patches from the entire mammogram was fed as input. Training methodology was as described in Sect. 2.3. The patches (200 × 200) which were classified as positive by the model acts as the predicted ROI in the large sized mammogram image (around 4000 × 4000 in dimension). The model was assessed on CBIS-DDSM dataset [14]. A total of 5218 training images (2017 abnormal and 3201 Normal) were split into train and validation in the ratio of 12:5. A model trained on 1:3 ratio of annotated images, was evaluated on a test set of 1298 images (709 abnormal and 589 Normal). The AUC/sensitivity(SN)%/specificity (SP)% attained was 0.98/90/93 respectively. Three sample images with ground truth regions and model-predicted ROIs (bounding boxes) are shown in Fig. 5. A baseline model (M_b) was also trained and tested. It achieved a AUC/SN/SP of 0.972/88.2/ 91.3 respectively. A recent method [15] that does normal/cancerous classification also reports on [14]. It is based on transfer learning with a Resnet50 and reports AUC/SN/SP to be 0.91/86%/80.1%.

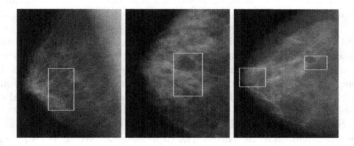

Fig. 5. Breast cancer detection. Green - Predicted ROI, Red - Ground Truth. (Color figure online)

5   Conclusion and Discussion

The need to make decisions of CAD systems for disease detection more explainable was addressed in this paper. It is worth emphasising that the primary problem here is not segmentation and that the evidence provided are an added benefit to the main classification task. Using a data set with very limited local annotations, a lightweight network design was proposed and trained using a novel methodology to provide classification and an explanation via heatmaps. The approach has been illustrated for DME and breast cancer detection, both employing different modality. The proposed solution also serves to draw the attention of the image reader to the areas deemed to be suspect. Results of extensive experiments indicate that a model trained with labeled images where only a third have basic bounding box type of local annotation, can achieve above 90% classification accuracy and provide explanation in the form of heatmaps with over

75% accuracy; the accuracy of heatmaps do improve with more accurate and abundant local annotation. The proposed solution thus enables an explainable CAD design with a *flexible* use of available annotations.

References

1. Sahlsten, J., et al.: Deep learning fundus image analysis for diabetic retinopathy and macular edema grading. Sci. Rep. **9**(1), 1–11 (2019)
2. Chen, X., Xu, Y., Wong, D.W.K., Wong, T.Y., Liu, J.: Glaucoma detection based on deep convolutional neural network. In: 37th Annual International Conference of the IEEE Engineering in Medicine and Biology Society (EMBC), pp. 715–718. IEEE (2015)
3. Dekhil, O., et al.: A novel cad system for autism diagnosis using structural and functional MRI. In: IEEE ISBI, pp. 995–998. IEEE (2017)
4. Holzinger, A., Biemann, C., Pattichis, C.S., Kell, D.B.: What do we need to build explainable AI systems for the medical domain? arXiv preprint arXiv:1712.09923 (2017)
5. Lee, H., et al.: An explainable deep-learning algorithm for the detection of acute intracranial haemorrhage from small datasets. Nature Biomed. Eng. **3**(3), 173 (2019)
6. Wang, X., Peng, Y., Lu, L., Lu, Z., Bagheri, M., Summers, R.M.: Chestx-ray8: hospital-scale chest x-ray database and benchmarks on weakly-supervised classification and localization of common thorax diseases. In: Proceedings of the IEEE Conference on Computer Vision and Pattern Recognition, pp. 2097–2106 (2017)
7. Kermany, D., Zhang, K., Goldbaum, M.: Large dataset of labeled optical coherence tomography (OCT) and chest x-ray images (2018)
8. Ronneberger, O., Fischer, P., Brox, T.: U-Net: convolutional networks for biomedical image segmentation. In: Navab, N., Hornegger, J., Wells, W.M., Frangi, A.F. (eds.) MICCAI 2015. LNCS, vol. 9351, pp. 234–241. Springer, Cham (2015). https://doi.org/10.1007/978-3-319-24574-4_28
9. Chiu, S.J., Allingham, M.J., Mettu, P.S., Cousins, S.W., Izatt, J.A., Farsiu, S.: Kernel regression based segmentation of optical coherence tomography images with diabetic macular edema. Biomed. Opt. Express **6**(4), 1172–1194 (2015)
10. Bogunović, H., Venhuizen, F., Klimscha, S., Apostolopoulos, S., Bab-Hadiashar, A., Bagci, U., et al.: RETOUCH - the retinal OCT fluid detection and segmentation benchmark and challenge. IEEE Trans. Med. Imaging **38**(8), 1858–1874 (2019)
11. Zhou, B., Khosla, A., Lapedriza, A., Oliva, A., Torralba, A.: Learning deep features for discriminative localization. In: Proceedings of the IEEE Conference on CVPR, pp. 2921–2929 (2016)
12. Kermany, D.S., et al.: Identifying medical diagnoses and treatable diseases by image-based deep learning. Cell **172**(5), 1122–1131 (2018)
13. Srinivasan, P., et al.: Fully automated detection of diabetic macular edema and dry age-related macular degeneration from optical coherence tomography images. Biomed. Opt. Express **5**, 10 (2014)
14. Lee, R.S., Gimenez, F., Hoogi, A., Miyake, K.K., Gorovoy, M., Rubin, D.L.: A curated mammography data set for use in computer-aided detection and diagnosis research. Sci. Data **4**, 170177 (2017)
15. Shen, L., Margolies, L.R., Rothstein, J.H., Fluder, E., McBride, R., Sieh, W.: Deep learning to improve breast cancer detection on screening mammography. Sci. Rep. **9**(1), 1–12 (2019)

Reliable Saliency Maps for Weakly-Supervised Localization of Disease Patterns

Maximilian Möller[1,2](✉), Matthias Kohl[1], Stefan Braunewell[1],
Florian Kofler[2,3], Benedikt Wiestler[3], Jan S. Kirschke[3], Björn H. Menze[2],
and Marie Piraud[1,2]

[1] Konica Minolta Laboratory Europe, München, Germany
maximilian.moeller@tum.de
[2] Department of Informatics, Technical University of Munich, München, Germany
[3] Department of Neuroradiology, Klinikum rechts der Isar, München, Germany

Abstract. Training convolutional neural networks with image-based labels leads to black-box image classification results. Saliency maps offer localization cues of class-relevant patterns, without requiring costly pixel-based labels. We show a failure mode for recently proposed weakly supervised localization models, e.g., models highlight the wrong input region, but classify correctly across all samples. Subsequently, we tested multiple architecture modifications, and propose two simple, but effective training approaches based on two-stage-learning and optional bounding box guidance, that avoid such misleading projections. Our saliency maps localize pneumonia patterns reliably and significantly better than gradCAM in terms of localization scores and expert radiologist's ratings.

Keywords: Weakly-supervised · Visualization · Saliency maps

1 Introduction

Deep Convolutional Neural Networks (DCNNs) reach expert performance on difficult classification tasks in the medical domain [6]. Unfortunately, we can not yet understand how DCNNs reach their decisions [5]. However, various visualization approaches create saliency maps that localize class-relevant patterns in the input image, exploiting deep filter activation maps [15]. Post-hoc methods, like gradCAM [12], create saliency maps, using filter activations in the final convolutional layer, their gradients, and converged model parameters. A popular application of GradCAMs and CAMs [16] is to provide a visual argument that the classifier learned essential features [9,13]. Though their saliency maps are of low resolution, do not capture the extent of objects, have difficulties detecting multiple instances, in particular small object instances, and

---

Electronic supplementary material The online version of this chapter (https:// doi.org/10.1007/978-3-030-61166-8_7) contains supplementary material, which is available to authorized users.

J. Cardoso et al. (Eds.): iMIMIC 2020/MIL3ID 2020/LABELS 2020, LNCS 12446, pp. 63–72, 2020.
https://doi.org/10.1007/978-3-030-61166-8_7

sometimes also highlight class-unrelated regions [12]. Despite these shortcomings, gradCAM is the state of the art post-hoc decision visualization technique that passes saliency map sanity checks [1]. Recently, weakly supervised localization approaches allow explicitly learning saliency maps during training instead of creating saliency maps post-hoc [4,7,14,17]. An encoder-decoder architecture uses filter activation maps from multiple resolutions to project a class saliency map. Subsequently, a spatial pooling function reduces the saliency map to a class score. Yao et al. [14] argue that this learning approach makes the saliency maps meaningful. We hypothesize that the weak localization constraint in an end-to-end trained weakly-supervised approach does not strictly enforce meaningful saliency maps. To the best of our knowledge, no previous work compares weakly supervised localization approaches with gradCAM. We systematically compare localization performance and stability of saliency maps from different approaches on the task of pneumonia detection on a large public dataset. We compare our assessment with the localization ratings of expert radiologists, to prove their clinical usefulness. Our contribution is twofold:

- We confirm our hypothesis and show a failure mode for end-to-end weakly supervised localization models. Their saliency maps can systematically highlight class-irrelevant patterns across all samples, without a drop in classification performance. Such saliency maps are useless for a radiologist and misleading for non-experts.
- Based on this observation, we test modifications of saliency map creation methods with stricter constraints, namely intermediate supervision using global labels and optional bounding box guidance. Our approaches outperform gradCAM, do not exhibit the failure mode, and receive the highest localization ratings by expert radiologists.

2   Methods

Our methods leverage different training strategies to use global labels, with optional bounding box guidance to project class-indicative saliency maps. They share the same architecture shown in Fig. 1, but use different training procedures and loss functions. The first approach, "Ours-end-to-end" uses end-to-end training like Yao et al. [14]. Our novel approaches ("Ours-two-stage", "Ours-oracle") offer a weakly supervised, and a bounding-box-guided approach to project stable and meaningful saliency maps.

In all approaches, we use a truncated U-net [10] (left box) and a localization-classification head (center box) with a class-specific spatial pooling function, as proposed by Yao et al. [14]. The encoder is a modified ResNet50 with six instead of five downsampling operations, that forwards k filter activation maps F_r^k at four isotropic resolutions $r = \{8, 16, 32, 64\}$. At the bottleneck, we add an optional classifier with an average pooling layer (AP), fully connected layer (FCL), softmax, and negative log-likelihood loss (NLLLoss) to allow for two-stage-training. The decoder uses 2D-convolutions with ReLU non-linearities and skip connections as in the standard U-net. The final decoded filter activation map space $D_{r=64}^{k=256}$ is passed to the localization classification head. There, a 2D-convolution with a kernel size of 1, followed by a sigmoid activation function, projects the 256 filter maps to C class saliency maps $S_{r=64}^{k=C}$, where C denotes the number of classes. Optionally, $S_{r=64}^{k=C}$ are passed to the bounding box guidance block, to allow for supervised saliency map creation (oracle). Then we extend the lower-bounded-adaption of log-sum exp pooling function (LSELBA) [8,14] to class-specific versions $LSELBA_c$, that pool each class saliency map S^c to its class score $\tilde{y}_c$, preserving the range $[0,1]$, where $c \in \{1...C\}$. Normalizing class scores $\tilde{y}_c$ creates class probabilities $\hat{y}_c$, which are passed to the classification loss function. The $LSELBA_c$ pooling function (see Eq. 1) learns β_c for each class, allowing the network to interpolate between max ($\beta_c \to \infty$) and average ($\beta_c \to -\infty$) pooling for each class, thereby reducing the size bias. The lower-bound parameter r_l is set to 10 for best performance [13]. w, h denote the spatial extent of all saliency maps $S_{r=h=w=64}^C$.

$$\tilde{y}_c(S,r) = \frac{1}{r_l + exp(\beta_c)} log(\frac{1}{w \times h} \sum_{i=1}^{w} \sum_{j=1}^{h} exp([r_l + exp(\beta_c)] \times S^c(i,j))) \quad (1)$$

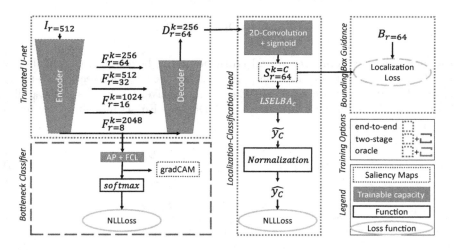

Fig. 1. Our training approaches use the core architecture in the blue-dotted frames. Optionally, it can be extended by the bottleneck classifier for "Ours-two-stage" training, or the bounding box guidance for "Ours-oracle" training. (Color figure online)

Ours-end-to-end approach trains the truncated U-net and the localization- classification head in end-to-end weakly supervised fashion, using only global labels and NLLLoss. Additionally, we reproduced the architecture of Yao et al. and trained it equally for comparison. These end-to-end approaches can fail to localize disease patterns while classifying correctly. We believe that the decoder can manipulate encoder features to highlight any region and magnify their amplitude to convey classification relevant information.

Ours-two-stage follows a two-stage training approach to improve the stability of predicted saliency maps. It removes the influence of decoder and localization-classification head on the encoder capacity. We first train the encoder until convergence, using only the bottleneck classifier. Secondly, the encoder is frozen and the decoder and localization-classification head are trained, using only the NLLLoss at the tip of the localization-classification head. We further use the trained encoder to compute gradCAM saliency maps for comparison with our extracted saliency maps.

Ours-oracle uses bounding boxes annotations $B_{r=64}$, downsampled to a resolution of 64×64 during training unlike all other approaches. The saliency maps are guided inside the bounding boxes, assuming that essential underlying disease patterns are found there. An additional supervised localization loss l_{loc} is computed with $B_{r=64}$ and $S_{r=64}^{k=C}$ and optimized jointly with the classification loss. l_{loc} (Eq. 2) is the weighted sum of negative sIoP (see Eq. 3) and sDice (see Eq. 4) scores. "Ours-oracle(0.01)" and "Ours-oracle(0.1)" denote that $\alpha = 0.01$ and $\alpha = 0.1$ are used for multitask training. Larger α can lead to over-segmentation due to imperfect bounding box labels during multi-task training.

$$l_{loc} = -\alpha(sDice + sIoP) \tag{2}$$

Evaluation metrics compare ground truth bounding boxes $B_{r=64}$ and saliency maps $S_{r=64}$ to quantify localization performance. The soft intersection over predicted saliency map score (sIoP see Eq. 3) qualitatively indicates if $S_{r=64}$ lies within $B_{r=64}$ [17]. The soft Dice score [2] (sDice see Eq. 4) indicates how much $S_{r=64}$ and $B_{r=64}$ match.

$$sIoP(S, B) = \frac{\sum_{i=1}^{h} \sum_{j=1}^{w} S_{i,j} B_{i,j}}{\sum_{i=1}^{h} \sum_{j=1}^{w} S_{i,j}} \tag{3}$$

$$sDice(S, B) = \frac{2 \sum_{i=1}^{h} \sum_{j=1}^{w} S_{i,j} B_{i,j}}{\sum_{i=1}^{h} \sum_{j=1}^{w} S_{i,j}^{2} + B_{i,j}^{2}} \tag{4}$$

We aim at maximizing sIoP. sDice is a control metric, which should reach a range between 0.1 and 0.4. Higher sDice scores indicate that saliency maps become similar to bounding boxes instead of the underlying pneumonia patterns. Lower scores indicate that models do not capture the extent of underlying pneumonia patterns. Further, ROC AUC averaged over all classes measures classification performance. For completeness, we report RSNA detection score in the supplementary material.

2.1   Dataset, Preprocessing, Training Parameters

We compare localization approaches on pneumonia detection using the RSNA Pneumonia Detection dataset [11]. It is a revisited subset of the ChestX-ray14 dataset [13], containing 25684 frontal chest X-ray images with three classes: 31% pneumonia, 40% no pneumonia/other diseases, 29% normal, where bounding boxes for all pneumonia samples are available. Images are down-sampled to 512×512 resolution ($I_{r=512}$). We created five train/validation splits with 1000 images in the validation set using a stratified shuffle split on unique patient IDs. Classification scores across the five validation sets reveal no difference. For localization performance evaluation in Sect. 3, we use a randomly selected train/validation split and report validation set performance. We apply data augmentation in all training runs. Random horizontal flips, random rotation ($\pm 5°$), translation ($\pm 5\%$) , zoom ($\pm 10\%$), and normalization with ImageNet statistics are applied during training. We use Adam optimizer [3] with learning rate of 10^{-3} and set weight decay to 10^{-5}. We train models for 100 epochs and set the batch size to 32, except for Yao et al. [14], due to large network size.

2.2   Radiologist Survey

For further evaluation of our methods, three expert radiologists rate the localization ability of saliency maps for 20 randomly selected images of pneumonia cases. We pick representative models for gradCAM, "Ours-two-stage" and "Ours-oracle(0.1)" to test superiority of our methods over gradCAM. Secondly, we pick the best and worst model of "Ours-end-to-end" to test if models with failure mode provide meaningful information to radiologists. Image and approach are presented in a randomized order. Saliency maps are overlayed on the original X-ray image and can be toggled on or off. Radiologists rate the usefulness of saliency maps, given a four-star scale: The predicted saliency map does not (1) /partially (2) /mostly (3) /perfectly (4) indicate underlying pneumonia patterns.

3   Experiments

The experiments investigate the localization performance and stability of saliency maps of gradCAM, Yao et al., and our approaches. We train each approach at least ten times, with random parameter initialization and sampling order. Every training run leads to one model. We compare mean classification and localization metrics, and their standard deviation across the models in Table 1. We use one sided heteroscedastic t-test for significance testing. Then, in Fig. 2 we plot each trained model as a dot on a 2D-scatter plot, to compare mean sIoP and mean sDice score and identify models with poor localization behavior, indicative for the failure mode. We evaluate our findings against the perception of a radiologist in Fig. 3. Lastly, in Fig. 4 we present some pneumonia cases with saliency maps that are part of the radiologist survey.

Table 1. Comparison of gradCAM, Yao et al., "Ours-end-to-end", "Ours-two-stage", and "Ours-oracle". Mean scores across models are reported with their standard deviations in brackets. Best scores without bounding box guidance are in bold. Fails denotes the fraction of models with dominantly misleading saliency maps.

Approach	Fails	ROCAUC	sDice	sIoP
gradCAM [12]	No	0.884 (±0.002)	**0.334 (±0.043)**	0.367 (±0.052)
Yao et al. [14]	54%	0.87 (±0.005)	0.115 (±0.064)	0.392(± 0.183)
Ours-end-to-end	30%	0.886 (±0.005)	0.168 (±0.066)	0.464 (±0.18)
Ours-two-stage	No	**0.886 (±0.002)**	0.309 (±0.053)	**0.485 (±0.061)**
Ours-oracle(0.01)	No	0.88 (±0.002)	0.248 (±0.069)	0.614 (±0.04)
Ours-oracle(0.1)	No	0.881 (±0.002)	0.474 (±0.024)	0.714 (±0.014)

3.1 Results

In Table 1 "Ours-two-stage" approach reaches superior performance over other previous weakly supervised methods in sIoP, and performs significantly better than gradCAM ($p \leq 5 \times 10^{-5}$). As expected "Ours-oracle" improves localization performance when α increases from 1% to 10%, and significantly outperforms all weakly supervised sIoP scores ($p \leq 6 \times 10^{-4}$). However, using bounding box guidance significantly degrades classification performance ($p \leq 2 \times 10^{-3}$).

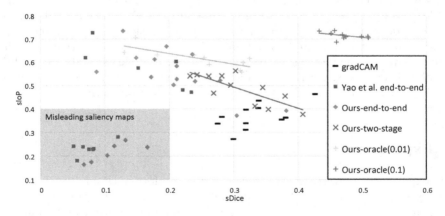

Fig. 2. The 2D-Scatter plot compares the average localization scores (sIoP and sDice) for each trained model of the six approaches. The red box indicates models with misleading saliency maps, e.g. failure mode. Trendlines are plotted for visual guidance. (Color figure online)

In Fig. 2, models, represented by different dots, that lie inside the red box exhibit poor average sIoP and sDice scores and consistently produce pneumonia unrelated saliency maps across the majority of all samples (see row 5 in Fig. 4 for examples). Notably, only models from Yao et al. (end-to-end) and "Ours-end-to-end" exhibit this failure mode. The column "fails" in Table 1 denotes the fraction of models that exhibit the failure mode.

Figure 3 summarizes the results of our expert survey with three participants. The survey results support our hypothesis; we use the Wilcoxon signed-rank test for significance testing and report a mixed effect model for further analysis in the supplemental material.

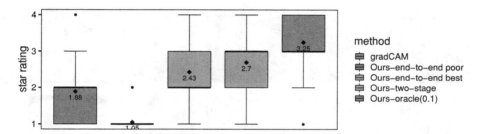

Fig. 3. The boxplot summarizes the results from our expert-survey with sixty points per box plot. Differences in expert star ratings are illustrated per method, and diamonds depict mean values. "Ours-end-to-end poor" represents a model that exhibits the failure mode. Yao et al. [14] was not evaluated, because "Ours-end-to-end" performs similar.

Firstly, saliency maps from models that exhibit the failure mode provide no meaningful information to radiologists (see "Ours-end-to-end poor"). Secondly, similar to the results in Table 1 our stable weakly supervised approach "Ours-two-stage" significantly outperforms gradCAM ($p \leq 2 \times 10^{-4}$). Additionally, our bounding box guidance extension, "Ours-oracle(0.1)" receives the highest localization rating.

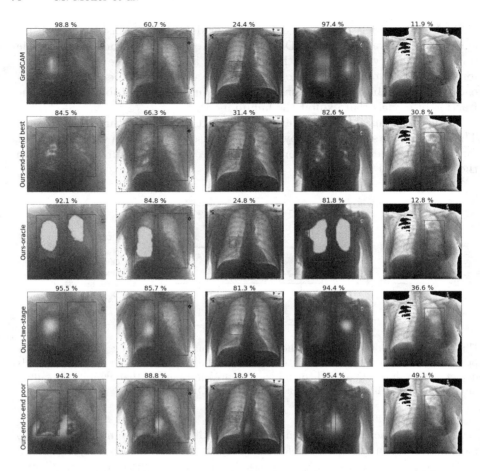

Fig. 4. Columns one to five contain randomly selected chest x-ray images with pneumonia. Each row overlays saliency maps indicative for pneumonia, using the denoted training approach. These saliency maps were evaluated by radiologists. For visualization, saliency maps are interpolated bilinearly and normalized across channels. Each image's title states the predicted pneumonia probability. Row five shows saliency maps from "Ours-end-to-end poor" model representing the failure mode, that can appear in end-to-end training. "Ours-oracle" may lead to over-segmentation.

4   Discussion and Conclusion

The results support our hypothesis. Indeed, end-to-end trained weakly supervised models show a failure mode that leads to misleading saliency maps. Such models seem to learn to project class-relevant activations onto class-irrelevant background patterns, that we call projection surfaces. While we do not claim that we fully understand the misleading projective behavior yet, we offer two avoidance strategies. Two-stage training reduces the incentive to learn such a misleading projection by separating the learning objective into two parts. First, the encoder learns classification-relevant features. Second, the decoder learns to reconstruct spatial context for classification-relevant features and predicts saliency maps used for classification. "Ours-oracle" uses bounding box annotations to guide the saliency map creation towards meaningful regions in a multi-task fashion, with the risk of potentially oversegmenting underlying pneumonia patterns, in case of imperfect labels. We explain the small drop in classification score when training with ground-truth bounding boxes with imperfect bounding box labels that miss some less relevant features for humans. Both approaches do not provide guarantees for reliable saliency maps, though the extensive testing is conclusive. Future work will analyze why end-to-end weakly supervised training can project saliency maps that highlight disease-unrelated patterns. Further, we hope that "Ours-2-stage" approach is helpful for a variety of advanced classification topics such as attention learning and active learning approaches.

References

1. Adebayo, J., Gilmer, J., Muelly, M., Goodfellow, I., Hardt, M., Kim, B.: Sanity checks for saliency maps. In: Advances in Neural Information Processing Systems, pp. 9505–9515 (2018)
2. Fidon, L., et al.: Generalised Wasserstein dice score for imbalanced multi-class segmentation using holistic convolutional networks. In: Crimi, A., Bakas, S., Kuijf, H., Menze, B., Reyes, M. (eds.) BrainLes 2017. LNCS, vol. 10670, pp. 64–76. Springer, Cham (2018). https://doi.org/10.1007/978-3-319-75238-9_6
3. Kingma, D.P., Ba, J.: Adam: a method for stochastic optimization. arXiv preprint arXiv:1412.6980 (2014)
4. Lee, S., Lee, J., Lee, J., Park, C.K., Yoon, S.: Robust tumor localization with pyramid grad-cam. arXiv preprint arXiv:1805.11393 (2018)
5. Lipton, Z.C.: The mythos of model interpretability. Queue **16**(3), 31–57 (2018)
6. Litjens, G., et al.: A survey on deep learning in medical image analysis. Med. Image Anal. **42**, 60–88 (2017)
7. Oquab, M., Bottou, L., Laptev, I., Sivic, J.: Is object localization for free?-weakly-supervised learning with convolutional neural networks. In: Proceedings of the IEEE Conference on Computer Vision and Pattern Recognition, pp. 685–694 (2015)
8. Pinheiro, P.O., Collobert, R.: From image-level to pixel-level labeling with convolutional networks. In: Proceedings of the IEEE Conference on Computer Vision and Pattern Recognition, pp. 1713–1721 (2015)
9. Rajpurkar, P., et al.: CheXNet: radiologist-level pneumonia detection on chest x-rays with deep learning. arXiv preprint arXiv:1711.05225 (2017)
10. Ronneberger, O., Fischer, P., Brox, T.: U-Net: convolutional networks for biomedical image segmentation. In: Navab, N., Hornegger, J., Wells, W.M., Frangi, A.F. (eds.) MICCAI 2015. LNCS, vol. 9351, pp. 234–241. Springer, Cham (2015). https://doi.org/10.1007/978-3-319-24574-4_28
11. RSNA: RSNA detection challenge kernel description (2018). https://www.kaggle.com/c/rsna-pneumonia-detection-challenge/overview
12. Selvaraju, R.R., Cogswell, M., Das, A., Vedantam, R., Parikh, D., Batra, D.: Grad-CAM: visual explanations from deep networks via gradient-based localization. In: Proceedings of the IEEE International Conference on Computer Vision, pp. 618–626 (2017)

13. Wang, X., Peng, Y., Lu, L., Lu, Z., Bagheri, M., Summers, R.M.: ChestX-ray8: hospital-scale chest x-ray database and benchmarks on weakly-supervised classification and localization of common thorax diseases. In: Proceedings of the IEEE conference on computer vision and pattern recognition, pp. 2097–2106 (2017)
14. Yao, L., Prosky, J., Poblenz, E., Covington, B., Lyman, K.: Weakly supervised medical diagnosis and localization from multiple resolutions. arXiv preprint arXiv:1803.07703 (2018)
15. Zhou, B., Khosla, A., Lapedriza, A., Oliva, A., Torralba, A.: Object detectors emerge in deep scene CNNs. arXiv preprint arXiv:1412.6856 (2014)
16. Zhou, B., Khosla, A., Lapedriza, A., Oliva, A., Torralba, A.: Learning deep features for discriminative localization. In: Proceedings of the IEEE Conference on Computer Vision and Pattern Recognition, pp. 2921–2929 (2016)
17. Zhou, Y., Zhu, Y., Ye, Q., Qiu, Q., Jiao, J.: Weakly supervised instance segmentation using class peak response. In: Proceedings of the IEEE Conference on Computer Vision and Pattern Recognition, pp. 3791–3800 (2018)

Explainability for Regression CNN in Fetal Head Circumference Estimation from Ultrasound Images

Jing Zhang[1]([✉]), Caroline Petitjean[1], Florian Yger[2], and Samia Ainouz[1]

[1] Normandie Univ, INSA Rouen, UNIROUEN, UNIHAVRE, LITIS, Rouen, France
{jing.zhang,samia.ainouz}@insa-rouen.fr, caroline.petitjean@univ-rouen.fr
[2] LAMSADE, Université Paris-Dauphine, Paris, France
florian.yger@dauphine.fr

Abstract. The measurement of fetal head circumference (HC) is performed throughout the pregnancy to monitor fetus growth using ultrasound (US) images. Recently, methods that directly predict biometric from images, instead of resorting to segmentation, have emerged. In our previous work, we have proposed such method, based on a regression convolutional neural network (CNN). If deep learning methods are the gold standard in most image processing tasks, they are often considered as black boxes and fail to provide interpretable decisions. In this paper, we investigate various saliency maps methods, to leverage their ability at explaining the predicted value of the regression CNN. Since saliency maps methods have been developed for classification CNN mostly, we provide an interpretation for regression saliency maps, as well as an adaptation of a perturbation-based quantitative evaluation of explanation methods. Results obtained on a public dataset of ultrasound images show that some saliency maps indeed exhibit the head contour as the most relevant features to assess the head circumference and also that the map quality depends on the backbone architecture and whether the prediction error is low or high.

Keywords: Saliency maps · Explanation evaluation · Regression CNN · Biometric prediction · Medical imaging

1 Introduction

The measurement of fetal head circumference (HC) is performed throughout the pregnancy as a key biometric to monitor fetus growth and estimate gestational age. In clinical routine, this measurement is performed on ultrasound (US) images (Fig. 1), via manually tracing the skull contour and fitting it into an ellipse. Automated segmentation approaches have been proposed, lately based on CNN in order to solve this tedious task, but these models require large dataset of manually segmented data. In our previous work [21], we departed from the mainstream approach of segmentation and instead proposed a regression network, in order to directly predict the head circumference.

© Springer Nature Switzerland AG 2020
J. Cardoso et al. (Eds.): iMIMIC 2020/MIL3ID 2020/LABELS 2020, LNCS 12446, pp. 73–82, 2020.
https://doi.org/10.1007/978-3-030-61166-8_8

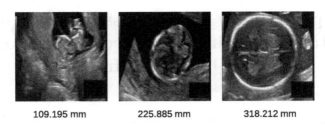

109.195 mm   225.885 mm   318.212 mm

Fig. 1. Ultrasound images of fetus head with head circumference in millimeters

Compared to a classification model, the last layer of a regression CNN model is a linear or sigmoid activation function, instead of the softmax layer. Also, the regression loss function is metric-inspired, for instance, it can be the Mean Absolute Error (MAE) or the Mean Squared Error (MSE). It is known that the high accuracy of deep learning methods comes at the cost of a low interpretability, i.e. the model is seen as a black box, which does not provide explanations along with the prediction. In this paper, our goal is to investigate how explanation methods can help us to get some insights on the regression network and to appreciate its behavior [12]. In classification networks, explanations may take the form of saliency or sensitivity maps [10], highlighting the areas that particularly contributed to a decision. The saliency maps have been applied on different neural networks such as CNN, LSTM, and in various tasks, for example classification, detection and image segmentation [16]. To the best of our knowledge, this paper is the first interpretation of a regression CNN that is dedicated to the estimation of biometric from medical images.

In this paper, our contributions are the following: we adapt explanation methods in regression CNN and provide an interpretation of what a saliency map is, in the regression case. We are thus able to gain insight into the CNN regression model for our HC prediction problem, and see what pixels contribute the most to the estimation of the HC: we expect them to be those of the head contour. We also address the problem of evaluating the explanation methods, in the regression case. Adebayo's sanity checks consist in performing randomization tests, in the data or in the model, and evaluate the changes in the produced saliency maps [1]. Another example is Samek's proposal, that has particularly inspired us [11], to compare and assess different explanation methods. The principle is to inject noise gradually in the image, in locations that have been highlighted by the saliency maps, and see how the prediction is affected by this perturbation. However, the method is designed for classification networks and requires some adaptation.

In Sect. 2 we briefly recall the state-of-the-art in saliency maps algorithm for classification CNN and their meaning in case of a regression network; we also presented the evaluation methodology used to assess the explanation methods. Experimental results are presented in Sect. 3 and conclusions are drawn in Sect. 4.

2   Saliency Map Methods for Regression CNN

In this section, we briefly describe 8 explanation methods from the state-of-the-art that are used to produce saliency maps in classification CNN [12,16,22]. Then, we present the evaluation method of perturbation analysis [11] and adapt it to the regression CNN to evaluate the performance of these methods.

2.1   State-of-the-Art Saliency Maps in CNN

Two categories of saliency maps are generally considered, perturbations-based or propagation-based. In perturbation-based approaches, the goal is to estimate how perturbation applied to the input image, such as blurring or injecting noise, changes the predicted class [5,22]. In propagation-based techniques, the idea is to backpropagate a relevance signal from the output to the input. In this paper, we will focus on the latter category of methods that actually encompass (i) sensitivity (or gradient-based) analysis, (ii) deconvolution methods, and (iii) Layer-wise Relevance Propagation (LRP) variants.

The sensitivity analysers include the **Gradient** [14] method, that simply computes the gradient of the output w.r.t. input image, and expresses how much the output value changes w.r.t. a small change in input; the **SmoothGrad** [17], that averages the gradient over random samples in a neighborhood of the input with added noise, and which is an improvement of Gradient method that can sharpen the saliency map; the **Input*Gradient** [13] technique, that strengthens the saliency map by multiplying Gradient with input information; and the **Integrated Gradients** [19], that computes the integration of the gradient along a path from the input to a baseline black image.

Deconvolution methods are the **DeConvNet** [20] that acts equivalently as a decoder of CNN models, which reverses the CNN layers, and the **Guided BackProp** [18] that combines backpropagation and DeConvNet.

The core idea of **Layer-wise Relevance Propagation (LRP)** [3] is to compute a relevance score for each input pixel layer by layer in backward direction. It first forward-passes the image so as to collect activation maps and backpropagates the error taking into account the network weights and activations. The **DeepTaylor** [9] method identifies the contribution of input features as the first-order of a Taylor expansion, through Taylor decomposition, then it can estimate the attribution of each neuron one by one.

In the classification setting, a saliency map provides an estimation of how much each pixel contributes to the class prediction. In the regression setting, the saliency map will provide an estimation of how much each pixel is impacting the model, and is contributing to decrease the prediction error, as measured by the loss function, that is in general the MAE or MSE.

2.2   Evaluation of Explanation Methods Based on Perturbation

Explanation methods (also called analyzers) perform differently depending on the model, the task at hand, the data, etc. In order to quantitatively evaluate those analyzers, we build upon the perturbation analysis of [11], originally designed to assess explainability methods in classification networks. Let us first describe the perturbation process and then the evaluation metric.

First, the input image to be analyzed is subsampled by a grid. Each subwindow of the grid is ranked according to its importance w.r.t. to the pixel-wise saliency scores assigned by the analyzers. Then, the information content of the image is gradually corrupted by adding perturbation (Gaussian noise) to each subwindow, starting with the most relevant subwindow, w.r.t. the ranking just mentioned. The effect of this perturbation on the model performance is measured with the prediction error. This procedure is repeated for each subwindow. Generally, the accuracy of model will drop quickly when important information is removed and remains largely unaffected when perturbing unimportant regions. Thus, the analyzers can be compared by measuring how quickly their performance drops. That is to say, the quicker the model performance drops after introducing perturbation, the better the analyzer is capable of identifying the input components responsible for the output of the model.

The quantitative evaluation proposed in [11] for classification network, consists in computing the difference between the score $f(x)$ indicating the certainty of the presence of an object in the image x, in the presence and in the absence of perturbation. This difference is called Area over Perturbation Curve (AOPC) and defined more precisely defined in [11] as:

$$\text{AOPC}_{Analyzer} = \frac{1}{N} \sum_{n=0}^{N} (f(x_n)^{(0)} - \frac{1}{K} \sum_{k=0}^{K} f(x_n)^{(k)}) \tag{1}$$

where N is the number of images, K is the number of perturbation steps, x is the input image.

Here, we propose to adapt the AOPC to the regression case, and if we denote by $\epsilon(x)^{(0)}$ the prediction error of initial image evaluated by the analyzer and $\epsilon(x_n)^{(k)}(1 \leq k \leq K)$ the prediction error of the perturbed image $(x_n)^{(k)}$ at step k, we can define the $\text{AOPC}_{Analyzer}^{regression}$ as:

$$\text{AOPC}_{Analyzer}^{regression} = \frac{1}{N} \sum_{n=0}^{N} (\epsilon(x_n)^{(0)} - \frac{1}{K} \sum_{k=0}^{K} \epsilon(x_n)^{(k)}) \tag{2}$$

A larger AOPC in absolute value means that an analyzer has a steep decrease while the perturbation steps is increasing.

3 Experiments

3.1 Experimental Setup

We analyse two regression models that we proposed in our previous work [21], namely the regression ResNet50 and regression VGG16 (implemented using Keras). As their names show, the backbone architectures are ResNet50 [6] and VGG16 [15] resp., and the loss is the mean absolute error. Both models are pre-trained on ImageNet; subsequently the last (softmax) layer is replaced by a linear layer and the network is fully retrained on a public dataset of ultrasound fetal head images called HC18 [7]. The HC18 dataset contains 999 US images, along with the corresponding head circumference, that we randomly split into a training (600), a validation (200) and a test set (199). We augment the data of the training set to 1800 images, and perform resizing of the images to the size 128 × 128 pixels. With a 5-fold cross validation, the mean absolute errors (MAE) that we obtained on the test set were 37.34 ± 37.46 pixels (4.78 ± 4.41 mm) in reg-ResNet50 and 40.17 ± 40.99 pixels (5.46 ± 5.99 mm) in reg-VGG16.

In the following, we will compute the saliency maps on the test set images. We first show the saliency maps of various explanation methods for our regression problem, for both architectures Reg-ResNet50 and Reg-VGG16, the quantitative evaluation of explanation methods, and a more in-depth study of prediction results, with the best ranked methods, namely Input*Gradient and LRP. We have used the iNNvestigate toolbox to perform our experiments [2].

Visualization of Explanation Methods. We visualize the saliency maps provided by the 8 selected explanation methods in Fig. 2. From these images, we can barely see the features retrieved by explanation method DeConvNet and Gradient in both models, that is to say these two methods seem somehow insensitive to the models. This may be explained by the gradient shattering problem [4] for the gradient method. Regarding DeConvNet's saliency map, it may be due to the architecture of deconvolution network which reconstructs the convolution networks reversely. In addition, for Reg-ResNet50, methods Gradient, GuidedBackprop and SmoothGrad fail to highlight the head contour. We will see that these observations are confirmed by the quantitative evaluation.

Quantitative Evaluation of Explanation Methods Based on Perturbation. Here, we compare the explanation methods through perturbation analysis. In this experiment, the input image of size 128 × 128 pixels is divided into a grid of 4 ×4 subwindows of size 32 × 32 pixels. Gaussian noise with mean value 0 and standard deviation 0.3 is added to each subwindow, according to their importance assigned by analyzers during the 16 steps. Figure 3 is an example of the perturbation process of Gradient analyzer.

In Fig. 4, we show the evolution of the prediction error w.r.t. the quantity of noise added at each perturbation steps, on first the most significant subwindow in the analyzer's sense, to the least significant one. One can observe that consistently, the prediction error is increasing, as the level of noise increases. Methods with the steepest curve, LRP and Input*gradient, exhibit the largest sensitivity to perturbations, and as such, should highlight the contributing pixels, in the sense of this criterion. Interestingly the Integrated gradient analyzer seems to be relevant for VGG16, but not for Reg-ResNet50. In the future, it will be interesting to vary the subwindow size to see if results are affected. We expect that a finer grid will be better suited to a thin structure like the head skull.

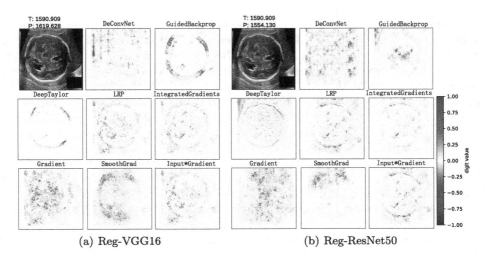

(a) Reg-VGG16   (b) Reg-ResNet50

Fig. 2. Comparison of different saliency maps with Reg-VGG16 and Reg-ResNet50. P: predicted HC value, T: ground truth HC value (in pixels).

Table 1. Performance (AOPC scores) of different analysis methods after perturbation, with two regression models. G: Gradient, SG: SmoothGrad, DCN: DeConvNet, DT: DeepTaylor, GB: GuidedBackprop, I*G: Input*Gradient, IG: IntegratedGradients. Lower is better. Best scores in bold.

Model	G	SG	DCN	DT	GB	I*G	IG	LRP
Reg_VGG16	−7.312	−7.398	−2.869	−7.401	−1.663	**−9.189**	−9.490	−9.175
Reg_ResNet50	−11.533	−11.841	−9.249	−9.890	−9.717	**−14.748**	−5.603	−14.577

In Table 1, we compared AOPC scores on regression VGG16 and regression ResNet50 models respectively. Since the AOPC is the difference between the prediction error with and without perturbation, we expect that the analyzer that are indeed perturbed by the noise will return a large AOPC score, in absolute value. We can see that the regression ResNet50 has higher AOPC score than regression VGG16 model. Again we can gather from this table that both the LRP and Input*Gradient methods perform well in those two models.

Note that other explanation methods have inconsistent performance depending on the model. This highlights the necessity to choose the proper explanation method before analyzing a specific model.

Comparison of Regression Models. As shown in Fig. 2, both regression VGG16 and regression ResNet50 are successful in learning the features from ultrasound images to assess the HC. From Table 1, we can gather that the regression ResNet50 has slight better performance on the whole, since AOPC values are larger in absolute value.

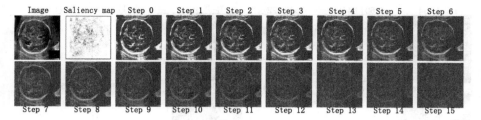

Fig. 3. Perturbation process for the saliency map produced by the Gradient method. Step 0 is the original input image. From step 1 to step 15, Gaussian noise is added gradually on the image subwindows. The perturbation order of these subwindows corresponds to the saliency scores assigned by the Gradient method analysis, i.e. the most contributing pixels are perturbed first. Red: noise, blue: original image pixels. (Color figure online)

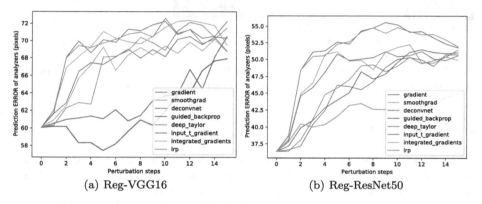

Fig. 4. Prediction error (in pixels) of different analyzers during each perturbation step based on Regression VGG16 and Regression ResNet50 model. The horizontal axis is the perturbation steps.

Comparison of Saliency Maps for Correct vs Incorrect Prediction. In this experiment, we arbitrarily pick one of the best performing methods from the previous results, and thus the use Input*Gradient explanation method to generate saliency maps from images with small prediction error (Fig. 5(a)), and with large prediction error (Fig. 5(b)). We can see that the well predicted images have obvious head contour, at least in the 2 last rows of Fig. 5(a). The models are able to learn the features from these images, therefore the saliency maps show key features. However, it is not always the case: the first row shows a small prediction error, and the head contour are not specifically highlighted. For the badly predicted images, the saliency maps highlight features that are spread and not localized into meaningful segments. The models can not learn the features from these images.

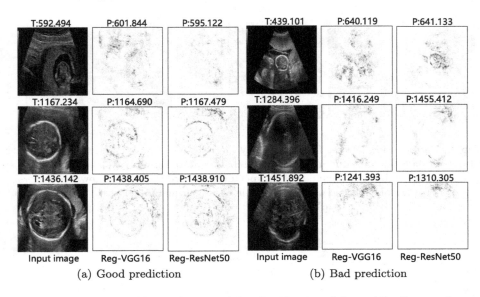

Fig. 5. Saliency map of Reg-VGG16 and Reg-ResNet50 with Input*Gradient explanation method. P and T: resp. predicted and ground truth HC values (pixels).

4   Conclusion

Understanding whether the model can learn the relevant features in images and taking the right decision is crucial in the medical domain. Whereas there have been a wealth of works in classification networks, there is a void for interpreting regression networks. In this paper, we address the problem of estimating the head circumference in fetal head directly from US images. We use several post-hoc explanation techniques that produce saliency maps and adapt a perturbation based quantitative evaluation method, to assess the relevance of the saliency maps. The experimental results proved that the regression CNN models are able to learn the key features from the input ultrasound fetus images, and in particular, the head circumference. One finding is that for this application, Gradient and DeConvNet method are particularly insensitive to different CNN models or data, and that ResNet50 seem to have better learnt the head features. Thus so far, we have extended the model property from classification to regression and explored a specific regression task. Future works also include investigating the explainability of other regression losses: in this paper, we used the MAE, but the mean square error or the Huber loss are alternatives, and there is no heuristic yet to decide which loss is better [8]. This will allow us to adapt or design new loss functions, that can account for an enhanced learnability of the regression CNN, to further improve the HC prediction. In addition to investigate individual image-wise explanations, we also intend to explore the generation of meta-explanations by aggregating individual explanations, to gain additional insight into the model behavior. Other regression applications will also be interesting to explore.

References

1. Adebayo, J., Gilmer, J., Muelly, M., Goodfellow, I., Hardt, M., Kim, B.: Sanity checks for saliency maps. In: Advances in Neural Information Processing Systems, pp. 9505–9515 (2018)
2. Alber, M., et al.: Investigate neural networks! CoRR abs/1808.04260 (2018). http://arxiv.org/abs/1808.04260
3. Bach, S., Binder, A., Montavon, G., Klauschen, F., Müller, K.R., Samek, W.: On pixel-wise explanations for non-linear classifier decisions by layer-wise relevance propagation. PloS One **10**(7), e0130140 (2015)
4. Balduzzi, D., McWilliams, B., Butler-Yeoman, T.: Neural Taylor approximations: convergence and exploration in rectifier networks. In: Proceedings of the 34th International Conference on Machine Learning, vol. 70, pp. 351–360. JMLR.org (2017)
5. Fong, R., Vedaldi, A.: Interpretable explanations of black boxes by meaningful perturbation. In: IEEE International Conference on Computer Vision (ICCV), pp. 3449–3457, October 2017. https://doi.org/10.1109/ICCV.2017.371. http://arxiv.org/abs/1704.03296, arXiv: 1704.03296
6. He, K., Zhang, X., Ren, S., Sun, J.: Deep residual learning for image recognition. In: Proceedings of the IEEE CVPR, pp. 770–778 (2016)
7. van den Heuvel, T.L.A., de Bruijn, D., de Korte, C.L., Ginneken, B.V.: Automated measurement of fetal head circumference using 2D ultrasound images. PlOs One **13**(8), 1–20 (2018). https://doi.org/10.1371/journal.pone.0200412
8. Lathuilière, S., Mesejo, P., Alameda-Pineda, X., Horaud, R.: A comprehensive analysis of deep regression. IEEE Trans. Pattern Anal. Mach. Intell. **41**, 1–17 (2019)
9. Montavon, G., Lapuschkin, S., Binder, A., Samek, W., Müller, K.R.: Explaining nonlinear classification decisions with deep Taylor decomposition. Pattern Recognit. **65**, 211–222 (2017)
10. Morch, N.J., et al.: Visualization of neural networks using saliency maps. In: Proceedings of IEEE International Conference on Neural Networks, vol. 4, pp. 2085–2090 (1995)
11. Samek, W., Binder, A., Montavon, G., Lapuschkin, S., Müller, K.R.: Evaluating the visualization of what a deep neural network has learned. IEEE Trans. Neural Netw. Learn. Syst. **28**(11), 2660–2673 (2016)
12. Samek, W., Müller, K.-R.: Towards explainable artificial intelligence. In: Samek, W., Montavon, G., Vedaldi, A., Hansen, L.K., Müller, K.-R. (eds.) Explainable AI: Interpreting, Explaining and Visualizing Deep Learning. LNCS (LNAI), vol. 11700, pp. 5–22. Springer, Cham (2019). https://doi.org/10.1007/978-3-030-28954-6_1
13. Shrikumar, A., Greenside, P., Shcherbina, A., Kundaje, A.: Not just a black box: learning important features through propagating activation differences. CoRR abs/1605.01713 (2016). http://arxiv.org/abs/1605.01713
14. Simonyan, K., Vedaldi, A., Zisserman, A.: Deep inside convolutional networks: visualising image classification models and saliency maps. CoRR abs/1312.6034 (2014)
15. Simonyan, K., Zisserman, A.: Very deep convolutional networks for large-scale image recognition. In: ICLR (2015)

16. Singh, A., Sengupta, S., Lakshminarayanan, V.: Explainable deep learning models in medical image analysis. J. Imaging **6**(6), 52 (2020)
17. Smilkov, D., Thorat, N., Kim, B., Viégas, F.B., Wattenberg, M.: SmoothGrad: removing noise by adding noise. In: Workshop on Visualization for Deep Learning, ICML (2017)
18. Springenberg, J., Dosovitskiy, A., Brox, T., Riedmiller, M.: Striving for simplicity: the all convolutional net. In: ICLR (Workshop Track) (2015). http://lmb. informatik.uni-freiburg.de/Publications/2015/DB15a
19. Sundararajan, M., Taly, A., Yan, Q.: Axiomatic attribution for deep networks. In: Proceedings of the 34th International Conference on Machine Learning, vol. 70, pp. 3319–3328. JMLR.org (2017)
20. Zeiler, M.D., Fergus, R.: Visualizing and understanding convolutional networks. In: Fleet, D., Pajdla, T., Schiele, B., Tuytelaars, T. (eds.) ECCV 2014. LNCS, vol. 8689, pp. 818–833. Springer, Cham (2014). https://doi.org/10.1007/978-3-319-10590-1_53
21. Zhang, J., Petitjean, C., Lopez, P., Ainouz, S.: Direct estimation of fetal head circumference from ultrasound images based on regression CNN. In: Medical Imaging with Deep Learning (2020)
22. Zintgraf, L.M., Cohen, T.S., Adel, T., Welling, M.: Visualizing deep neural network decisions: prediction difference analysis (2017). http://eprints.gla.ac.uk/214152/

MIL3ID 2020

Recovering the Imperfect: Cell Segmentation in the Presence of Dynamically Localized Proteins

Özgün Çiçek[1]([✉]), Yassine Marrakchi[1,2], Enoch Boasiako Antwi[1,2,3], Barbara Di Ventura[1,2], and Thomas Brox[1,2]

[1] University of Freiburg, Freiburg, Germany
{cicek,marrakch}@cs.uni-freiburg.de
[2] Signalling Research Centres BIOSS and CIBSS, Freiburg, Germany
[3] Heidelberg Biosciences International Graduate School (HBIGS), Heidelberg, Germany

Abstract. Deploying off-the-shelf segmentation networks on biomedical data has become common practice, yet if structures of interest in an image sequence are visible only temporarily, existing frame-by-frame methods fail. In this paper, we provide a solution to segmentation of imperfect data through time based on temporal propagation and uncertainty estimation. We integrate uncertainty estimation into Mask R-CNN network and propagate motion-corrected segmentation masks from frames with low uncertainty to those frames with high uncertainty to handle temporary loss of signal for segmentation. We demonstrate the value of this approach over frame-by-frame segmentation and regular temporal propagation on data from human embryonic kidney (HEK293T) cells transiently transfected with a fluorescent protein that moves in and out of the nucleus over time. The method presented here will empower microscopic experiments aimed at understanding molecular and cellular function.

1 Introduction

In the past decades, it has become evident that proteins are very dynamic and their localization within the cell dictates which function they perform [3]. Cell biologists carry out time-lapse fluorescence microscopy experiments to study protein localization and unravel how this affects the protein's function. Optogenetics can be used to control the localization of the protein of interest [18] to observe the effects of dynamic localization patterns. Take Fig. 1 for an example: HEK293T cells expressing mCherry fused to the optogenetic tool LINuS [15]

Ö. Çiçek and Y. Marrakchi—Equal contribution.

Electronic supplementary material The online version of this chapter (https:// doi.org/10.1007/978-3-030-61166-8_9) contains supplementary material, which is available to authorized users.

J. Cardoso et al. (Eds.): iMIMIC 2020/MIL3ID 2020/LABELS 2020, LNCS 12446, pp. 85–93, 2020.
https://doi.org/10.1007/978-3-030-61166-8_9

were observed over time to analyze the effect of blue light on the nuclear import of the protein. In the absence of blue light, the fusion protein localizes predominantly in the cytosol. Hence, the nucleus appears dark. However, in presence of blue light, the fusion protein enters the nucleus making it appear bright and difficult to distinguish from the cytosol. Giving light repeatedly creates oscillations of the protein in and out of the nucleus in time.

Learning-based methods, such as U-Net [19] and Mask R-CNN [6], succeed at segmenting structures in data with clearly visible patterns, but fail when the visibility deteriorates. When the signal in the nucleus is similar to that in the cytosol, nuclei segmentation from any network is not reliable. While in a single image it is not possible to improve these segmentations, past and future frames in a video provide additional information for refinement. An expert can play the video back and forth to infer the segmentation of ambiguous nuclei. An automated segmentation method, too, must (1) automatically identify critical frames and (2) propagate predictions from neighbouring frames.

In this paper, we address both challenges by (1) equipping Mask R-CNN with uncertainty estimation to identify erroneous predictions and (2) incorporating optical flow to improve the identified erroneous predictions by propagating certain predictions from neighboring frames. Doing so, we introduce the most recent uncertainty estimation methods in biomedical instance segmentation and solve a real task commonly experienced in signalling studies which is not yet addressed. So far, nuclear markers have been employed in the experiments so that the available automated segmentation tools can be used [4]. However, additional markers cause unreliable quantification since different proteins bleed-through and interfere with each other. They also limit the channel space needed for other proteins of interest. The presented method makes the use of a nuclear marker dispensable.

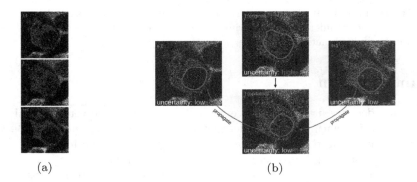

(a)   (b)

Fig. 1. (a) Exemplary time-lapse images of HEK293T cells depicting an oscillatory nuclear signal. (b) Oscillation at time t causing bad nuclei segmentation (up) and the corrected segmentation of it using our propagation method (down).

2   Related Work

We are the first to address instance segmentation of structures with an oscillatory fluorescent signal in biomedical videos. Our work is related to methods that benefit from temporal features in their design. Milan et al. [14] and Payer et al. [17] used Recurrent Neural Network (RNN) to aggregate temporal features. Paul et al. [16] incorporated temporal cues for segmentation via optical flow. Similarly, Jain et al. [9] propagated features instead of segmentation with flow from key frames. A similar idea was applied to instance segmentation by Bertasius and Torresani [2]. Although these methods seem close, the task is different: (1) they heavily rely on dense annotation in time to learn interpolations explicitly while we cannot afford it due to high cost of expert annotation and difficulty in fine-grained annotation of imperfect frames and (2) they benchmark only on visible objects, while we are solely interested in objects with limited visibility.

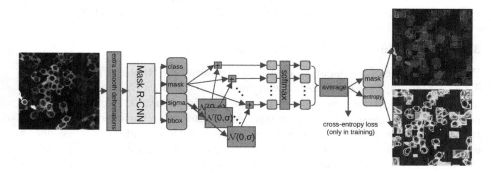

Fig. 2. Overview of Mask R-CNN with added data uncertainty. Changes to the original architecture are shown in red (operations) and green (outputs). (Colof figure online)

3 Methods

3.1 Instance-Aware Segmentation with Uncertainty Estimation

We base our model on Mask R-CNN [6] equipped with elastic deformations of U-Net [19] to create additional biomedically plausible images on-the-fly for better generalization. We incorporate uncertainty estimation in our Mask R-CNN architecture to detect erroneous predictions; see Fig. 2. We consider data uncertainty (aleatoric), model uncertainty (epistemic), and their combination.

Data Uncertainty. For data uncertainty, we use the modified cross-entropy loss [10] in the mask branch of Mask R-CNN. This models the data uncertainty as the learned noise scale from the data. To learn both the class scores and their noise scale, the negative expected log-likelihood is minimized for pixel i as:

$$L(s) = - \sum_i \log \left[\frac{1}{T} \sum_t \exp(\hat{s}_{i,t,c'} - \log \sum_c \exp(\hat{s}_{i,t,c})) \right], \tag{1}$$

where c' is the correct class among all classes (c) and $\hat{s}$ are predicted logits corrupted by Gaussian noise with standard deviation σ. σ is also learned by the network alongside with the logits. Figure 2 shows where this loss is embedded in the network. At test time, the uncertainty is computed as the entropy of the class pseudo-probabilities for each pixel as: $\mathcal{U}(p) = -\sum_c p(c) \log p(c)$.

Model Uncertainty. Model uncertainty requires sampling from the model. We experimented with the latest sampling strategies. **Dropout** is one of the common techniques to sample from networks [10]. **Ensemble** replaces the sampling by dropout by individually trained networks [11]. **SGDR ensemble** replaces the individual trained networks by the pre-converged snapshots of the same model [7]. Pre-converged models are obtained at the end of each SGDR (SGD with warm restarts) [12] cycle. **WTA** (Winner-Takes-All) [8] aims for training a single network with multiple heads where at each iteration only the head with the best prediction gets penalized. **EWTA** (Evolving WTA) [13] is a variant of WTA which improves the trade-off between diversity and consistency.

3.2   Uncertainty-Based Nuclei Propagation

We base the nuclei propagation on the cell tracker[1] by Ronneberger et al. [19]. This overlapping-based tracker is well suited for the data at hand, which does not show large motion over time. Despite its simplicity, it is one of the top performing cell trackers in the challenge. After computing the overlapping-based tracks, we go over all the frames of each track with ascending average nuclei uncertainty order (without nuclei, uncertainty is set to infinity) and update the identified uncertain predictions by propagating masks from more certain neighbours as depicted in Fig. 1. A prediction is marked uncertain if its average uncertainty is higher than a threshold θ. Neighbours are more certain if they meet a relative threshold of α in case of a single-side propagation and β in case of a two-sided propagation as detailed in Algorithm 1. This resembles the procedure that experts follow to find non-visible nuclei by sliding over time.

3.3   Motion Estimation for Biomedical Videos

Propagation of predictions over time requires motion estimation between frames to warp the certain predictions onto the less certain ones. One simple way is to warp by the shift and scaling parameters computed between the not yet updated and neighbouring nuclei predictions. This approach assumes the shape of the nuclei does not change over time; however, slight deformations can occur. Optical flow can provide fine-grained motion. Recent optical flow methods for natural images are networks [5] trained on synthetic datasets. These methods perform well on real images, but their performance deteriorates as the gap between real and synthetic images grows. Our data is very different from existing synthetic datasets and no synthetic data exists for biomedical data. Therefore, with the prior knowledge that expected flow in our videos can be well explained by smooth deformations, and being able to generate them on the fly via deformation augmentations of U-Net, we explicitly train a network to predict these deformations

---

[1] http://celltrackingchallenge.net/.

Algorithm 1: Uncertainty-Based Nuclei Propagation

Input: instance segmentation masks $\mathcal{S}$, uncertainty maps $\mathcal{U}$, motion per consecutive frame pairs $\mathcal{M}$, hard threshold θ, relative thresholds: α,β

Output: tracks

1   tracks$\leftarrow$ IoU-BasedCellTracker($\mathcal{S}$);
2   **for** track t in tracks **do**
3   **for** frame f in track t **do**
4       $\bar{u}_{t,f} \leftarrow$ mean_over_nuclei($\mathcal{U}_{t,f}, S$);
5   order$\leftarrow$argsort($\bar{u}_t$);
6   **for** frame id f in order **do**
7   **if** $\bar{u}_{t,f} \geq \theta$ and $\beta \times \bar{u}_{t,f} \geq \bar{u}_{t,f-1}$ and $\beta \times \bar{u}_{t,f} \geq \bar{u}_{t,f+1}$ **then**
8   mask$_{\text{prev}} \leftarrow$ warp($\mathcal{S}_{t,f}, \mathcal{S}_{t,f-1}, \mathcal{M}$);
9   mask$_{\text{next}} \leftarrow$ warp($\mathcal{S}_{t,f}, \mathcal{S}_{t,f+1}, \mathcal{M}$);
10        $\mathcal{S}_{t,f} \leftarrow$ union(mask$_{\text{prev}}$, mask$_{\text{next}}$);
11   **else if** $\bar{u}_{t,f} \geq \theta$ and $\alpha \times \bar{u}_{t,f} \geq \bar{u}_{t,f-1}$ **then**
12         $\mathcal{S}_{t,f} \leftarrow$ warp($\mathcal{S}_{t,f}, \mathcal{S}_{t,f-1}, \mathcal{M}$);
13   **else if** $\bar{u}_{t,f} \geq \theta$ and $\alpha \times \bar{u}_{t,f} \geq \bar{u}_{t,f+1}$ **then**
14         $\mathcal{S}_{t,f} \leftarrow$ warp($\mathcal{S}_{t,f}, \mathcal{S}_{t,f+1}, \mathcal{M}$);

---

as explored by Sokooti et al. [20]. We warp an image (I_{t+1}) backward in time with randomly generated smooth deformations ($f_{t \to t+1}$) to obtain the previous image (I_t). Then we train a FlowNet on the on-the-fly-generated image pairs and ground-truth flows. Even though the fluorescent signal in the nuclei regularly disappears making the optical flow challenging to estimate, the motion of the cytosol helps infer the motion of the nuclei.

4   Experiments

Implementation Details. We based our implementation on the Mask R-CNN by Abdulla [1]. We incorporated publicly available elastic deformations[2] for U-Net [19] as additional augmentation. We used ResNet50-FPN as backbone. We used softmax cross entropy for the mask head with 3 classes (background, cytosol and nuclei). We trained the networks from scratch for 12k epochs except the SGDR ensemble, which was trained for 15k with 3k cycles. We used the last pre-converged models. The WTA and EWTA models were trained for 11k and then merged with a network trained for 4k. We used 4 as ensemble size for all methods. Ensembling was performed over bounding boxes. For nuclei propagation, we used $\theta = 0.5$, $\alpha = 0.7$ and $\beta = 0.85$. For Mask R-CNN and FlowNet (variant C) training we used deformations with 3 and 10 control points, respectively and deformation magnitude of 10. Please see the original papers for more details. Our code and examplar dataset is publicly available[3].

---

[2] https://github.com/fcalvet/image_tools.
[3] https://lmb.informatik.uni-freiburg.de/Publications/2020/CMB20/.

Data and Annotation. While the challenge we address is common in signalling studies, there is no public dataset for this purpose. Our data was generated after 24 h of transient transfection of human embryonic kidney (HEK293T) cells with a construct expressing the fusion protein mCherry-LINuS with a ZEISS LSM780 confocal microscope. MaskRCNN was trained on 82 cropped images from original 1800×1800 images with pixel size 0.11×0.11 and 6 cells on average in each and tested on 2 full-sized images with 73 cells in total to validate instance segmentation and uncertainty estimation. We evaluated our propagation algorithm on an unseen video with 35 full-sized frames with randomly selected 117 cells annotated by experts. In cases where the nuclei appeared ambiguous experts interpolated the annotations from past and future frames.

4.1 Instance Segmentation and Uncertainty Estimation Evaluation

We evaluated our instance segmentation by mean average precision (mAP). In Table 1 we refer to standard mAP as mAP (sm) since it is based on softmax scores. To further evaluate the quality of the uncertainty estimation, we replaced the softmax scores by average entropy over the cell and re-computed mAP (mAP (ent)). This simulates our approach as we rely on averaged uncertainties in our nuclei propagation. This measure shows how reliable the predicted uncertainties are at ranking the prediction quality in the precision-recall curves, which are used commonly to evaluate uncertainty estimation for classification tasks. In Table 1 we see that WTA merged is the best at 0.5 IoU (Intersection over Union) threshold and at 0.75 threshold ensemble is the best. In the rest of our experiments we used the WTA merged with data uncertainty since it is computationally more efficient than ensemble.

Table 1. Quantitative evaluation for instance segmentation and uncertainty estimation in mAP (@0.5/@0.75 IoU).

	Model uncertainty		Combined uncertainty	
	mAP (sm)	mAP (ent)	mAP (sm)	mAP (ent)
Single	0.77/0.48	0.80/0.49	0.74/0.60	0.83/0.69
Dropout	0.74/0.61	0.78/0.65	0.77/0.61	0.83/0.67
Ensemble	0.82/**0.64**	0.78/0.61	0.78/0.63	0.83/**0.70**
SGDR ensemble	0.75/0.54	0.72/0.51	0.71/0.49	0.63/0.44
WTA merged	0.74/0.47	0.82/0.49	**0.83**/0.56	**0.85**/0.64
EWTA merged	0.64/0.51	0.73/0.58	0.80/**0.64**	0.77/0.59

4.2 Nuclei Propagation Evaluation

We report the mean IoU of the nuclei segmentation for all experiments in Table 2. We use *interpolated* for nuclei completely missed before our improvement, *updated* for the nuclei that were segmented but our method decided to improve, *non-updated* for the nuclei that were chosen not to be improved by our method and *all* for all the mentioned cases. The first row shows the results before any propagation algorithm. We present 3 variants of motion used in propagation (column: *warped with*). We also explored using the certainties as pixel-wise weights in fusing segmentations from candidate neighbors and computed the weighted average to find the final mask. To isolate the gain by our uncertainty-based error detection, we created a baseline which is identical to our mean-flow variant, but performs the propagation on all the nuclei over a track. The significant improvement obtained by all our variants, especially flow variants without fusion, shows that our method can effectively improve erroneous nuclei predictions. The baseline that propagates to all frames (*all*) had a lower performance on the *non-updated* frames, showing that propagation independent of an uncertainty measure harms nuclei with high confidence (approx. 90% of all nuclei). For a better understanding of the challenge we tackle and the usefulness of our proposed method, we provide more qualitative results in the supplementary material.

Table 2. Quantitative evaluation of uncertainty-based propagation in mean IoU for all/updated/interpolated/non-updated nuclei with respective nuclei counts.

Update	Warp with	Mask fusion	All (117)	Updated (51)	Extrapolated (11)	Non-updated (55)
None	None	No	0.62	0.55	0.00	**0.80**
Uncertain	Shift + Scale	No	0.71	0.68	0.39	**0.80**
Uncertain	Mean nuclei flow	No	**0.73**	0.71	**0.45**	**0.80**
All	Mean nuclei flow	No	0.69	0.70	0.40	0.74
Uncertain	Pixel-wise flow	No	**0.73**	**0.72**	0.44	**0.80**
Uncertain	Pixel-wise flow	Yes	0.72	0.70	0.40	**0.80**

5   Conclusion

We addressed automated segmentation of image sequences, which cannot be analyzed frame-by-frame due to temporary uncertainties causing errors in predictions. First, we estimate uncertainty from Mask R-CNN to identify unreliable predictions. Second, we improve the less reliable predictions by propagating the more certain ones from neighbouring frames. We evaluated our method on HEK293T cells expressing a protein that oscillates in and out of the nucleus over time making the nucleus invisible temporarily. Our method improves nuclei segmentation over several baselines while keeping the cytosol and background segmentation untouched. We believe that our method will facilitate further time-series analysis for quantitative biology to understand the effect that the dynamic localization of the protein has on the cell without additional markers.

Acknowledgments. This project was funded by the German Research Foundation (DFG) and the German Ministry of Education and Science (BMBF). Gefördert durch die Deutsche Forschungsgemeinschaft (DFG) im Rahmen der Exzellenzstrategie des Bundes und der Länder - EXC-2189 - Projektnummer 390939984 und durch das Bundesministerium für Bildung und Forschung (BMBF) Projektnummer 01IS18042B und 031L0079.

References

1. Abdulla, W.: Mask R-CNN for object detection and instance segmentation on Keras and tensorflow. https://github.com/matterport/Mask_RCNN (2017)
2. Bertasius, G., Torresani, L.: Classifying, segmenting, and tracking object instances in video with mask propagation. Technical Report 1912.04573, arXiv (2019)
3. Brami-cherrier, K., et al.: Mechanisms of site-specific functions of focal adhesion kinase. Biophys. J. **104**, 609 (2013)
4. Chen, S.Y., et al.: Optogenetic control reveals differential promoter interpretation of transcription factor nuclear translocation dynamics. bioRxiv (2019)
5. Dosovitskiy, A., et al.: FlowNet: learning optical flow with convolutional networks. In: ICCV (2015)
6. He, K., Gkioxari, G., Dollár, P., Girshick, R.B.: Mask R-CNN. In: ICCV (2017)
7. Huang, G., Li, Y., Pleiss, G.: Snapshot ensembles: train 1, get M for free. In: ICLR (2017)
8. Ilg, E., et al.: Uncertainty estimates and multi-hypotheses networks for optical flow. In: ECCV (2018)
9. Jain, S., Wang, X., Gonzalez, J.E.: Accel: A corrective fusion network for efficient semantic segmentation on video. In: CVPR (2019)
10. Kendall, A., Gal, Y.: What uncertainties do we need in Bayesian deep learning for computer vision? In: NIPS (2017)
11. Lakshminarayanan, B., Pritzel, A., Blundell, C.: Simple and scalable predictive uncertainty estimation using deep ensembles. In: NIPS Workshop (2016)
12. Loshchilov, I., Hutter, F.: SGDR: stochastic gradient descent with warm restarts. In: ICLR (2017)
13. Makansi, O., Ilg, E., Çiçek, Ö., Brox, T.: Overcoming limitations of mixture density networks: a sampling and fitting framework for multimodal future prediction. In: CVPR (2019)
14. Milan, A., Rezatofighi, S.H., Dick, A., Reid, I., Schindler, K.: Online multi-target tracking using recurrent neural networks. In: AAAI (2017)
15. Niopek, D., et al.: Engineering light-inducible nuclear localization signals for precise spatiotemporal control of protein dynamics in living cells. Nature Commun. **5**, 4404 (2014)
16. Paul, M., Mayer, C., Gool, L.V., Timofte, R.: Efficient video semantic segmentation with labels propagation and refinement. Technical Report 1912.11844, arXiv (2019)
17. Payer, C., Štern, D., Neff, T., Bischof, H., Urschler, M.: Instance segmentation and tracking with cosine embeddings and recurrent hourglass networks. In: Frangi, A.F., Schnabel, J.A., Davatzikos, C., Alberola-López, C., Fichtinger, G. (eds.) MICCAI 2018. LNCS, vol. 11071, pp. 3–11. Springer, Cham (2018). https://doi.org/10.1007/978-3-030-00934-2_1
18. Repina, N.A., Rosenbloom, A., Mukherjee, A., Schaffer, D.V., Kane, R.S.: At light speed: Advances in optogenetic systems for regulating cell signaling and behavior. Ann. Rev. Chem. Biomol. Eng. **8**(1), 13–39 (2017)

19. Ronneberger, O., Fischer, P., Brox, T.: U-Net: convolutional networks for biomedical image segmentation. In: Navab, N., Hornegger, J., Wells, W.M., Frangi, A.F. (eds.) MICCAI 2015. LNCS, vol. 9351, pp. 234–241. Springer, Cham (2015). https://doi.org/10.1007/978-3-319-24574-4_28

20. Sokooti, H., de Vos, B., Berendsen, F., Lelieveldt, B.P.F., Išgum, I., Staring, M.: Nonrigid image registration using multi-scale 3D convolutional neural networks. In: Descoteaux, M., Maier-Hein, L., Franz, A., Jannin, P., Collins, D.L., Duchesne, S. (eds.) MICCAI 2017. LNCS, vol. 10433, pp. 232–239. Springer, Cham (2017). https://doi.org/10.1007/978-3-319-66182-7_27

Semi-supervised Instance Segmentation with a Learned Shape Prior

Long Chen(✉), Weiwen Zhang, Yuli Wu, Martin Strauch,
and Dorit Merhof

Institute of Imaging and Computer Vision, RWTH Aachen University,
Aachen, Germany
{long.chen,martin.strauch,dorit.merhof}@lfb.rwth-aachen.de
https://www.lfb.rwth-aachen.de/

Abstract. To date, most instance segmentation approaches are based on supervised learning that requires a considerable amount of annotated object contours as training ground truth. Here, we propose a framework that searches for the target object based on a shape prior. The shape prior model is learned with a variational autoencoder that requires only a very limited amount of training data: In our experiments, a few dozens of object shape patches from the target dataset, as well as purely synthetic shapes, were sufficient to achieve results en par with supervised methods with full access to training data on two out of three cell segmentation datasets. Our method with a synthetic shape prior was superior to pretrained supervised models with access to limited domain-specific training data on all three datasets. Since the learning of prior models requires shape patches, whether real or synthetic data, we call this framework semi-supervised learning. The code is available to the public (https://github.com/looooongChen/shape_prior_seg).

Keywords: Semi-supervised · Instance segmentation · Shape prior · Variational autoencoder · Edge loss

1 Introduction

Instance segmentation, where many instances of an object have to be segmented in one image, is the basis of several practically relevant applications of computer vision, such as cell tracking [1]. Many approaches [2–4] have been proposed for instance segmentation, the majority of which are based on supervised learning. The practical applicability of these methods is often limited by the lack of a large training dataset with manually outlined objects. Here, we introduce an instance segmentation approach that only relies on a shape prior which can be learned from a considerably smaller number of training samples or even synthetic data.

This work was supported by the Deutsche Forschungsgemeinschaft (Research Training Group 2416 MultiSenses-MultiScales).

J. Cardoso et al. (Eds.): iMIMIC 2020/MIL3ID 2020/LABELS 2020, LNCS 12446, pp. 94–102, 2020.
https://doi.org/10.1007/978-3-030-61166-8_10

The shape is one of the most informative cues in object segmentation and detection tasks. Anatomically constrained neural networks (ACNNs) [5] improve segmentation results by including a shape prior for model regularization. For segmentation refinement, a shape prior has been used by [6] as a separate post-processing step. Segmentations generated by the shape prior model are reconstructed to the original MRI images through several convolutional layers in [7]. By minimizing the reconstruction error, the segmentation model can be trained in an unsupervised fashion. All these works report promising results, but are limited to cases where object position and extent are roughly the same in all images, such as for the cardiac images in [5], the lung X-ray images in [6] and the brain MRI scans in [7]. To our knowledge, this is the first work considering instance segmentation based on a shape prior, i.e. we detect and segment multiple, scattered object instances. Similar to [8], we use the spatial transformer [9] to localize objects. The main advantage of using the spatial transformer lies in its differentiability, making the whole framework end-to-end trainable.

The main contributions of this work are: We propose (1) an semi-supervised instance segmentation approach that searches for target objects based a shape prior, and (2) a novel loss computing the difference between two gradient maps. This framework provides a way to achieve instance segmentation with a small amount of manual annotations, or by utilizing unpaired annotations (where the correspondence between annotations and images is unknown). We compared our approach to the state-of-the-art supervised method, Mask R-CNN [2], in different training scenarios. On three experimental datasets, our approach is proved to be en par with a Mask R-CNN with full access to training data, while it outperforms a pre-trained Mask R-CNN with limited access to domain-specific training data.

2   Approach

As shown in Fig. 1, our framework consists of three main parts: 1) the localization network, 2) the spatial transformer [9], and 3) the patch segmentation network. Based on the localization prediction, the spatial transformer crops local patches and feeds them to the patch segmentation network. The gradient maps of segmented patches are then stitched together. The entire model is trained by minimizing the reconstruction error of the gradient map.

During training, the model learns to predict the object position and to find the correspondence between the image patch and the segmentation. The shape prior model (gray part in Fig. 1; fixed during training) is guaranteed to output a plausible shape, but the correspondence has to be learned by the model itself.

2.1   Localization Network

The localization network consists of 8 convolutional layers and 4 max pooling layers after every 2 convolutional layers. Given an image of size (H_{img}, W_{img}), the localization network will spatially divide the image into an $(H_{img}/S_{cell}, W_{img}/S_{cell})$ grid of cells, where S_{cell} is the cell size and also the

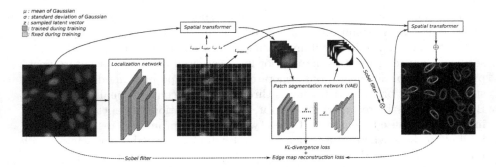

Fig. 1. Architecture of our framework: the localization network predicts the object position and a presence score, based on which object patches are cropped by a spatial transformer. A variational autoencoder with the decoder part fixed (shape prior) is responsible for the patch segmentation. At last, the gradient maps of segmented patches are stitched together. The model is trained by minimizing the reconstruction loss of the gradient map with the KL-divergence loss as regularization.

downsampling rate. Since 4 pooling layers with stride 2 are used, we have $S_{cell} = 16$.

Each cell is responsible to predict the presence of an object $L_{presence} \in [0, 1]$, its range described by the bounding box size (H_{obj}, W_{obj}) and the offset with respect to the cell center (O_x, O_y) (Fig. 2(a)), with the implementation:

$$L_{presence} = sigmoid(f_{presence})$$
$$L_{scale} = sigmoid(f_{scale}) \cdot (S_{max} - S_{min}) + S_{min}$$
$$L_{ratio} = \exp(tanh(f_{ratio}) \cdot \log(R_{max}))$$
$$(L_x, L_y) = (0.5 \cdot tanh(f_x), 0.5 \cdot tanh(f_y))$$

where $f_{[\cdot]}$ is the corresponding input feature map. $sigmoid(\cdot)$ and $tanh(\cdot)$ denote the sigmoid and tanh activation function. S_{min}, S_{max} and R_{max} are hyperparameters, which are the minimal scale, the maximal scale and the maximal aspect ratio, respectively. The position is parameterized according to:

$$(H_{obj}, W_{obj}) = (L_{scale} \cdot S_{cell}/\sqrt{L_{ratio}}, \ L_{scale} \cdot S_{cell} \cdot \sqrt{L_{ratio}})$$
$$(O_x, O_y) = (L_x \cdot S_{cell}, \ L_y \cdot S_{cell})$$

It is worth mentioning that the maximal offset is $0.5 \cdot S_{cell}$, which means that an object will be detected by the cell in which its center lies.

2.2　Patch Crop and Stitch

Given the location parameters obtained from the localization network, we use a spatial transformer to crop local patches. The spatial transformer implements the

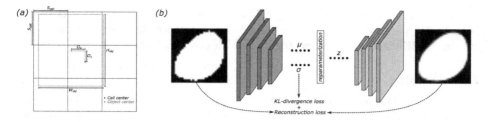

Fig. 2. (a) Demonstration of parameters of a bounding box. (b) Architecture of the patch segmentation network, which is firstly trained with shape patches. During the detector training, the decoder part is fixed and plays the role of shape prior.

crop by sampling transformed grid points, which is differentiable, enabling end-to-end training. The patch crop of the i-th cell can be described by transform:

$$
T^i_{crop} = \begin{bmatrix} W_{img}/W^i_{obj} & 0 & W_{img} \cdot (X^i_{cell} + O^i_y)/W^i_{obj} \\ 0 & H_{img}/H^i_{obj} & H_{img} \cdot (Y^i_{cell} + O^i_x)/H^i_{obj} \\ 0 & 0 & 1 \end{bmatrix}
$$

where (X^i_{cell}, Y^i_{cell}) is the cell center. (O^i_x, O^i_y) and (H^i_{obj}, W^i_{obj}) are the predicted offset and size of the object. All cropped patches will be rescaled to size $S_{patch} \times S_{patch}$ ($S_{patch} = 32$ in this work) and segmented by the patch segmentation network, as described in Sect. 2.3. After that, the gradient map of segmented objects will be stitched together by adding up back transformed patches through:

$$
T^i_{stitch} = \begin{bmatrix} W^i_{obj}/S_{patch} & 0 & X^i_{cell} + O^i_y \\ 0 & H^i_{obj}/S_{patch} & Y^i_{cell} + O^i_x \\ 0 & 0 & 1 \end{bmatrix}
$$

The gradient map is computed by applying the x- and y-directional Sobel filter to the image and taking the square root of the summed square. The gradient map is normalized to range 0 to 1. In this work, we use an input size of 256×256 for all experiments. Considering $S_{cell} = 16$, 256 patches are cropped in total.

2.3 Shape Prior and Patch Segmentation Network

Similar to [5–7], we employ a variational autoencoder (VAE) as our shape model. As shown in Fig. 2(b), the model is trained to reconstruct plausible patch segmentation masks with the KL-divergence loss as regularization. Compared to a standard autoencoder, a VAE learns a more continuous latent space, which is expected to generate plausible new shapes that do not appear in training data.

In this work, the VAE is trained with 32×32 patches. The encoder and decoder consist of 6 convolutional layers and 3 pooling/upsampling layers, respectively. Based on our experiments, model training requires only a small amount of data, especially when the shape variation is small. We train the shape prior with either annotations from a single image or synthetic data (Sect. 3).

After training, the decoder part will be used as the shape prior in the detector (Fig. 1). Its parameters will be fixed during the detector training. The encoder will be reinitialized and trained together with the localization network.

2.4 Training

The model is trained end-to-end by minimizing the gradient map reconstruction error with the KL-divergence loss as regularization. In initial experiments, we found the mean absolute/squared error (MAE/MSE) to be very unstable during training: The shape prior model tends to generate distorted shapes or degenerates into empty output. Thus, we propose the following novel loss:

$$L_{edge} = 1 - \frac{\frac{1}{N}\sum_i min^2(G^i_{image}, G^i_{reconstruction})}{\frac{1}{N}\sum_i G^i_{reconstruction} + \alpha} \tag{1}$$

where G_{image} and $G_{reconstruction}$ indicate the gradient map of the image and the reconstructed gradient map. N is the number of pixels. The $min()$ operation are conducted pixelwise. The parameter α prevents the model from pushing $G_{reconstruction}$ to zero and is set to 0.01 empirically.

Instead of optimizing the value of each pixel, as MSE and MAE, this loss maximizes the proportion of the reconstructed gradient map under the image gradient map. In addition, the square operator in the numerator is proved to be crucial for stable training in our experiments. Our interpretation is that the square operator modulates the back-propagated gradient with the reconstructed gradient map, giving more emphasis to positions around the edge.

2.5 Pre- and Post-processing

To reduce the influence of extreme values on the loss, we equalized the image and the gradient map by clipping and stretching. For all datasets, we truncated the gradient map at 0.8 times the maximum and normalized the value to the range 0 to 1. In addition, we also performed image equalization for the Fluo-N2DH-SIM+ dataset due to the bright spots inside the cell (Fig. 3). The clip value was set to 1.2 times the image mean.

As post-processing, we first filtered out predictions with $L_{presence}$ smaller than 0.1. Non-max suppression is then performed to eliminate duplicate predictions: An instance mask is compared with another mask, when the overlapping area is larger than $p_{non\_max} = 0.1$ with respect to its own area. A mask is only retained if its score is the highest in all comparisons.

3   Experiments and Results

3.1 Datasets and Experiments

We evaluate our approach on three datasets: the BBBC006 dataset[1] and two datasets Fluo-N2DH-SIM+ and PhC-C2DL-PSC from the cell tracking challenge [1]. In the following, we use BBBC, FLUO and PHC as abbreviations. The

---

[1] https://data.broadinstitute.org/bbbc.

BBBC dataset contains 768 microscopic images of human U2OS cells, while the FLUO (HL60 cells with Hoechst staining) and PHC (pancreatic stem cells on a polystyrene substrate) datasets are smaller with 215 and 202 annotated images.

For comparison, we also report the performance of the supervised method Mask R-CNN. The following experiments are performed:

Ours-Annotation: We first evaluate our approach with the shape prior learned from manual annotations. We only took segmentation patches from one image. Specifically, 67, 8 and 138 object patch masks were used for the BBBC, FLUO and PHC shape model training. To model small shape changes and object rotation, we performed rotation (in steps of $30°$) and elastic deformation [11] to augment the training set. The scale range and maximal aspect ratio was set to $2-3/3$, $1-2/1.5$ and $1-2/3$, respectively.

Ours-Synthetic: Since the objects are approximately circular, especially for the BBBC and FLUO datasets, we could train the shape prior model with synthetic data consisting of elastically deformed ellipses [11] with random angle and major-minor axis ratio. The maximal major-minor axis ratio was 2, 1.5 and 3 for the BBBC, FLUO and PHC dataset, respectively.

MRCNN-Scratch-One/Full: We trained a Mask R-CNN from scratch using ResNet-50 backbone. The anchor box scale, aspect ratio and non-maximum suppression (NMS) threshold were set to values equivalent to those used in our approach. Since the Ours-annotation scenario can be considered as one image training, we also trained a Mask R-CNN with one image for comparison.

MRCNN-Finetune-One/Full: Since the dataset in our experiments is small, especially for FLUO and PHC, we pretrained the Mask R-CNN on the MS COCO dataset[2]. Afterwards, we finetuned the model, with only the head layers trainable, on the actual target dataset.

For the BBBC and PHC dataset, we cropped images to 256×256 and 128×128 for training and test. All images were resized to 256×256 for the network input. For the scenarios using one training image (Ours-annotation, MRCNN-scratch-one, MRCNN-finetune-one), the images a01_s1, 02/t000, 02/t150 were used for BBBC, FLUO and PHC, respectively. MRCNN-scratch-full and MRCNN-finetune-full used a01_s1-b24_s2, 02/t000-t149, 02/t150-t250 for training. Ours-synthetic requires no manual annotations. All remaining images were kept for testing.

3.2   Results and Discussion

We report the *average precision*[3] (AP) over a range of IoU (intersection over union) thresholds from 0.3 to 0.9 as the evaluation score (Table 1). Our approach, including the evaluation scenarios where the shape prior is learned from one image annotation and synthetic data, outperforms the Mask R-CNN trained

---

[2] https://cocodataset.org/.

[3] https://www.kaggle.com/c/data-science-bowl-2018.

or finetuned with one image, which shows the advantage of our approach in cases where few or no annotations are available. Furthermore, our approach achieves comparable results with the Mask R-CNN trained/finetuned with the full training set on the BBBC and FLUO dataset, while the performance gap is apparent for the PHC dataset.

While Mask R-CNN achieved the best mean AP (mAP) on the BBBC dataset, our approach outperformed Mask R-CNN on the FLUO dataset by a relatively large margin. The main reason is that the FLUO dataset is indeed a very small one for Mask R-CNN training, even with finetuning. This again illustrates the advantage of our method on small datasets.

| Ground Truth | Ours-synthetic | Ours-annotation | MRCNN-finetune-one | MRCNN-finetune-full |

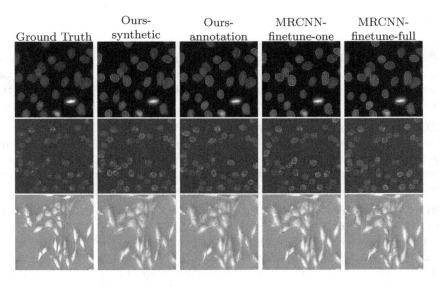

Fig. 3. Qualitative results: from top to bottom, the rows show the results on the BBBC006, Fluo-N2DH-SIM+ and PhC-C2DL-PSC datasets, respectively.

On the PHC dataset, neither method performed particularly well. Both methods tended to detect nearby objects as one if there was no clearly visible edge between them. The average precision of our method in the low IoU range was close to or better than that of Mask R-CNN. Figure 3 shows that our method could detect most objects as well as the Mask R-CNN. However, our method has been designed to heavily rely on the edge clue, so that the segmentation will converge to strong edges. For the PHC dataset, the object boundaries do not generally correspond to the strongest edges. This explains why objects were undersegmented by our approach (Fig. 3) and why the average precision decreased rapidly with increasing IoU (Table 1).

The performance improvement through training the shape prior with manually outlined shapes depends on the nature of the shape. On the FLUO dataset, annotated data and synthetic data shape priors performed almost equally well, while training with manual annotations was superior on the other two datasets, even though only a few dozen shapes were used.

Table 1. Average precision (AP) over different IoU for different datasets (the best two scores in bold). Experiments and abbreviations are introduced in Sect. 3.1.

Dataset	IoU	0.3	0.4	0.5	0.6	0.7	0.8	0.9	mAP
BBBC	Ours-annotation	**.8345**	**.8260**	**.7977**	**.7632**	.7083	.6100	.2660	**.6865**
	Ours-synthetic	.8171	.8012	.7641	.7170	.6525	.5247	.2042	.6401
	MRCNN-scratch-one	.6386	.5934	.5459	.4769	.3543	.1759	.0294	.4020
	MRCNN-scratch-full	.7901	.7851	.7708	.7473	**.7128**	**.6296**	**.3374**	.6817
	MRCNN-finetune-one	.7672	.7524	.7277	.7020	.6608	.5492	.1250	.6121
	MRCNN-finetune-full	**.7997**	**.7949**	**.7851**	**.7720**	**.7521**	**.6923**	**.3485**	**.7064**
FLUO	Ours-annotation	**.9605**	**.9538**	**.9312**	**.8999**	.8228	**.6777**	.1332	.7685
	Ours-synthetic	**.9600**	**.9497**	**.9336**	**.8986**	**.8324**	**.6768**	.1378	**.7698**
	MRCNN-scratch-one	.0458	.0324	.0156	.0018	.0000	.0000	.0000	.0014
	MRCNN-scratch-full	.9333	.9144	.8703	.7605	.5765	.2556	.01073	.6173
	MRCNN-finetune-one	.8224	.8133	.7905	.7389	.5909	.2404	.0049	.5716
	MRCNN-finetune-full	.9361	.9252	.8955	.8467	.7265	.4115	.0197	**.6802**
PHC	Ours-annotation	**.6840**	**.6034**	.4035	.1468	.0233	.0028	.0000	.2662
	Ours-synthetic	.6471	.5611	.3605	.1326	.0219	.0027	.0000	.2466
	MRCNN-scratch-one	.1124	.0991	.0847	.0668	.0353	.0049	.0000	.0576
	MRCNN-scratch-full	.6332	.6001	**.5226**	**.4467**	**.2981**	**.1079**	**.0023**	**.3730**
	MRCNN-finetune-one	.1647	.1602	.1460	.1146	.0633	.0108	.0000	.0942
	MRCNN-finetune-full	**.6551**	**.6380**	**.5855**	**.5014**	**.3425**	**.1144**	**.0007**	**.4053**

4   Conclusion and Outlook

We have proposed an instance segmentation framework which searches for target objects in images based on a shape prior model. In practice, this allows segmenting instances with a very limited amount of annotations, segmenting synthesizable shapes without any annotation, as well as reusing object annotations from other datasets.

The main limitation of our approach lies in the dependency on the edge cues. Images should have a relatively clear background, which is, however, the case for many biomedical datasets(see Footnote 3). Future work will focus on including area-based information, which will make our approach applicable to further datasets, e.g. in cases where edges and object boundaries do not always coincide.

References

1. Ulman, V., et al.: An objective comparison of cell-tracking algorithms. Nat. Methods **14**, 1141–1152 (2017)

2. He, K., Gkioxari, G., Dollár, P., Girshick, R.: Mask R-CNN. In: ICCV 2017, pp. 2980–2988 (2017)
3. Schmidt, U., Weigert, M., Broaddus, C., Myers, G.: Cell detection with star-convex polygons. In: Frangi, A., Schnabel, J., Davatzikos, C., Alberola-López, C., Fichtinger, G. (eds.) MICCAI 2018. LNCS, vol. 11071, pp. 265–273. Springer, Cham (2018). https://doi.org/10.1007/978-3-030-00934-2_30
4. Chen, L., Strauch, M., Merhof, D.: Instance segmentation of biomedical images with an object-aware embedding learned with local constraints. In: Shen, D., et al. (eds.) MICCAI 2019. LNCS, vol. 11764, pp. 451–459. Springer, Cham (2019). https://doi.org/10.1007/978-3-030-32239-7_50
5. Oktay, O., et al.: Anatomically constrained neural networks (ACNNs): application to cardiac image enhancement and segmentation. IEEE Trans. Med. Imaging **37**(2), 384–395 (2018)
6. Larrazabal, A.J., Martinez, C., Ferrante, E.: Anatomical priors for image segmentation via post-processing with denoising autoencoders. In: Shen, D., et al. (eds.) MICCAI 2019. LNCS, vol. 11769, pp. 585–593. Springer, Cham (2019). https://doi.org/10.1007/978-3-030-32226-7_65
7. Dalca, A.V., Guttag, J., Sabuncu, M.R.: Anatomical priors in convolutional networks for unsupervised biomedical segmentation. In: CVPR 2018, pp. 9290–9299 (2018)
8. Crawford, E., Pineau, J.: Spatially invariant unsupervised object detection with convolutional neural networks. In: AAAI 2019, pp. 3412–3420 (2019)
9. Jaderberg, M., Simonyan, K., Zisserman, A., Kavukcuoglu, K.: Spatial transformer networks. In: NIPS 2015, pp. 2017–2025 (2015)
10. Kingma, D.P., Welling, M.: Auto-encoding variational Bayes. In: ICLR 2014 (2014)
11. Simard, P.Y., Steinkraus, D., Platt, J.C.: Best practices for convolutional neural networks applied to visual document analysis. In: Proceedings of the Seventh International Conference on Document Analysis and Recognition, p. 958. IEEE (2003)

COMe-SEE: Cross-modality Semantic Embedding Ensemble for Generalized Zero-Shot Diagnosis of Chest Radiographs

Angshuman Paul[1]([⊠]), Thomas C. Shen[1], Niranjan Balachandar[1],
Yuxing Tang[1], Yifan Peng[2], Zhiyong Lu[2], and Ronald M. Summers[1]

[1] Imaging Biomarkers and Computer-Aided Diagnosis Laboratory, Radiology
and Imaging Sciences, National Institutes of Health Clinical Center,
Bethesda, MD 20892, USA
paul.angshuman@nih.gov
[2] Biomedical Text Mining Group, National Center for Biotechnology Information,
National Institutes of Health, Bethesda, MD 20894, USA

Abstract. Zero-shot learning, in spite of its recent popularity, remains an unexplored area for medical image analysis. We introduce a first-of-its-kind generalized zero-shot learning (GZSL) framework that utilizes information from two different imaging modalities (CT and x-ray) for the diagnosis of chest radiographs. Our model makes use of CT radiology reports to create a semantic space consisting of signatures corresponding to different chest diseases and conditions. We introduce a CrOss-Modality Semantic Embedding Ensemble (COMe-SEE) for zero-shot diagnosis of chest x-rays by relating an input x-ray to a signature in the semantic space. The ensemble, designed using a novel semantic saliency preserving autoencoder, utilizes the visual and the semantic saliency to facilitate GZSL. The use of an ensemble not only helps in dealing with noise but also makes our model useful across different datasets. Experiments on two publicly available datasets show that the proposed model can be trained using one dataset and still be applied to data from another source for zero-shot diagnosis of chest x-rays.

Keywords: Zero-shot learning · Chest x-ray · Semantic saliency · Autoencoder · Ensemble

1 Introduction

In radiology diagnosis, it is often difficult to find image instances for all possible diseases, especially for the rare ones. Diagnosis of such diseases from radiology images poses a great challenge in clinical practice. If a disease is less common, radiologists may have to rely on the clinical features available from the descriptions of the disease for its diagnosis. In such a situation, a radiologist may need

Electronic supplementary material The online version of this chapter (https://doi.org/10.1007/978-3-030-61166-8_11) contains supplementary material, which is available to authorized users.

J. Cardoso et al. (Eds.): iMIMIC 2020/MIL3ID 2020/LABELS 2020, LNCS 12446, pp. 103–111, 2020.
https://doi.org/10.1007/978-3-030-61166-8_11

to map the visual information from the radiology images to the semantic information embedded in the descriptions of the diseases.

Zero-shot learning (ZSL) is the branch of machine learning that attempts to mimic the human ability to make predictions about the data belonging to the classes not seen during training. ZSL methods use the knowledge learned from the classes seen during training (*seen* classes) to make predictions about classes, not seen during training (*unseen* classes) by utilizing some form of auxiliary information about the classes. ZSL has achieved impressive results for natural images [5,6,8,11,13].

Generalized zero-shot learning (GZSL) is the branch of ZSL in which test datasets may contain data from both the seen and the unseen classes. GZSL may play a key role for automated identification of rare diseases. However, applying GZSL techniques for medical image diagnosis presents significant challenges due to noise in the data and labels, multiple regions of interest in each image and the relative scarcity of auxiliary information to aid the ZSL methods.

We propose a GZSL framework for the diagnosis of chest x-rays by addressing the above challenges. We use computed tomography (CT) radiology reports to mine auxiliary information about different diseases in the form of semantic signatures. CT reports are chosen because CT scans and reports contain richer information than x-ray images and reports, and may provide higher diagnostic accuracy for some diseases [3]. To relate x-ray images to these CT-based semantic signatures, we introduce a cross-modality semantic embedding ensemble (COMe-SEE) which is composed of semantic saliency preserving autoencoders (SSP-AEs). We utilize visual saliency by letting each SSP-AE in the ensemble to explore a different visual subspace and learn a unique mapping to the semantic space. The use of an ensemble also helps to mitigate the effects of noise and makes the model robust across datasets. In this work, our contributions are as follows:

- Design of a first-of-its-kind GZSL method for the diagnosis of chest x-rays.
- Introduction of a semantic saliency preserving autoencoder (SSP-AE).
- Design of an ensemble of SSP-AEs that utilizes visual saliency for visual-to-semantic mapping across two different imaging modalities (x-ray and CT).

2   Methods

To design the GZSL framework for the diagnosis of chest x-rays, we divide the diseases and conditions of interest into seen classes and unseen classes. During training, we have access to auxiliary information about both the seen and the unseen classes in the form of disease-specific semantic signatures alongside the x-ray images of seen classes. The signatures are generated by a *signature generator* from CT radiology reports. We use Intelligent Word Embedding [1] (IWE), a Word2Vec [12] model trained on CT reports, to generate signatures for the chest diseases and conditions of interest. Subsequently, we design a semantic embedding ensemble that performs a cross-modality semantic embedding between the visual characteristics derived from x-ray images of different diseases and the corresponding semantic signatures. A diagram of the proposed framework is presented in Fig. 1.

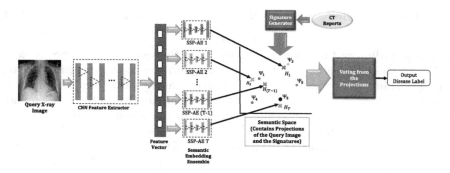

Fig. 1. The block diagram of the proposed model. Ψ_1 to Ψ_5: signatures in the semantic space; $H_1, ..., H_T$: projections for the input query x-ray image from different SSP-AEs.

2.1   Cross-modality Semantic Embedding

We take a two-step approach to design the semantic embedding model. In the first step, we extract features from x-ray images using a feature extractor composed of DenseNet-121 [10]. The output from the penultimate layer of the network is taken as the feature vector for an input x-ray image. Since we train the feature extractor with x-ray images only from the seen classes, the extracted features are likely to be noisy especially for the images of unseen classes at inference time. In the second step, we perform semantic embedding through the utilization of visual saliency using autoencoders, well-known for extracting saliency from noisy feature vectors [9]. Consider an autoencoder that takes an input feature vector $\mathbf{X}$ and reconstructs it ($\widehat{\mathbf{X}}$) at the output. A vanilla autoencoder is trained by minimizing reconstruction loss $L_{re} = \|\widehat{\mathbf{X}} - \mathbf{X}\|$. However, this training does not guarantee a good semantic embedding. Therefore, we introduce the following loss terms to design a novel semantic saliency preserving autoencoder (SSP-AE) for a meaningful semantic embedding.

Semantic Embedding Loss: Let Ψ_c be the signature of the cth disease that belongs to the set of seen classes. Also, assume $\mathbf{H}(n)$ to be the hidden space representation of training feature vector $\mathbf{X}(n)$ and the true class label of $\mathbf{X}(n)$ be k. Then, we train the hidden space representation $\mathbf{H}(n)$ to be close to Ψ_k, the semantic signature of its class label. To do this, we use an indicator function $I_c(n)$ that takes value 1 when c is the true class label of training data $\mathbf{X}(n)$ and -1 otherwise. Consider N_{tr} to be the number of training data points and C_S to be the set of seen classes. Then we introduce the following semantic embedding loss

$$L_{se} = \sum_{n=1}^{N_{tr}} \sum_{\forall c \in C_S} I_c(n)\|H(n) - \Psi_c\|, \tag{1}$$

Minimization of L_{se} forces the training data of the seen classes to be projected close to their corresponding signatures in the hidden space (which is also the semantic space), helping the autoencoder to learn a visual-to-semantic embedding.

Semantic Saliency Loss: Feature vectors from the seen classes are projected in the hidden space through the autoencoder, and form a cluster for each seen class. During training, these clusters should be created as far away from each other as possible; the farther the clusters are from each other, the better the semantic saliency is preserved. We define a loss component that exploits this. Let $\mathbf{G}_i$ and $\mathbf{G}_j$ be the cluster centers corresponding to seen classes i and j, respectively, in the hidden (semantic) space of the autoencoder. Then we construct a vector $\mathbf{V}_{ij}$ by vector subtraction of $\mathbf{G}_j$ from $\mathbf{G}_i$, given by $\mathbf{V}_{ij} = \mathbf{G}_i - \mathbf{G}_j$. We want to maximize $\mathbf{V}_{ij}$. Hence, we formulate the semantic saliency loss as:

$$L_{sal} = \sum_{\forall i,j \in \mathcal{C}_S} \frac{1}{(\|\mathbf{V}_{ij}\|^2 + \epsilon)}, \tag{2}$$

where ϵ is a constant of small value to avoid division by zero. The proposed SSP-AE preserves semantic saliency by minimizing L_{sal}. Taking into account the above loss components, the overall loss function for the SSP-AE is:

$$\mathcal{L} = L_{re} + \beta_1 L_{se} + \beta_2 L_{sal}, \tag{3}$$

where β_1 and β_2 are pre-defined constants. For an useful semantic embedding, we need to utilize the salient features from the noisy feature vectors. Towards that end, we design an ensemble of SSP-AEs.

2.2   Semantic Embedding Ensemble

We need the ensemble of SSP-AEs to generalize well for the unseen classes at the test time. This is possible if each SSP-AE is exposed to different training data [2]. For this, we use bootstrap sampling [2] where a random subset of training data is sampled to train each SSP-AE. Thus each autoencoder is exposed to different training data. Let $B(t)$ be the bootstrap sample for autoencoder t. Using $B(t)$, we find a subspace rich in visually salient features for the tth SSP-AE to explore. Thus each SSP-AE in the ensemble explores a different subspace and learns a unique semantic mapping. This is achieved through subspace sampling.

Subspace Sampling: The subspace sampling process is semi-deterministic. Let the dimension of the feature vectors obtained from the feature extractor be d. For the tth SSP-AE, we first randomly sample M number of d' dimensional subspaces from this d dimensional feature space. Consider subspace m. Let $B_m(t)$ be the projection of $B(t)$ on this subspace. We find the average of inter-cluster distances (for the clusters corresponding to different seen classes) in this subspace. The subspace m^* having the highest average of inter-cluster distances is chosen for the tth SSP-AE which is trained by $B_{m^*}(t)$ (projection of $B(t)$ on subspace m^*). We repeat this process for each SSP-AE in the ensemble.

Ensemble Training: Training the ensemble involves training each of the SSP-AEs in the ensemble. Let us consider an ensemble composed of T number of SSP-AEs. For each SSP-AE, we do the following steps. First, for the tth SSP-AE, we find the effective training data $B_{m^*}(t)$. Subsequently, the tth SSP-AE is trained by minimizing the loss of (3) on $B_{m^*}(t)$. Note that all the autoencoders in the ensemble can be trained in parallel. Therefore, in a parallel environment, the training time of the ensemble is equal to the training time of an individual autoencoder, leading to faster training. Due to the randomness involved in subspace sampling, the semantic mapping learned by different SSP-AEs during training are nonuniform in their usefulness. Therefore, we measure the usefulness of the mapping learned by an SSP-AE, and assign a weight to the SSP-AE based on the usefulness.

Weight Assignment: Let $\mathbf{G}_c(t)$ be the cluster center corresponding to seen class c in the hidden space of the tth SSP-AE after training. For class c in the set of seen classes C_S, a perfect training would cause the cluster center $\mathbf{G}_c(t)$ to fall on the semantic signature of class c (i.e. Ψ_c). The farther $\mathbf{G}_c(t)$ is from Ψ_c, the less useful the visual-to-semantic mapping learned by the tth SSP-AE. Based on this, we assign weights to each trained SSP-AE and use the weights for decision making during inference. The weight of the tth SSP-AE is:

$$w(t) = \exp\left(-\sum_{\forall i \in \mathcal{C}_S} \|\Psi_c - \mathbf{G}_c(t)\|\right). \tag{4}$$

2.3 Generalized Zero-Shot Diagnosis of Chest Radiographs

Let a query x-ray image be I. First, the feature extractor extracts a feature vector $\mathbf{X}(I)$ from I. We apply $\mathbf{X}(I)$ to the semantic embedding ensemble. Let the hidden space projection of $\mathbf{X}(I)$ by the tth SSP-AE be $\mathbf{H}_t(I)$. If the semantic signature closest to $\mathbf{H}_t(I)$ is Ψ_k, the tth SSP-AE assigns class label k to the query image I. In this way, every other SSP-AE also assigns a class label to the query image I. After a weighted voting with the weights of the SSP-AEs (obtained from (4)), the class label with the highest vote is assigned to the query image I.

2.4 Implementation Details

We use 25, 20, and 20 SSP-AEs to construct the ensembles for NIH, NIH-900, and Open-i datasets, respectively. The signature generator produces 160-dimensional signatures for each of the diseases and conditions. Each SSP-AE explores a subspace of dimension $d' = 40000$ from the feature space of dimension 50176. An SSP-AE consists of an encoder and a decoder. The encoder of an SSP-AE contains linear layers (each followed by a ReLU activation) mapping the input to 2048, 512 and 160-dimensions (hidden space) respectively. The decoder has the same linear layers (each followed by a ReLU activation) in the exact opposite order. Adam optimizer [7] is used with a mini-batch size of 16 and a learning rate of 0.001 to train each SSP-AE.

Table 1. Performances of different methods in terms of seen recall (Re_S), unseen recall (Re_U) and harmonic mean of seen and unseen recall (Re_H) (**bold** fonts: best values in each column). The proposed method and its variants are indicated in *italics*.

Methods	NIH			NIH-900			Open-i		
	Re_S	Re_U	Re_H	Re_S	Re_U	Re_H	Re_S	Re_U	Re_H
GMN [13]	35.07	5.33	9.25	26.26	1.56	2.94	30.48	1.6	3.4
GDAN [6]	32.14	8.95	14	29.8	6.24	10.32	31.51	4.57	7.98
SAE [8]	29.17	8.63	13.32	29.66	9.61	14.52	29.45	11.42	16.46
ESZSL [11]	34.16	1.80	3.42	**32.67**	0.00	0.00	**45.10**	0.00	0.00
DeViSe [5]	**41.55**	0.03	0.07	14.33	0.00	0.00	21.23	0.00	0.00
Sng. SSP-AE	34.32	13.32	**19.19**	26.00	5.28	8.78	24.65	8.31	12.43
RandSub	8.21	**23.04**	12.11	10.33	**18.75**	13.32	13.01	16.10	14.39
Proposed	12.03	23.01	15.80	17.66	18.26	**17.96**	26.71	**16.10**	**20.09**

3   Experiments and Results

To demonstrate the robustness of our method, we perform experiments on chest x-ray datasets from two different sources: the NIH chest x-ray dataset [14] and the Open-i dataset [4]. Note that we train our model only using the training data from NIH dataset [14] and test the model on both the test sets of NIH [14] and Open-i. Since the labels for images in the above datasets are extracted using an automated rule-based approach, the labels may be noisy in some instances. To see the performance on noiseless labels, we separately test our model on a manually labeled 900-image subset [15] (NIH-900) from the NIH dataset.

Based on the availability of semantic signatures from CT reports, we consider nine chest diseases and conditions for our experiments. Those are cardiomegaly, consolidation, edema, effusion, emphysema, infiltration, nodule, pneumonia and pneumothorax. Out of these, we randomly select cardiomegaly, edema and emphysema as unseen classes and take the rest as seen classes.

3.1   Performance Measures and Comparisons

Following the usual practice for ZSL methods [13], we report the results in terms of recall on seen classes (Re_S), recall on unseen classes (Re_U) and the harmonic mean of the recall values in seen and unseen classes (Re_H) [13]. Since our datasets are multi-label, if the algorithm output matches with one of the ground truth labels of a test image, we consider the output to be a true positive.

We compare the performance of the proposed method with a number of state-of-the-art ZSL techniques including GMN [13], GDAN [6], SAE [8], ESZSL [11] and DeViSe [5]. We use the feature vectors from the feature extractor as inputs to each of the above competing methods. We compare the performance of a single SSP-AE (abbreviated as Sng. SSP-AE) to that of the proposed ensemble as well. The performances of the above methods are presented in Table 1. The effect of the proposed method with random subspace sampling (abbreviated as RandSub) instead of the proposed semi-deterministic sampling is also shown in Table 1. Results of our method on some example chest x-ray images are presented in Fig. 2. We run each experiment five times to look into the repeatability of our method. We find that for every dataset, the standard deviation of $Re_\mathcal{H}$ is <0.7. We further perform several ablation studies which are discussed in the supplementary materials.

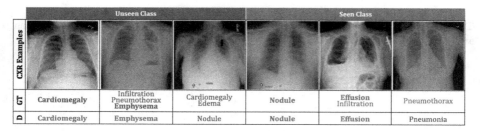

Fig. 2. Examples of chest x-ray (CXR) images showing the performance of the proposed method with ground truth (GT) and detected (D) labels. Correct (green) detection: D matches with one of the GT labels (**bold**); incorrect (red) detection: otherwise. CXR images of unseen classes are the ones having at least one unseen class as a GT label. (Color figure online)

3.2 Discussion on Performance

In reviewing Table 1, we first note that the performance of the proposed method is better than all of its state-of-the-art competitors for both the NIH-900 and Open-i datasets in terms of the harmonic mean of recall $Re_\mathcal{H}$. The single SSP-AE yields the best performance for NIH dataset. Our proposed method does not obtain the best results on the seen classes. However, the methods that have the best performance on seen classes (DeViSe in NIH dataset and ESZSL on the NIH-900 and Open-i datasets) almost completely fail for the unseen classes.

Second, the consistent values of harmonic mean establishes the robustness of our method across datasets from different sources. Since our model is trained using only the NIH dataset, the consistency of results across different test datasets indicates the generalizability of the training. This fact makes our model potentially useful in a clinical setup. Although SAE outperforms the proposed method for the seen classes, the best results in terms of the unseen recall and harmonic mean of recall for all the datasets are always obtained by either the proposed method or one of its variants. Furthermore, notice that single SSP-AE (Sng. SSP-AE) performs quite well for the seen classes. However, it shows poor performance for the unseen classes. In contrast, our method performs consistently for the unseen classes across all the datasets. This indicates the utility of the proposed ensemble in providing generalization ability for the unseen classes.

However, our method does not perform equally well for each of the classes (see Table 2). For example, the performance of our method in the case of pneumothorax is poor. This may be due to the small size of pneumothorax-affected regions in x-ray images, making it difficult to find visually salient features. Consolidation, infiltration and pneumonia, on the other hand, may cause similar lung opacity in chest x-rays. Therefore, it is difficult to find visually distinguishable features from the chest x-ray for these three classes as well leading to poor diagnosis.

Table 2. Performance of the proposed method in terms of recall for different unseen and seen classes from different datasets.

Dataset	Unseen classes			Seen classes					
	Cardiomegaly	Edema	Emphysema	Consolidation	Effusion	Infiltration	Nodule	Pneumonia	Pneumothorax
NIH	8.98	56.07	6.17	1.78	11.86	3.91	18.85	4.79	5.54
NIH-900	16.67	40.74	6.45	2.78	13.41	3.53	25.00	4.65	2.56
Open-i	20.99	10.53	3.53	0.00	12.50	4.55	30.00	6.45	0.00

4   Conclusions

We propose a method for zero-shot diagnosis of chest x-ray images. Through rigorous experiments on different datasets, we have shown that classification of both unseen and seen disease classes is possible from chest x-ray images, with the help of signatures generated from CT reports. To manage the effects of noise, we pioneer an ensemble-based approach that performs semantic embedding using visually salient features. Experiments show the robustness of our algorithm across datasets from different sources, making it potentially applicable for the diagnosis of rare diseases in a clinical setup. In the future, we will look into integrating the signature generation as part of training so that the models for visual-to-semantic embedding and signature generation may be trained in concert.

Acknowledgment. This project was supported by the Intramural Research Programs of the National Institutes of Health, Clinical Center and National Library of Medicine. We thank NVIDIA for GPU card donation.

References

1. Banerjee, I., Madhavan, S., Goldman, R.E., Rubin, D.L.: Intelligent word embeddings of free-text radiology reports. In: AMIA Annual Symposium Proceedings, vol. 2017, p. 411. American Medical Informatics Association (2017)
2. Breiman, L.: Random forests. Mach. Learn. **45**(1), 5–32 (2001)
3. Dajac, J., Kamdar, J., Moats, A., Nguyen, B.: To screen or not to screen: low dose computed tomography in comparison to chest radiography or usual care in reducing morbidity and mortality from lung cancer. Cureus **8**(4), e589 (2016). https://doi.org/10.7759/cureus.589

4. Demner-Fushman, D., et al.: Preparing a collection of radiology examinations for distribution and retrieval. J. Am. Med. Inform. Assoc. **23**(2), 304–310 (2015)

5. Frome, A., et al.: Devise: a deep visual-semantic embedding model. In: Advances in Neural Information Processing Systems, pp. 2121–2129 (2013)

6. Huang, H., Wang, C., Yu, P.S., Wang, C.D.: Generative dual adversarial network for generalized zero-shot learning. In: Proceedings of the IEEE Conference on Computer Vision and Pattern Recognition, pp. 801–810 (2019)

7. Kingma, D.P., Ba, J.: Adam: a method for stochastic optimization. arXiv preprint arXiv:1412.6980 (2014)

8. Kodirov, E., Xiang, T., Gong, S.: Semantic autoencoder for zero-shot learning. In: Proceedings of the IEEE Conference on Computer Vision and Pattern Recognition, pp. 3174–3183 (2017)

9. Krizhevsky, A., Hinton, G.E.: Using very deep autoencoders for content-based image retrieval. In: ESANN, vol. 1, p. 2 (2011)

10. Rajpurkar, P., et al.: CheXNet: radiologist-level pneumonia detection on chest x-rays with deep learning. arXiv preprint arXiv:1711.05225 (2017)

11. Romera-Paredes, B., Torr, P.: An embarrassingly simple approach to zero-shot learning. In: International Conference on Machine Learning, pp. 2152–2161 (2015)

12. Rong, X.: word2vec parameter learning explained. arXiv preprint arXiv:1411.2738 (2014)

13. Sariyildiz, M.B., Cinbis, R.G.: Gradient matching generative networks for zero-shot learning. In: Proceedings of the IEEE Conference on Computer Vision and Pattern Recognition, pp. 2168–2178 (2019)

14. Wang, X., Peng, Y., Lu, L., Lu, Z., Bagheri, M., Summers, R.M.: ChestX-ray8: Hospital-scale chest X-ray database and benchmarks on weakly-supervised classification and localization of common thorax diseases. In: Proceedings of the IEEE Conference on Computer Vision and Pattern Recognition, pp. 2097–2106 (2017)

15. Wang, X., Peng, Y., Lu, L., Lu, Z., Summers, R.M.: TieNet: text-image embedding network for common thorax disease classification and reporting in chest X-rays. In: Proceedings of the IEEE Conference on Computer Vision and Pattern Recognition, pp. 9049–9058 (2018)

Semi-supervised Machine Learning with MixMatch and Equivalence Classes

Colin B. Hansen[1]([✉]), Vishwesh Nath[1], Riqiang Gao[1], Camilo Bermudez[1], Yuankai Huo[1], Kim L. Sandler[2], Pierre P. Massion[2], Jeffrey D. Blume[3], Thomas A. Lasko[3], and Bennett A. Landman[1,3]

[1] Computer Science, Vanderbilt University, Nashville, TN 37235, USA
colin.b.hansen@vanderbilt.edu
[2] Vanderbilt University Medical Center, Nashville, TN 37235, USA
[3] Department of Biomedical Informatics, Vanderbilt University, Nashville, TN 37235, USA

Abstract. Semi-supervised methods have an increasing impact on computer vision tasks to make use of scarce labels on large datasets, yet these approaches have not been well translated to medical imaging. Of particular interest, the Mix-Match method achieves significant performance improvement over popular semi-supervised learning methods with scarce labels in the CIFAR-10 dataset. In a complementary approach, Nullspace Tuning on equivalence classes offers the potential to leverage multiple subject scans when the ground truth for the subject is unknown. This work is the first to (1) explore MixMatch with Nullspace Tuning in the context of medical imaging and (2) characterize the impacts of the methods with diminishing labels. We consider two distinct medical imaging domains: skin lesion diagnosis and lung cancer prediction. In both cases we evaluate models trained with diminishing labeled data using supervised, MixMatch, and Nullspace Tuning methods as well as MixMatch with Nullspace Tuning together. MixMatch with Nullspace Tuning together is able to achieve an AUC of 0.755 in lung cancer diagnosis with only 200 labeled subjects on the National Lung Screening Trial and a balanced multi-class accuracy of 77% with only 779 labeled examples on HAM10000. This performance is similar to that of the fully supervised methods when all labels are available. In advancing data driven methods in medical imaging, it is important to consider the use of current state-of-the-art semi-supervised learning methods from the greater machine learning community and their impact on the limitations of data acquisition and annotation.

Keywords: Semi-supervised learning · Skin lesion · Lung cancer

1 Introduction

Semi-supervised learning methods seek to leverage performance in models using information extracted from both labeled and unlabeled data [1]. Many forms of semi-supervised learning and regularization rely on data augmentation as well as the stochasticity of deep learning models. In data augmentation, a sample is transformed to introduce new example variations to which a model should be robust without altering the label of

© Springer Nature Switzerland AG 2020
J. Cardoso et al. (Eds.): iMIMIC 2020/MIL3ID 2020/LABELS 2020, LNCS 12446, pp. 112–121, 2020.
https://doi.org/10.1007/978-3-030-61166-8_12

the sample. An effective semi-supervised approach is to encourage models to make the same prediction for two different variants of the same sample [2, 3]. Recent success in the CIFAR-10 classification task with limited labeled training data has been achieved through applying Mixup [4] to both labeled and unlabeled data in an algorithm called MixMatch [3]. However, the variations introduced by data augmentation are typically dataset specific. This is especially true for medical imaging tasks in which data augmentation must not alter the image outside of what is possible, considering the anatomy involved and the type of acquisition.

In some tasks, pairs or groups of unlabeled examples can be identified as having the same label even if the label itself is unknown. This is an advantage in medical imaging as many studies typically have repeat acquisitions of the same subject. Assuming the time between acquisitions is not large enough that the anatomy or diagnosis should change, then we know these same subject acquisitions have the same label. We call this knowledge *partial label information*. In prior work, partial label information has been used to predict fiber orientation distributions in diffusion weighted magnetic resonance imaging [5] and to detect coronary calcium in non-contrast computer tomography (CT) [6].

We use the term *equivalence class* to indicate a subset of unlabeled examples for which the label is known to be the same. Formally, an equivalence class Q of examples x in a data subset D under a true but unknown labeling function f is defined as:

$$Q = \{x \in D | f(x) = c\} \tag{1}$$

where c is a constant. We use the expression $x_1 \sim x_2$ to indicate a pair of samples such that $x_1, x_2 \in Q$. If the labeling function f is a linear function, the difference between a pair of examples $x_1 \sim x_2$ from Q lies in the nullspace of f:

$$f(x_1) = f(x_2) \Leftrightarrow f(x_1 - x_2) = 0 \tag{2}$$

We abuse the term *nullspace* by using it to conceptually refer to comparisons between elements in an equivalence class, even though (2) does not hold for nonlinear functions. Using the equivalence classes that can be found naturally in medical imaging, we can help tune a model by encouraging it to make the same predictions for x_1 and x_2 when $x_1 \sim x_2$ in a process we call *Nullspace Tuning*.

The purpose of this work is to show the effectiveness of recent methods MixMatch and Nullspace Tuning in medical imaging tasks and characterize their performance with diminishing labeled data. Additionally, we explore how these methods can be used in tandem to leverage aspects from both methods in training models. We do this for natural images in the task of skin lesion diagnosis using the HAM10000 skin lesion dataset [7] and for CT in the task of lung cancer diagnosis using data from the National Lung Screening Trial (NLST) with follow up confirmed diagnoses [8].

2 Related Work

2.1 Data Augmentation

Data augmentation artificially expands a training dataset by modifying examples using transformations that are believed not to affect the label. Image deformation and additive noise are common examples of such transformations [2, 9, 10]. Natural images can be effectively augmented using random cropping, mirroring, and color shifting [11]. In CT, data augmentation can consist of spatial deformations, translations, rotations, and non-rigid deformations [12]. Effective data augmentation policies can be automatically selected from a search space of image transformations [2]. Generative adversarial networks are also being used to generate anatomically informed data augmentations as well as completely new data to supplement training [13, 14].

2.2 Equivalence Classes in Labeled Data

Some tasks exist in which the equivalence classes describe the label completely. Signature verification and facial recognition are two examples. The verification model tunes the nullspace through minimizing the distance between different signatures or images of the same person [15, 16]. Contrastive loss extends this concept to learn from the contrast of two samples whether they are from the same or different classes [17, 18]. Triplet networks [19] use a similar concept to learn from tuples (x, x^+, x^-), where $x \sim x^+$ and $x \nsim x^-$, and the predicted class probability pairs are encouraged to be near or far, respectively.

2.3 Semi-supervised Learning

Recent semi-supervised learning methods constrain the model through an additional term in the loss function that is computed over unlabeled data. The goal of these methods is to extract useful features from unlabeled data that will allow the model to generalize more effectively to unseen data. This can be done by penalizing the distance in predictions for two perturbations of the same sample [20, 21], by stabilizing the target for unlabeled data through obtaining predictions from a moving average of model weights during training [22], or by using the prediction function to update a guessed label for the unlabeled data periodically during training [23]. Virtual Adversarial Training approximates a small perturbation which, if added to x, would most significantly change the resulting prediction without altering the underlying class [24]. Of particular interest is the method called MixMatch which was developed by taking key aspects of dominant semi-supervised methods and incorporating them in to a single algorithm [3]. The key steps are augmenting all examples, guessing low-entropy labels for unlabeled data, and then applying MixUp to provide more interpolated examples between labeled, unlabeled, and augmented data [4].

3 Methods

Nullspace Tuning is a form of contrastive learning, but unlike some semi-supervised contrastive methods [25], Nullspace Tuning does not rely on data augmentation. Rather it relies on the natural augmentations that exist between samples that can be identified as being equivalent in class. This section describes the use of partial labels in Nullspace Tuning. First, it is described as a standalone method. Second, we illustrate how to combine Nullspace Tuning with MixMatch.

3.1 Nullspace Tuning

To perform Nullspace Tuning, we add a penalty on the distance between predicted probabilities for known equivalence class pairs to a standard loss function $\mathcal{L}_s$. If we have labeled data $\{x_i, y_i\} \in D$ and unlabeled data $\{x_i^*\} \in D^*$, the new loss function can be defined using the model's vector-valued prediction function h and a known equivalence class paring $x_j^* \sim x_k^*$ as:

$$\mathcal{L} = \mathcal{L}_s(h(x_i), y_i) + \lambda \|h\left(x_j^*\right) - h(x_k^*)\|_2^2 \tag{3}$$

where λ is a hyperparameter weighting the contribution of the nullspace loss term. It is not necessary to make any assumptions about the relationship between the labeled data x_i and the unlabeled data x_j^* and x_k^*. Additionally, in cases where the equivalency class has more than two elements, the randomization of chosen pairs within the equivalency class can provide further data augmentation. We choose cross entropy as the standard loss function $\mathcal{L}_s$ in all experiments contained in this work.

3.2 MixMatchNST

The original MixMatch algorithm uses two forms of data augmentation. The first is a set of dataset-specific transformations. By averaging the predicted class distribution function across K augmentations, a guessed label distribution q is assigned to each unlabeled sample x^*. To reduce entropy, temperature sharpening is applied to q [26]. The second form of data augmentation applies MixUp [4] to the labeled data $\{x_i, y_i\}$ and the unlabeled data $\left\{x_j^*, q_j\right\}$ to produce interpolated data $\{\tilde{x}_i, \tilde{y}_i\}$ and $\left\{\tilde{x}_j^*, \tilde{q}_j\right\}$. A hyperparameter α controls how much the examples are altered during MixUp. To prevent overfitting, weight decay is applied using an exponential moving average during training [27, 28].

MixMatchNST modifies the MixMatch loss function with the addition of a Nullspace-Tuning term. The loss function then becomes a combination of the standard loss $\mathcal{L}_s$ calculated using labeled data, the unlabeled loss term weighted by hyperparameter λ_U, and the Nullspace Tuning term weighted by hyperparameter λ_E:

$$\mathcal{L} = \mathcal{L}_s(h(\tilde{x}_i), \tilde{y}_i) + \lambda_U \|h\left(\tilde{x}_j^*\right) - \tilde{q}_j\|_2^2 + \lambda_E \|q_j - q_k\|_2^2 \tag{4}$$

where x_j^* is chosen such that $x_j^* \sim x_k^*$. The Nullspace Tuning term is calculated using the guessed labels q_j and q_k before the MixUp step, whereas the labeled and unlabeled MixMatch terms are calculated using MixUp interpolated examples.

4 Experiments

We evaluate the benefit of Nullspace Tuning over partial label information as well as the benefit of MixMatch over unlabeled data in two medical imaging examples. The first is skin lesion diagnosis in natural images, and the second is lung cancer diagnosis in CT. We follow the precedent of simulating randomly unlabeled data in these datasets to characterize these methods as the amount of labeled data diminishes while the amount of unlabeled data increases [29].

4.1 Implementation Details

All experiments were implemented in PyTorch 1.0.0 [30] and trained on Nvidia 2080Ti GPUs. In both datasets there is a class imbalance which must be considered in both the labeled and in unlabeled data. For the supervised loss, we sample evenly from each class in the labeled data. For the semi-supervised loss, the average prediction for each equivalence class is used as a guessed label, and the unlabeled or paired data are sampled evenly across the guessed labels. Additionally, for each fold, a balanced validation set is created to evaluate the model during training. The class imbalance is kept in proportion when splitting the data in to test sets for each fold, so we report balanced multi-class accuracy and AUC in our evaluation for diminishing amount of labeled. For each method, we perform a hyperparameter search on the λ loss hyperparameters.

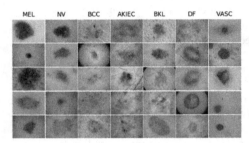

Fig. 1. The difficulty in the skin lesion diagnosis task is the similarity between classes and the variation within classes. This can be seen as especially true for melanoma.

Experiment 1: Using the HAM10000 skin lesion dataset, we train supervised, Nullspace Tuning, MixMatch, and MixMatchNST models, using varying numbers of labeled examples. The supervised model ignores unlabeled examples. The dataset consists of 10,015 color photographs (RGB format, 600 × 450 pixels) of skin lesions categorized as: melanoma (MEL) (1113 images); melanocytic nevus (NV) (6705 images); basal cell carcinoma (BCC) (514 images); actinic keratosis and intraepithelial carcinoma (AKIEC) (327 images); benign keratosis, solar lentigo, and lichen-planus (BKL) (1099 images); dermatofibroma (DF) (115 images); or vascular lesions (VASC) (142 images) [7]. Figure 1 shows examples of each class. For the network architecture, we use a DenseNet [31] which was the top performing single model in the ISIC 2018 challenge which did not use external data [32]. The method is defined by Li et al. and serves as our baseline. The weights of this model are initialized from a model pretrained on Imagenet [11]. Unlike the lung cancer data, HAM10000 does not have natural equivalence classes. We simulate these by randomly pairing the unlabeled data once at the beginning of training such that there are many unchanging equivalence classes of size two, where each example in the pair has the same known but withheld label. Random data affine transforms, mirroring, and color shifting is applied as data augmentation strategies. Validation is performed using k-fold cross validation (k = 5).

Experiment 2: Here we use the NLST as well as a pretrained model from the top performing method in the 2017 Data Science Bowl lung cancer diagnosis challenge [33]. The pretrained model is defined by Liao et al. and was pretrained on a dataset provided by the National Cancer Institute which included some of the NLST data. From the NLST, data used consists of 5710 subjects and a total of 16,053 CT scans with follow-up confirmed diagnoses that successfully passed the preprocessing of Liao et al. There are 1055 subjects with a positive final diagnosis and 4655 with a negative final diagnosis. Most subjects have multiple longitudinal scans which are used as natural equivalence classes for Nullspace Tuning when a subject is simulated as unlabeled data. When splitting the data into training, validation, and testing sets as well as into labeled and unlabeled data, we keep subject data together to avoid bias in the model. We obtain the feature vectors of the five most likely nodule patches just before the final fully connected layer in the pretrained model and train a fully connected neural network on the NLST data as described in Fig. 2. This is similar to the method used by Gao et al. [34]. For data augmentation, a small amount of random Gaussian noise is applied to the feature vectors obtained from the pretrained model. Validation is done by repeating 100 rounds of training and testing under 80/20 random splits. The training data is further split into sets of labeled, unlabeled, and validation data.

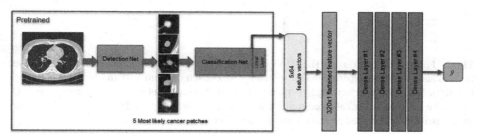

Fig. 2. The feature vectors extracted from the top five most likely cancer patches from the Liao pretrained model are used to train a four-layer FCNN with approximately 300,000 total parameters.

4.2   Results

Experiment 1: For the skin lesion data, balanced multiclass accuracy is reported for models trained using from 779 to 6998 labeled examples (Fig. 3). Both MixMatch and Nullspace tuning show large performance gains over the standard supervised model. When only 779 samples are labeled in the training set, both methods achieve an increase in balance multiclass accuracy of over 20%. At the same point, MixMatch achieves an increase of approximately 7% over the next best method and achieves the best performance at all amounts of labeled data (Fig. 3). At 3888 labeled data or approximately 40% of the original challenge training set, MixMatchNST achieves comparable performance to the that achieved by the Li et al. in the withheld challenge test set, and comparable performance to using all 6998 labeled examples in a supervised model. For both methods, a larger λ which controls the contribution of the loss term generally achieves better performance when less data is available (Fig. 3).

Experiment 2: The AUC is reported after training each model with between 40 and 400 labeled subjects (Fig. 4). Here, the baseline represents the AUC from applying the Liao et al. pretrained model. All other methods train a small fully connected network using pretrained feature vectors as described by Fig. 2. Other than the baseline, Nullspace Tuning and MixMatchNST achieve the highest AUC at 200 and 400 labeled data whereas

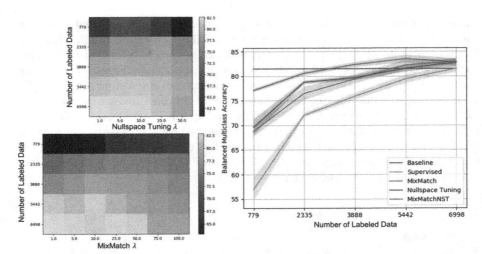

Fig. 3. For experiment 1 using HAM10000, the mean balanced multiclass accuracy across five folds is shown for the hyperparameter search for MixMatch (bottom left) and Nullspace Tuning (top left). The highest performing hyperparameter is used in reporting the final performance (right) where the baseline is the balanced multiclass accuracy reported by the ISIC 2018 challenge for the Li method. The shaded region represents the standard error of the mean.

MixMatch achieves nearly the same AUC as the standard supervised approach. In general, a λ of 5 achieves the best Nullspace Tuning performance and a λ of 0.1 achieves the best MixMatch performance.

5   Discussion

Experiment 1 depicts the full extent of the semi-supervised methods' ability to regularize the model. Even though MixMatch and Nullspace Tuning appear to have similar performance, the high performance of MixMatchNST suggests that features extracted or constrained by each method is additive to the generalizability of the model. In experiment 2, we see that even in fine tuning a pretrained model, the scarcity in labeled data has a large impact on the performance of the model. Here, the semi-supervised learning methods have a small but distinct advantage when labeled data is limited. It is possible the MixMatch algorithm is at a disadvantage when data augmentation is limited to the addition of noise rather than a full suite of randomized transforms. Additionally, the choice of using longitudinal scans as equivalence classes introduces noise due to only fine-tuning the diagnosis model without training the detection model at all. Two sets of patches each from different scans of the same subject then may not belong to the same class. While this work does not show this method is clinically applicable, it does show the added value of these semi-supervised methods in medical imaging tasks.

Conclusion: The use of semi-supervised learning methods such as MixMatch can greatly benefit tasks in which labeled data is scarce or annotations are expensive to

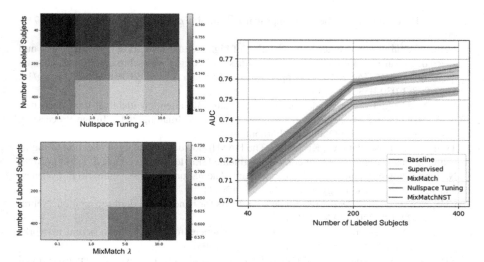

Fig. 4. For **experiment 2** using the NLST, the mean AUC across five folds is shown for the hyperparameter search for MixMatch (bottom left) and Nullspace Tuning (top left). The highest performing hyperparameter is used in reporting the final performance (right) where the baseline is the AUC reported from directly applying the Liao model to the NLST dataset. The shaded region represents the standard error of the mean.

obtain. We advocate for the adoption of these methods to medical image processing especially when domain specific data augmentations are available. Additionally, the ability to acquire partial label information such as equivalence classes should be considered when full labels are impractical. Incorporating partial label information and unlabeled information in semi-supervised learning paradigms can largely benefit models used in medical image processing domains.

Acknowledgements. This work was supported by the National Institutes of Health under award numbers R01EB017230, and T32EB001628, and in part by the National Center for Research Resources, Grant UL1 RR024975-01. The content is solely the responsibility of the authors and does not necessarily represent the official views of the NIH. This study was in part using the resources of the Advanced Computing Center for Research and Education (ACCRE) at Vanderbilt University, Nashville, TN which is supported by NIH S10 RR031634.

References

1. Chapelle, O., Scholkopf, B., Zien, A.: Semi-supervised learning (chapelle, o. et al., eds.; 2006) [book reviews]. IEEE Trans. Neural Netw. **20**(3), 542–542 (2009)
2. Cubuk, E.D., et al.: AutoAugment: learning augmentation strategies from data. In: Proceedings of the IEEE Conference on Computer Vision and Pattern Recognition (2019)
3. Berthelot, D., et al.: MixMatch: a holistic approach to semi-supervised learning. In: Advances in Neural Information Processing Systems (2019)

4. Zhang, H., et al.: mixup: beyond empirical risk minimization. arXiv preprint arXiv:1710. 09412 (2017)
5. Nath, V., et al.: Inter-scanner harmonization of high angular resolution DW-MRI using null space deep learning. In: Bonet-Carne, E., Grussu, F., Ning, L., Sepehrband, F., Tax, Chantal M.W. (eds.) MICCAI 2019. MV, pp. 193–201. Springer, Cham (2019). https://doi.org/10. 1007/978-3-030-05831-9_16
6. Huo, Y., et al.: Coronary calcium detection using 3D attention identical dual deep network based on weakly supervised learning. In: Medical Imaging 2019: Image Processing. International Society for Optics and Photonics (2019)
7. Tschandl, P., Rosendahl, C., Kittler, H.: The HAM10000 dataset, a large collection of multi-source dermatoscopic images of common pigmented skin lesions. Sci. Data **5**, 180161 (2018)
8. National Lung Screening Trial Research Team: The national lung screening trial: overview and study design. Radiology **258**(1), 243–253 (2011)
9. Cireşan, D.C., et al.: Deep, big, simple neural nets for handwritten digit recognition. Neural Comput. **22**(12), 3207–3220 (2010)
10. Simard, P.Y., Steinkraus, D., Platt, J.C.: Best practices for convolutional neural networks applied to visual document analysis. In: ICDAR (2003)
11. Krizhevsky, A., Sutskever, I., Hinton, G.E.: ImageNet classification with deep convolutional neural networks. In: Advances in Neural Information Processing Systems (2012)
12. Roth, H.R., et al.: Anatomy-specific classification of medical images using deep convolutional nets. In: 2015 IEEE 12th International Symposium on Biomedical Imaging (ISBI). IEEE (2015)
13. Frid-Adar, M., et al.: GAN-based synthetic medical image augmentation for increased CNN performance in liver lesion classification. Neurocomputing **321**, 321–331 (2018)
14. Frid-Adar, M., et al.: Synthetic data augmentation using GAN for improved liver lesion classification. In: 2018 IEEE 15th International Symposium on Biomedical Imaging (ISBI 2018). IEEE (2018)
15. Bromley, J., et al.: Signature verification using a "siamese" time delay neural network. In: Advances in Neural Information Processing Systems (1994)
16. Wen, Y., Zhang, K., Li, Z., Qiao, Yu.: A discriminative feature learning approach for deep face recognition. In: Leibe, B., Matas, J., Sebe, N., Welling, M. (eds.) ECCV 2016. LNCS, vol. 9911, pp. 499–515. Springer, Cham (2016). https://doi.org/10.1007/978-3-319-46478-7_31
17. Chopra, S., Hadsell, R., LeCun, Y.: Learning a similarity metric discriminatively, with application to face verification. In: 2005 IEEE Computer Society Conference on Computer Vision and Pattern Recognition (CVPR 2005). IEEE (2005)
18. Hadsell, R., Chopra, S., LeCun, Y.: Dimensionality reduction by learning an invariant mapping. In: 2006 IEEE Computer Society Conference on Computer Vision and Pattern Recognition (CVPR 2006). IEEE (2006)
19. Schroff, F., Kalenichenko, D., Philbin, J.: FaceNet: a unified embedding for face recognition and clustering. In: Proceedings of the IEEE Conference on Computer Vision and Pattern Recognition (2015)
20. Laine, S., Aila, T.: Temporal ensembling for semi-supervised learning. arXiv preprint arXiv: 1610.02242 (2016)
21. Sajjadi, M., Javanmardi, M., Tasdizen, T.: Regularization with stochastic transformations and perturbations for deep semi-supervised learning. In: Advances in Neural Information Processing Systems (2016)
22. Tarvainen, A., Valpola, H.: Mean teachers are better role models: weight-averaged consistency targets improve semi-supervised deep learning results. In: Advances in Neural Information Processing Systems (2017)

23. Lee, D.-H.: Pseudo-label: the simple and efficient semi-supervised learning method for deep neural networks. In: Workshop on Challenges in Representation Learning, ICML (2013)
24. Miyato, T., et al.: Virtual adversarial training: a regularization method for supervised and semi-supervised learning. IEEE Trans. Pattern Anal. Mach. Intell. **41**(8), 1979–1993 (2018)
25. Chen, T., et al.: A simple framework for contrastive learning of visual representations. arXiv preprint arXiv:2002.05709 (2020)
26. Goodfellow, I., Bengio, Y., Courville, A.: Deep Learning. MIT Press, Cambridge (2016)
27. Loshchilov, I., Hutter, F.: Fixing weight decay regularization in adam (2018)
28. Zhang, G., et al.: Three mechanisms of weight decay regularization. arXiv preprint arXiv:1810.12281 (2018)
29. Oliver, A., et al.: Realistic evaluation of deep semi-supervised learning algorithms. In: Advances in Neural Information Processing Systems (2018)
30. Paszke, A., et al.: Automatic differentiation in PyTorch (2017)
31. Iandola, F., et al.: DenseNet: implementing efficient ConvNet descriptor pyramids. arXiv preprint arXiv:1404.1869 (2014)
32. Li, K.M., Li, E.C.: Skin lesion analysis towards melanoma detection via end-to-end deep learning of convolutional neural networks. arXiv preprint arXiv:1807.08332 (2018)
33. Liao, F., et al.: Evaluate the malignancy of pulmonary nodules using the 3-D deep leaky noisy-or network. IEEE Trans. Neural Netw. Learn. Syst. **30**(11), 3484–3495 (2019)
34. Gao, R., et al.: Distanced LSTM: time-distanced gates in long short-term memory models for lung cancer detection. In: Suk, H.-I., Liu, M., Yan, P., Lian, C. (eds.) MLMI 2019. LNCS, vol. 11861, pp. 310–318. Springer, Cham (2019). https://doi.org/10.1007/978-3-030-32692-0_36

Non-contrast CT Liver Segmentation Using CycleGAN Data Augmentation from Contrast Enhanced CT

Chongchong Song[1,2], Baochun He[1,3], Hongyu Chen[1,3], Shuangfu Jia[4], Xiaoxia Chen[5], and Fucang Jia[1,3(✉)]

[1] Shenzhen Institutes of Advanced Technology, Chinese Academy of Sciences, Shenzhen, China
fc.jia@siat.ac.cn
[2] School of Sino-Dutch Biomedical and Information Engineering, Northeastern University, Shenyang, China
[3] Shenzhen College of Advanced Technology, University of Chinese Academy of Sciences, Shenzhen, China
[4] Hejian People's Hospital, Cangzhou, China
[5] The Third Medical Center of the General Hospital of PLA, Beijing, China

Abstract. Non-contrast CT is often preferred in clinical screening while segmentation of such CT data is more challenging due to the low contrast in tissue boundaries and scarce supervised training data than contrast-enhanced CT (CTce) segmentation. To alleviate manual labelling work of radiologists, we generate training samples for 3D U-Net segmentation network by transforming the existing CTce liver segmentation dataset to the non-contrast CT styled volumes with CycleGAN. We validated the performance of CycleGAN in both unsupervised and hybrid supervised training strategy. The results show that using CycleGAN in unsupervised segmentation can achieve higher mean Dice coefficients than fully supervised manner in liver segmentation. The hybrid training of generated samples and the target task samples can improve the generalization ability of segmentation.

Keywords: Non-contrast CT · Liver segmentation · CycleGAN · Data augmentation · 3D U-Net

1 Introduction

The accurate measurements from CT, including liver volume, shape, and location can assist doctors in decision making in diagnosis and treatment. In order to reduce the manual labeling work of radiologists, varieties of efficient and accurate methods have been proposed to segment the liver [1–3]. There also exits a lot of public contrast enhanced CT (CTce) liver segmentation dataset such as Sliver07 [4], LiTS [5]. Almost all of the existing liver segmentation were CTce data. In fact, non-contrast CT is often preferred in clinical screening of the patients as CTce images are scanned by injecting contrast

---

C. Song and B. He—Contributed equally to this work.

© Springer Nature Switzerland AG 2020
J. Cardoso et al. (Eds.): iMIMIC 2020/MIL3ID 2020/LABELS 2020, LNCS 12446, pp. 122–129, 2020.
https://doi.org/10.1007/978-3-030-61166-8_13

agent which may cause some adverse reactions include contrast-induced nephropathy, hyperthyroidism, and possibly metformin accumulation [6]. Therefore, the segmentation from non-contrast CT data has great significance in routine clinical applications.

However, the segmentation of liver from non-contrast CT is more challenging due to the low contrast between the liver and its surrounding tissues, compared to that from CTce volume. For the state of the art, few methods [7] have been focused on the segmentation of non-contrast CT data. Besides, there was only one public dataset that can be traced which named Anatomy3 [8] with only 20 non-contrast CT volumes. Therefore, an obvious idea is that whether we can segment the non-contrast data which has few labels by using the existing public CTce dataset.

In the image style transformation area, CycleGAN [9] is among the most popular method due to its great performance in image to image translation task with unpaired training set. It has been also commonly applied in multi-modality medical image transfer learning. For example, Jiang et al. [10] first exploited CycleGAN to translate CT images to MRI images in tumor segmentation and then used the generated MRI images and a few real MRI data for semi-supervised tumor segmentation. Liu et al. [11] proposed a novel unrolling mechanism that jointly optimizes a generative model and a detector, in this way it improves the accuracy of nodules' location. Chen et al. [12] also applied CycleGAN in image appearance transformation on unsupervised domain adaption task, and they trained an end-to-end CycleGAN and segmentation network which must be 2D network due to the memory limitation.

In this study, we also employed CycleGAN to generate more non-contrast CT training data from the public CTce dataset. To the best of our knowledge, this is the first paper applying CycleGAN in liver CTce to CT style transformation and cross-modality segmentation. The focus of this paper was to explore whether the transformation from CTce to non-contrast CT is useful for non-contrast CT segmentation. Hence, we just used the general 3D U-Net model as segmentation network. As U-Net model [13] was not only the basic segmentation network but also an efficient model which has shown stable performance by the famous nnU-Net (no new U-Net) proposed by Isensee et al. [14]. They just used the U-Net architecture and achieved the first place in 13 tasks of the MICCAI 2018 medical image decathlon segmentation challenge. In this paper, we also used the 3D low resolution version of the nnU-Net for liver segmentation network.

2 Method

We used a two-stage network consisted of CycleGAN and 3D U-Net. As shown in Fig. 1, in the first stage we trained the 2D CycleGAN to obtain more non-contrast CT training samples from contrast enhanced CT data. Then the generated slices were stacked to 3D volume as samples for training the non-contrast CT 3D U-Net model with the framework of nnU-Net. As we used the original CycleGAN and nnU-Net, in this section we only described the image processing and training details.

Image Cropping. In the original CT dataset, the whole body was scanned which would result a very coarse resolution input in z-axis for later 3D U-Net segmentation network. Therefore, we cropped all images according to the liver range in z-axis which were added

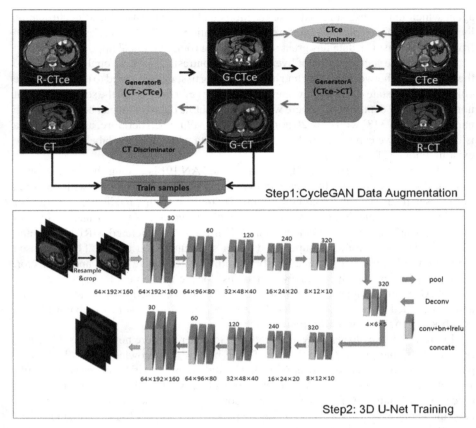

Fig. 1. The pipeline of our CT liver segmentation from public CTce dataset based on CycleGAN data augmentation

with 20 mm in each side for final cropping. The cropped images will be taken as final study samples.

Step1-CycleGAN Data Augmentation. Firstly, the gray scale of the selected data was clipped to the range of −100 HU to 300 HU. Then the intensities were linear mapped to the range of 0 to 255 that 8-bit PNG images can store. After preprocessing, 14 CT and 10 CTce volumes were random selected for CycleGAN training. Among them 753 CTce slices and 277 non-contrast CT slices with original size were random selected as training samples. The resting 241 CTce slices and 98 non-contrast CT slices were validation sets. The original CycleGAN used Adam optimizer with learning rate initially set as 2×10^{-4} and linearly attenuated. The whole training iterations were 200 times with the learning rate decayed after 100 iterations. The training GPU memory and time consumption were about 4 GB and 20 h (Nvidia Tesla V100), respectively. In the testing stage, all slices of volumes were input to the network to obtain a slice size of 256×256 output. Then bilinear interpolation is adopted to upsample them to 512×512. The grayscales of the up-sampled slices were remapped to -100–300. Finally, all post processed slices were stacked to the original 3D volume format. By doing this, the CTce volume was transformed to non-contrast CT style data to implement data augmentation.

Step2-3D U-Net Segmentation Network. We used nnU-Net framework which was a fully automatic image segmentation pipeline including adaptive spacing & intensity normalization and U-Net architecture generation. It resampled all images to the input size of the 3D U-Net. Then the median spacing of all resampled images was regarded as final resampling spacing on the original training images. Then all images were normalized by clipping to the [0.5, 99.5] percentiles of target organ intensity values, followed by a z-score normalization based on the mean and standard deviation of all collected organ intensity values. The 3D network shown in Fig. 1 was also automatically designed by the nnU-Net. The initial feature map and batch size was set as 30 and 2, respectively. The optimizer was Adam and the loss was the sum of Dice and cross entropy. The number of training epochs was about 550 with 250 iterations each. The training consumption of GPU memory was about 12 GB, and the time required for the whole training was about 50 h. Other parameters and training details can be referred to nnU-Net.

3   Experimental Results

3.1   Datasets

The contrast-enhanced CT segmentation dataset used in this study is from the challenge of LiTS-ISBI2017 [5] which aims for the liver and liver tumor segmentation. It contains 130 abdomen contrast enhanced CT volumes scanned. In this study, the two labels were mapped to one label for only the whole liver segmentation. The non-contrast CT dataset used is named Anatomy3 [8] which comes from the multi-organ segmentation challenge hold with ISBI 2015. It contains 20 non-contrast enhanced abdomen CT scans from real patient.

3.2   Experimental Settings

We testified the efficiency of CycleGAN by adopting three training schemas: 1) only using the original CT data; 2) only using the data from the CTce dataset; 3) using hybrid training of the original CT and CTce data. For the last two schemas we compared the training with and without CycleGAN style transformation. All the training models were testified on all images from non-contrast CT dataset (the test data were not used in experiment planning and training steps in the nnU-Net framework). For the first and third schema, we used 15 for training and the other 5 for testing in total of four folds training and testing. The evaluation metrics were the commonly used Dice similarity coefficient (DSC) and HD (Hausdorff distance). All experiments were named with the format of N_X_C/G_Y where N, C and G denotes the original CT dataset, the original CTce dataset without CycleGAN transformation, and the generated CT dataset from CTce data using CycleGAN respectively. X and Y denotes the number of the original CT dataset and the CTce dataset used for training respectively.

3.3 Results

Figure 2 shows the comparison result of experiments with training samples measured with DSC and Hausdorff distance. It can be obviously seen that the method using Cycle-GAN data augmentation is better than the result of directly using the original CTce data in both unsupervised or hybrid training group. In the unsupervised group (names started with N_0), directly using the model trained with the original CTce dataset to segment the non-contrast CT data only obtained a mean DSC value of 0.8826 and 0.9031 in N_0_C_15 and N_0_C_60, much lower than the value of 0.9272 and 0.9420 in N_0_G_15 and N_0_G_60. The segmentation accuracy can be improved by increasing the number of CycleGAN generated training samples while the trend is contrary when just using original CTce samples. The smaller standard division value of CycleGAN domain adaption experiments also indicated its good robustness in the non-contrast CT liver segmentation. Besides, the unsupervised experiment N_0_G_60 achieved a mean DSC of 0.9420 which is higher than the value 0.9379 of fully supervised experiment N_15. The same trends can also be seen from the Handoff distance comparison result. All of the above results testified the effectiveness of the CycleGAN image appearance transformation.

The generalization ability is very important as the manual refinement work after the automatic segmentation result also consumes a lot of time if the DSC is smaller than 0.95. It can be observed that the unsupervised experiment N_0_G_60 had an outlier case (bad result) was much higher than the outlier of fully supervised experiment N_15. Figure 2 also proved the good generalization ability of our method using hybrid training samples of both the target (non-contrast CT) dataset and CycleGAN data augmentation samples from the public CTce dataset. The experiment N_15_G_60 achieved the best result with a very small standard division and also no large outliers compared with the resting experiment training schemas.

Figure 3 shows the performance of CycleGAN on CTce to CT image style transformation. The generation from CTce to CT presented good intensity transformation while preserved the original anatomy structure. However, the transformation from CT to CTce generated CTce style intensities but also some wrong structures. The result conforms to the fact that CTce contains more information such as vessels and tumors which cannot be generated.

3.4 Discussions

All of the existing liver CTce segmentation grand challenges [4, 5] have shown that the automatic segmentation of liver CTce has achieved a very high DSC value and has good consistency with experts' manual labelling ground truth based on the deep learning techniques [2, 3]. In contrast, the segmentation of non-contrast CT gained few attention and investigation of such topic can only be traced to the literature of [7]. Although in clinical practice, non-contrast CT is more generally used in regular examination and the data is more easily accessed, supervised datasets are scarce. Recently, a lot of work have explored the domain adaption and data augmentation based on GAN especially Cycle-GAN and shown promising performance to alleviate the manual labelling work. Based on the exiting studies, we first investigated the liver segmentation of palin scan CT by using

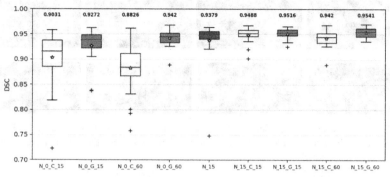

(A) Boxplot comparison result measured in DSC

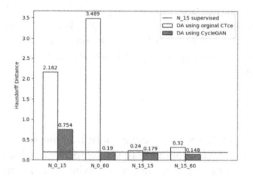

(B) Bar chart of comparison result measured in Hausdorff distance

Fig. 2. Comparison result of all experiments measured with DSC and Hausdorff distance in box-plot of A and bar chart of B. The first and second number of experiment name denotes the number of training samples from original non-contrast CT dataset and the CTce dataset respectively. C and G means data augmentation directly using the original and the generated non-contrast CT styled data by CycleGAN from the public CTce dataset, respectively, which also represented by the white and grey box. The red box or line denotes fully supervised result (baseline). (Color figure online)

the existing supervised liver CTce segmentation dataset through CycleGAN style transformation. We demonstrated the efficiency of CycleGAN by comparing the performance in an unsupervised/hybrid training strategy. The result shows that using CycleGAN in an unsupervised approach can achieved better performance than the supervised training. And the hybrid training can obtain more accurate and generalized segmentation model. This may assist researchers and experts in labeling work of segmentation task. The researchers may first use CycleGAN to generate the target task training samples from existed supervised source domain samples for segmentation in an unsupervised approach. Then partial segmentation results can be refined by the experts and then mixed up with more generated samples for training a more generalized model.

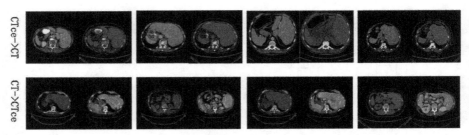

Fig. 3. Visualization result of CycleGAN transformation between contrast enhanced CT (CTce) and non-contrast CT. The left and right image of each group denotes the original image and the generated image respectively.

One limitation of our work is that we did not investigate the factor of image pre-processing in nnU-Net on the cross-modality segmentation from CTce to CT. The nnU-Net used statistical liver intensities of training samples for greyscale normalization. The intensity mean and standard deviation of CTce dataset may not fit for non-contrast CT, especially in the unsupervised training experiments. Another limitation is that we tested the method on only twenty patients.

4 Conclusions

We proposed to use CycleGAN to generate more training samples from supervised contrast enhanced CT dataset for non-contrast CT segmentation. We validate our method on two public datasets in unsupervised/supervised/hybrid training manner. Experiments results demonstrate the effectiveness of our method in alleviating the labelling work and improving the performance of non-contrast CT liver segmentation.

Acknowledgments. This work was partially supported by the National Key Research and Development Program (2019YFC0118100 and 2017YFC0110903), the Guangdong Key Area Research and Development Program (2020B010165004), the Shenzhen Key Basic Science Program (JCYJ20170413162213765 and JCYJ20180507182437217), the Shenzhen Key Laboratory Program (ZDSYS201707271637577), the NSFC-Shenzhen Union Program (U1613221).

References

1. Huang, C., et al.: Fully automatic liver segmentation using a probability atlas registration. In: O'Conner, L. (ed.) Proceedings of ICECC, pp. 126–129. IEEE Computer Society Press (2012)
2. Xu, L.: H-DenseUNet: hybrid densely connected UNet for liver and tumor segmentation from CT volumes. IEEE Trans. Med. Imaging **37**(12), 2663–2674 (2018)
3. Seo, H., Huang, C., Bassenne, M., Xiao, R., Xing, L.: Modified U-Net (mU-Net) with incorporation of object-dependent high level features for improved liver and liver-tumor segmentation in CT images. IEEE Trans. Med. Imaging **39**(5), 1316–1325 (2020)
4. Heimman, T., van Ginneken, B., Styner, M.A., et al.: Comparison and evaluation of methods for liver segmentation from CT datasets. IEEE Trans. Med. Imaging **28**(8), 1251–1265 (2009)

5. Bilic, P., Christ, P.R., Vorontsov, E., et al.: The liver tumor segmentation benchmark (LiTS). arXiv reprint. arXiv: 1901.040506 (2019)
6. Insideradiology Homepage. https://www.insideradiology.com.au/iodine-containing-contrast-medium-hp/
7. Tomoshige, S., Oost, E., Shimizu, A., et al.: A conditional statistical shape model with integrated error estimation of the conditions; application to liver segmentation in non-contrast CT images. Med. Image Anal. **18**(1), 130–143 (2014)
8. Goksel, O., Foncubierta-Rodríguez, A., Jimenez-del-Toro, O., et al.: Overview of the VISCERAL challenge at ISBI 2015. In: Goksel, O. (ed.) Proceedings of the VISCERAL Anatomy Grand Challenge at the 2015 IEEE International Symposium on Biomedical Imaging (ISBI), New York, NY, 16 April 2015
9. Zhu, J., Park. T., Isola, P., Efros, A.: Unpaired image-to-image translation using cycle-consistent adversarial networks. In: Proceedings of the IEEE International Conference on Computer Vision 2017, Venice, pp. 2242–2251 (2017)
10. Jiang, J., et al.: Tumor-aware, adversarial domain adaptation from CT to MRI for lung cancer segmentation. In: Frangi, A.F., Schnabel, J.A., Davatzikos, C., Alberola-López, C., Fichtinger, G. (eds.) MICCAI 2018. LNCS, vol. 11071, pp. 777–785. Springer, Cham (2018). https://doi.org/10.1007/978-3-030-00934-2_86
11. Liu, L., Muelly, M., Deng, J., Pfister, T., Li, L.: Generative modeling for small-data object detection. In: Proceedings of the IEEE International Conference on Computer Vision, 2019, Seoul, Korea (South), pp. 6072–6080 (2019)
12. Chen, C., Dou, Q., Chen, H., et al.: Unsupervised bidirectional cross-modality adaptation via deeply synergistic image and feature alignment for medical image segmentation. IEEE Trans. Med. Imaging **39**(7), 2494–2505 (2020)
13. Ronneberger, O., Fischer, P., Brox, T.: U-Net: convolutional networks for biomedical image segmentation. In: Navab, N., Hornegger, J., Wells, W.M., Frangi, A.F. (eds.) MICCAI 2015. LNCS, vol. 9351, pp. 234–241. Springer, Cham (2015). https://doi.org/10.1007/978-3-319-24574-4_28
14. Isensee, F., Petersen, J., Klein, A., et al.: nnU-Net: self-adapting framework for u-net-based medical image segmentation. arXiv preprint arXiv:1809.10486 (2018)

Uncertainty Estimation in Medical Image Localization: Towards Robust Anterior Thalamus Targeting for Deep Brain Stimulation

Han Liu[1]($\boxtimes$), Can Cui[1], Dario J. Englot[2], and Benoit M. Dawant[1]

[1] Department of Electrical Engineering and Computer Science, Vanderbilt University, Nashville, TN 37235, USA
han.liu@vanderbilt.edu
[2] Department of Neurosurgery, Vanderbilt University Medical Center, Nashville, TN 37235, USA

Abstract. Atlas-based methods are the standard approaches for automatic targeting of the Anterior Nucleus of the Thalamus (ANT) for Deep Brain Stimulation (DBS), but these are known to lack robustness when anatomic differences between atlases and subjects are large. To improve the localization robustness, we propose a novel two-stage deep learning (DL) framework, where the first stage identifies and crops the thalamus regions from the whole brain MRI and the second stage performs per-voxel regression on the cropped volume to localize the targets at the finest resolution scale. To address the issue of data scarcity, we train the models with the pseudo labels which are created based on the available labeled data using multi-atlas registration. To assess the performance of the proposed framework, we validate two sampling-based uncertainty estimation techniques namely Monte Carlo Dropout (MCDO) and Test-Time Augmentation (TTA) on the second-stage localization network. Moreover, we propose a novel uncertainty estimation metric called maximum activation dispersion (MAD) to estimate the image-wise uncertainty for localization tasks. Our results show that the proposed method achieved more robust localization performance than the traditional multi-atlas method and TTA could further improve the robustness. Moreover, the epistemic and hybrid uncertainty estimated by MAD could be used to detect the unreliable localizations and the magnitude of the uncertainty estimated by MAD could reflect the degree of unreliability for the rejected predictions.

Keywords: Deep Brain Stimulation · Anterior nucleus of thalamus · Medical image localization · Uncertainty estimation · Pseudo labels

1 Introduction

Epilepsy is one of the most common chronic neurological disorders characterized

---

The original version of this chapter was revised: the equation on page 5 was modified to improve its accuracy and readability. The correction to this chapter is available at https://doi.org/10.1007/978-3-030-61166-8_30

J. Cardoso et al. (Eds.): iMIMIC 2020/MIL3ID 2020/LABELS 2020, LNCS 12446, pp. 130–137, 2020.
https://doi.org/10.1007/978-3-030-61166-8_14

by spontaneous recurrent seizures and affects around 70 million patients worldwide [1]. Over 30% of the epilepsy patients have refractory seizures which may carry risks of structural damage to the brain and nervous system, comorbidities, and increased mortality [2]. Deep Brain Stimulation (DBS) is a recently FDA-approved neurostimulation therapy that can effectively reduce the occurrences of refractory seizures by delivering electric impulses to a deep brain structure called the anterior nucleus of the thalamus (ANT). Accurate localization of the ANT target is however difficult because of the well documented variability in the ANT size and shape and thalamic atrophy caused by persistent epileptic seizures [3]. Currently, the standard approach to automate this process is the atlas-based technique. While popular, atlas-based methods are known to lack robustness when anatomic differences between atlases and subjects are large. This is particularly acute for ANT-DBS targets that are close to the ventricles, which can be severely enlarged in some patients.

Over the past decade, DL-based techniques such as convolutional neural networks (CNN) have emerged as powerful tools and have achieved unprecedented performances in many medical imaging tasks. However, to train sufficiently robust and accurate models, deep learning methods typically require large amounts of labeled data, which is expensive to collect, especially in the medical domain. In the case of data scarcity and noisy labels, insufficiently trained models may fail catastrophically without any indication. Hence, it is extremely desirable for deep learning models to estimate the uncertainties regarding their outputs in these scenarios. The predictive uncertainty of neural networks can be categorized into two types: *epistemic* uncertainty and *aleatoric* uncertainty. Epistemic uncertainty, also known as model uncertainty, accounts for the uncertainty in the model and can be explained away by observing more training data. On the other hand, the aleatoric uncertainty is the input-dependent uncertainty that captures the noise and randomness inherent in observations. Recently, uncertainty estimation has also received increasing attention in medical image analysis. Ayhan et al. [4] proposed to estimate the heteroscedastic aleatoric uncertainty using TTA for classification task. Nair et al. [5] explored the uncertainty estimation for lesion detection and segmentation tasks based on MCDO. Wang et al. [6] proposed a theoretical formulation of TTA and demonstrated its effectiveness in uncertainty estimation for segmentation task. Nevertheless, uncertainty estimation for localization tasks has not been well studied.

In this work, we developed a novel two-stage deep learning framework aiming at robustly localizing the ANT targets. To the best of our knowledge, this is the first work to develop a learning-based approach for this task. To overcome the problem of data scarcity, we train the models with the pseudo labels which are created based on the available gold-standard annotations using multi-atlas registration. Moreover, we validate two sampling-based uncertainty estimation techniques to assess the localization performance of the developed method. We also propose a novel metric called MAD for sampling-based uncertainty estimation methods in localization tasks. Our experimental results show that the proposed method achieved more robust localization performance than the traditional multi-atlas method and TTA could further improve the robustness. Lastly, we show that the epistemic and hybrid uncertainty estimated by MAD can be used to detect the unreliable localizations and the magnitude of MAD can reflect the degree of unreliability when the predictions are rejected.

2   Materials and Methods

2.1   Data

Our own dataset consists of 230 T1-weighted MRI scans from a database of patients who underwent a DBS implantation for movement disorders, i.e., Parkinson Disease or Essential Tremor at Vanderbilt University. The resolution of the images varies from $0.4356 \times 0.4356 \times 1 \, mm^3$ to $1 \times 1 \times 6 \, mm^3$. The ground truth was manually annotated on a different dataset collected by an experienced neurosurgeon. In this dataset, the 3D coordinates of eight ANT targets (one on each side) on four MRI scans were identified and the thalamus mask on one of these volumes was delineated. With the available annotations, we generated the pseudo labels for the ANT targets and for the thalamus masks using multi and single-atlas registration [7]. In this study, 200 MRI scans were randomly selected for training and validation and the remaining 30 images were used for testing. For preprocessing, we use trilinear interpolation to resample all the images to isotropic voxel sizes of $1 \times 1 \times 1 \, mm^3$ and rescale the image intensities to $[0, 1]$.

2.2   Proposed Method

Typically, an MRI scan with original resolution cannot be fed to a 3D CNN directly due to the limitation of computational resource. A common approach to solve this problem, i.e., using downsampled images, is not appropriate here because the downsampling operation unavoidably leads to a loss in image resolution. This is a concern in our application because even a few-voxel shift in the deep brain can lead to target predictions that are unacceptable for clinical use. To address this issue, we propose a two-stage framework where the first stage coarsely identifies and crops the thalamus regions from the whole brain MRI and the second stage performs per-voxel regression on the cropped volume to localize the targets at the finest resolution scale.

The workflow of the proposed framework is shown in Fig. 1. In this first stage, we train a 3D U-net [8] using the downsampled $80 \times 80 \times 80$ MRI scans to coarsely segment the thalamus. The output layer of this network has three channels corresponding to background, left thalamus and right thalamus respectively. Once we obtain the segmentation results, we post-process the binary segmentation from each foreground channel by isolating the largest connected component and resample the results back to the original resolution. Thereafter, we compute the bounding box of each isolated component and crop a $64 \times 64 \times 64 \, mm^3$ volume around its center. The cropped volume fully encloses the entire left or right thalamus as well as its contextual information. Before passing the volumes to the second stage, we flip the left-thalamus volumes in the left-right direction so that the inputs of the second stage have consistent orientations. In the second stage, we employ another 3D U-net with the same architecture to localize

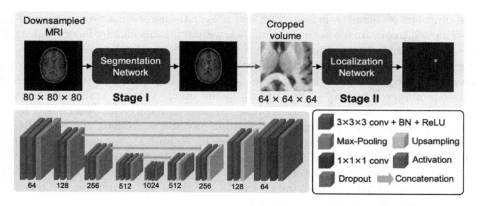

Fig. 1. The workflow of the proposed two-stage framework and the 3D U-net architecture we used. The segmentation and localization network share the same architecture. The number below each encode/decoder unit is the channel number of the $3 \times 3 \times 3$ convolution kernels.

the ANT target by performing per-voxel regression on the cropped volumes. Since the cropped volumes have the same resolution as the original MRI, there is no performance degradation in localization due to loss in resolution. To allow volume-to-volume mapping, we design the ground truth map to be a 3D Gaussian function centered at the pseudo label position with a standard deviation of 1.5 mm. The maximum value is scaled to 1 and any value below 0.05 is set to 0. During the testing phase, the left-thalamus localization maps are flipped back to the original orientation and the voxel with the **maximum** activation in each localization map is taken as the final prediction.

2.3 Uncertainty Estimation

Epistemic Uncertainty. We estimate the epistemic uncertainty of the localization task using the dropout variational inference. Specifically, we train the model with dropout (same as the baseline method) and during the testing phase we perform T stochastic forward passes with dropout to generate Monte Carlo samples from the approximate posterior. Let $y = f(x)$ be the network mapping from input x to output y. Let T be the number of Monte Carlo samples and $\hat{W}_t$ be the sampled model weights from MCDO. For regression tasks, the final prediction and the epistemic uncertainty can be estimated by calculating the predictive mean $E(y) \approx \frac{1}{T} \sum_{t=1}^{T} f^{\hat{W}_t}(x)$ and variance $Var(y) \approx \frac{1}{T} \sum_{t=1}^{T} f^{\hat{W}_t}(x)^T f^{\hat{W}_t}(x) - E(y)^T E(y)$ from these samples.

Aleatoric Uncertainty. To estimate the aleatoric uncertainty, we use the TTA technique which is a simple yet effective approach to study locality of testing samples. Recently, Wang et al. [6] provided a theoretical formulation for using TTA to estimate a distribution of prediction by Monte Carlo simulation with prior distributions of image transformation and noise parameters in an image acquisition model. In our image acquisition model, we extend this idea by incorporating both spatial transformations T_s and intensity transformation T_i to simulate the variations of spatial orientations and image brightness and contrast respectively. Our image acquisition model can thus be expressed as: $x = T_s(T_i(x_0))$, where x is our observed testing image and x_0 is the image without transformations in latent space. During the testing phase, we aim to reduce the bias caused by transformations in x by leveraging the latent variable x_0. Given the prior distributions of the transformation parameters in T_s and T_i, we can estimate y by generating N Monte Carlo samples and the n_{th} Monte Carlo sample can be inferred as: $y_n = T_{s_n}(y_{0_n}) = T_{s_n}(f(x_{0_n})) = T_{s_n}(f(T_{i_n}^{-1}(T_{s_n}^{-1}(x_n))))$. The final prediction and aleatoric uncertainty can be obtained by computing the mean and variance from the Monte Carlo samples.

Maximum Activation Dispersion. For regression tasks, the uncertainty maps are typically obtained by computing the voxel-wise variance from the Monte Carlo samples. However, this approach fails to generate useful uncertainty maps in our application. In our ground truth maps, the non-zero elements, i.e., foreground voxels within the Gaussian ball, are very sparse compared to the zero elements, and thus more difficult to localize and more likely to produce larger predictive variance. As a result, such uncertainty maps would display higher uncertainty at the predicted targets even if the targets are correctly localized (Fig. 2), and thus are not effective for uncertainty estimation regarding the localization performance. To address this issue, we propose a novel metric called Maximum Activation Dispersion (**MAD**) which can be directly applied to any sampling-based uncertainty estimation technique. This metric measures the consistency of the maximum activation positions of the Monte Carlo samples and ignores the activation variance at the same position. Note that MAD aims at estimating the image-wise uncertainty regarding the overall localization performance instead of voxel-wise uncertainty produced by uncertainty maps. Let N be the number of Monte Carlo samples and $\mathbf{p}_n = (x_n, y_n, z_n)$ be the maximum activation position of the n_{th} Monte Carlo sample. The maximum activation dispersion is computed as $\frac{1}{N}\sum_{n=1}^{N}\|\mathbf{p}_n - \bar{\mathbf{p}}\|$, where $\|\cdot\|$ is the L_2 norm and $\bar{\mathbf{p}} = \frac{1}{N}\sum_{n=1}^{N}\mathbf{p_n}$ is the geometric center of all maximum activation positions.

2.4   Implementation Details

Two five-level 3D U-nets with the same architecture were used in the proposed two-stage framework (Fig. 1). In the first stage (segmentation), optimization was performed using the Adam optimizer with a learning rate of 5×10^4, with Dice loss as the loss function, a batch size of 3, and early stopping based on validation loss with patience of 10 epochs. In the second stage (localization), dropout layers were added to allow MCDO. As suggested by Kendall et al. [9], applying dropout layers to all the encoders and decoders is too strong a regularizer. To avoid poor training fit, we followed the best dropout configuration

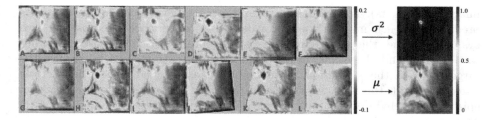

Fig. 2. Visualization of some Monte Carlo samples (A-L) by TTA, their mean (localization map) and variance (uncertainty map). The uncertainty map displays higher uncertainties at the correctly localized position and thus is not effective for localization uncertainty estimation. (Color figure online)

in [9] by adding dropout layers with a dropout rate of $p = 0.5$ only at the deepest half of encoders and decoders. During training, optimization was performed using the Adam optimizer with a learning rate of 2×10^4, with a batch size of 6, and early stopping based on validation loss with patience of 5 epochs. A weight decay of 5×10^4 was used. The weighted mean squared error (WMSE) was used as the loss function to alleviate the class imbalance issue by assigning higher weights to the sparse non-zero entries. The models with the smallest validation losses were selected for final evaluation.

During the testing phase, we forward passed the testing image once to the deterministic network with dropout turned off (**baseline**). With a given prior distribution of transformation parameters in image acquisition model, we forward passed the stochastically transformed testing image $N = 100$ times to the deterministic network with dropout turned off and transformed the predictions back to the original orientation. The mean and variance (aleatoric uncertainty) of the Monte Carlo samples were obtained (**baseline + TTA**). In the image acquisition model, the spatial transformations were modeled by translation and rotation along arbitrary axis. The intensity transformation was modeled by a smooth and monotonous function called Bézier Curve, which is generated using two end points P_0 and P_3 and two control points P_1 and P_2: $B(t) = (1-t)^3 P_0 + 3(1-t)^2 t P_1 + 3(1-t)t^2 P_2 + t^3 P_3, t \in [0, 1]$. In particular, we set $P_0 = (0,0)$ and $P_3 = (1,1)$ to obtain an increasing function to avoid invalid transformations. The prior distributions of the spatial and intensity transformation parameters were modeled by uniform distribution U as $s \sim U(s_0, s_1)$, $r \sim U(r_0, r_1)$ and $t \sim U(t_0, t_1)$, where s, r and t represent translation (voxels), rotation angle (degrees) and the fractional value for Bézier Curve. In our experiment, we set $s_0 = -10$, $s_1 = 10$, $r_0 = -20$, $r_1 = 20$, $t_0 = 0$ and $t_1 = 1$. Lastly, we forward passed the same testing image to the stochastic network $T = 100$ times with dropout turned on with a rate of $p = 0.5$ and obtained the mean and variance (epistemic uncertainty) of the Monte Carlo samples (**baseline + MCDO**).

In our experiment, we evaluated the localization performances of the multi-atlas, baseline, baseline + TTA and baseline + MCDO methods on 30 testing images (60 targets). The axial, sagittal and coronal views centered at the targets predicted by each method were provided to an experienced neurosurgeon to evaluate whether the predicted targets are acceptable for clinical use (the order was shuffled and the evaluator was blind to the method used to predict the target). When the predictions were evaluated as rejected, the evaluator was asked to provide the reasons for rejection. Furthermore, we analyze the aleatoric, epistemic and hybrid (aleatoric + epistemic) uncertainties estimated by MAD on the baseline rejected cases.

3   Experimental Results

Our results show that among a total number of 60 targets, 53, 55, 57 and 55 targets were evaluated as acceptable for the multi-atlas, baseline, baseline + TTA and baseline + MCDO respectively. In Fig. 3, we show the boxplots of aleatoric, epistemic and hybrid uncertainties estimated by MAD. It can be observed that when the rejected predictions are far away from the acceptable positions (red and blue), their estimated uncertainty correspond to the outliers above the upper whisker in the boxplots of epistemic and hybrid uncertainty. On the other hand, when the rejected predictions are close to the acceptable positions (cyan, green and magenta), their uncertainties fall in the range of upper quartile and the upper whisker (cyan and green) and the range of lower quartile and median (magenta), corresponding to their degree of unreliability.

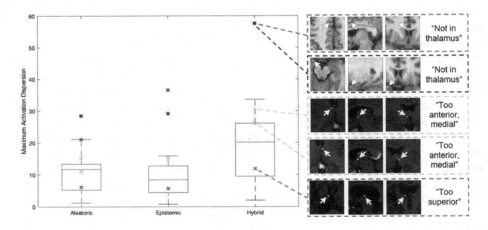

Fig. 3. Boxplots of aleatoric, epistemic and hybrid uncertainties estimated by MAD on the testing set (60 images). The rejected cases of the baseline method (5 cases) are shown in color. The axial, sagittal and coronal views of the rejected targets are shown with the reasons for rejection provided by the evaluator. (Color figure online)

It can be observed that the epistemic and hybrid uncertainty estimated by MAD could be used to detect unreliable localizations, i.e., the ones that not even in thalamus. Moreover, the magnitudes of MAD could reflect the degree of unreliability when the predictions were rejected. We also observe that even though the MCDO did not improve the localization robustness compared to the baseline method, the epistemic uncertainty obtained by this technique has great value for detecting the unreliable localizations, i.e., the outliers in the boxplot.

4 Conclusion

In this study, we present a two-stage deep learning framework to robustly localize the ANT-DBS targets in MRI scans. Results show that the proposed method achieved more robust localization performance than the traditional multi-atlas method and TTA-based aleatoric uncertainty estimation can further improve the localization robustness. We also show that the proposed MAD is a more effective uncertainty estimation metric for localization tasks.

References

1. Ngugi, A.K., Bottomley, C., Kleinschmidt, I., Sander, J.W., Newton, C.R.: Estimation of the burden of active and life-time epilepsy: a meta-analytic approach. Epilepsia $51(5)$, 883–890 (2010)
2. Laxer, K.D., et al.: The consequences of refractory epilepsy and its treatment. Epilepsy Behav. 37, 59–70 (2014)
3. Keller, S.S., et al.: Thalamotemporal alteration and postoperative seizures in temporal lobe epilepsy. Ann. Neurol. $77(5)$, 760–774 (2015)
4. Ayhan, M.S., Berens, P.: Test-time data augmentation for estimation of heteroscedastic aleatoric uncertainty in deep neural networks (2018)
5. Nair, T., Precup, D., Arnold, D.L., Arbel, T.: Exploring uncertainty measures in deep networks for multiple sclerosis lesion detection and segmentation. Med. Image Anal. 59, 101557 (2020)
6. Wang, G., Li, W., Aertsen, M., Deprest, J., Ourselin, S., Vercauteren, T.: Aleatoric uncertainty estimation with test-time augmentation for medical image segmentation with convolutional neural networks. Neurocomputing 338, 34–45 (2019)
7. Rohde, G.K., Aldroubi, A., Dawant, B.M.: The adaptive bases algorithm for intensity-based nonrigid image registration. IEEE Trans. Med. Imaging $22(11)$, 1470–1479 (2003)
8. Çiçek, Ö., Abdulkadir, A., Lienkamp, S.S., Brox, T., Ronneberger, O.: 3D U-Net: learning dense volumetric segmentation from sparse annotation. In: Ourselin, S., Joskowicz, L., Sabuncu, M.R., Unal, G., Wells, W. (eds.) MICCAI 2016. LNCS, vol. 9901, pp. 424–432. Springer, Cham (2016). https://doi.org/10.1007/978-3-319-46723-8_49
9. Kendall, A., Badrinarayanan, V., Cipolla, R.: Bayesian segnet: model uncertainty in deep convolutional encoder-decoder architectures for scene understanding. arXiv preprint arXiv:1511.02680 (2015)

A Case Study of Transfer of Lesion-Knowledge

Soundarya Krishnan[1]([✉]), Rishab Khincha[1], Lovekesh Vig[3], Tirtharaj Dash[1,2], and Ashwin Srinivasan[1,2]

[1] Department of CS and IS, BITS Pilani, K.K. Birla Goa Campus, Goa, India
`soundaryak4898@gmail.com`
[2] APP Centre for AI Research, BITS Pilani, K.K. Birla Goa Campus, Goa, India
[3] TCS Research, New Delhi, India

Abstract. All organs in the human body are susceptible to cancer, and we now have a growing store of images of lesions in different parts of the body. This, along with the acknowledged ability of neural-network methods to analyse image data, would suggest that accurate models for lesions can now be constructed by a deep neural network. However an important difficulty arises from the lack of annotated images from various parts of the body. Our proposed approach to address the issue of scarce training data for a target organ is to apply a form of *transfer learning*: that is, to adapt a model constructed for one organ to another for which there are minimal or no annotations. After consultation with medical specialists, we note that there are several discriminating visual features between malignant and benign lesions that occur consistently across organs. Therefore, in principle, these features boost the case for transfer learning on lesion images across organs. However, this has never been previously investigated. In this paper, we investigate whether lesion knowledge can be transferred across organs. Specifically, as a case study, we examine the transfer of a lesion model from the brain to lungs and lungs to the brain. We evaluate the efficacy of transfer of a brain-lesion model to the lung, and the transfer of a lung-lesion model to the brain by comparing against a model constructed: (a) without model-transfer (i.e. random weights); and (b) using model-transfer from a lesion-agnostic dataset (ImageNet). In all cases, our lesion models perform substantially better. These results point to the potential utility of transferring lesion-knowledge across organs other than those considered here.

Keywords: Transfer learning · Medical imaging · Tumour classification

1 Introduction

Cancer is one of the deadliest diseases in the world, with tens of millions diagnosed with some form of cancer annually. Early diagnosis is one of the most important factors in its control and prevention. Computer-aided detection (CAD) systems,

© Springer Nature Switzerland AG 2020
J. Cardoso et al. (Eds.): iMIMIC 2020/MIL3ID 2020/LABELS 2020, LNCS 12446, pp. 138–145, 2020.
https://doi.org/10.1007/978-3-030-61166-8_15

specifically deep learning models could potentially help radiologists by detecting features that can be missed even by the trained eye.

The performance of deep learning models is often heavily dependent on the data available. Given the extensive expertise required for generating annotated medical data [13], ethical and privacy concerns around sharing it [3], obtaining training data for deep models remains a bottleneck. For cancer, this problem is compounded by the fact that not some organs have sparse data. One approach to deal with the lack of data, is to draw on techniques of *transfer-learning* that exploit a commonality across datasets. If such a commonality exists, then the parts of the model constructed on a larger dataset (the "source" model) could be re-used to construct a model for the smaller dataset (the "target" model). In the case of lesions (tumours), there are clinical reasons to expect some common visual features in tumours across organs: (a) Malignant tumours across the body have irregular boundaries, and benign tumours have clear boundaries and sharp margins; (b) Malignant tumours are often found to have a thickening at the periphery; and (c) Malignant tumours are generally found to have inhomogenous attenuation, and benign tumours are characterised by homogeneous attenuation instead [1]. This suggests that source-models for detecting lesions that are constructed for one organ should be helpful in constructing a target-model for detecting tumours in a different organ. The usual approach to transfer-learning however focuses on the use of large, generic datasets for constructing a source model: it uses a source-model with pre-trained ImageNet weights. However, it has been pointed out that the nature of classification in the ImageNet dataset is far different from medical classification [12]. ImageNet weights are tuned for tasks in which the subject is prominent in the background, tasks such as cancer tumour classification involve local features, such as inhomogeneous intensities and irregular boundaries, and some subtle pattern-based features.

The principal motivation for this paper is provided by the work in [10], where it is shown that transfer of knowledge is possible across diseases. We are also inspired by the paper [8], in which the authors note that if the target task is localisation-sensitive (as it is in our case), the gains of using ImageNet weights for initialisation are limited to only a reduction in convergence time, without a big boost in performance.

In this paper, we study the effect of transfer learning from a different organ versus the standard practice of transfer learning from the ImageNet database. To the best of our knowledge, this kind of study has not been performed for lesion classification. We focus on transfer of lesion-knowledge from brain to lungs, and lungs to the brain. Our results using lesion-specific source data are promising: (1) Performance improves substantially (importantly, recall over the malignant class improves); (2) The time taken for model-construction decreases significantly; (3) Model performance has less variance; and (4) These effects are even more pronounced when target data is limited.

2   Note on Transfer Learning

Transfer learning is now a thriving area of application, especially using deep neural networks [14]. In essence, this consists of using a large *source* dataset to identify a deep network structure and weights. This model is then "transferred" to construct a model for *target* data, usually by re-estimating using the target data, the weights associating with higher layers of the network. Since this involves fewer estimates than the entire model, it can normally be done with lesser data than would be needed to estimate all weights in the model. For the purposes of this paper, we take the following high-level view of the transfer learning process:

- We will use "model m" to denote the pair (π, θ), where π denotes the structure of m (for example, of a deep network) and θ denotes the parameters in m (for example, the weights in the network). We will assume structures are drawn from some space Π and parameters from a space Θ, and $M = \Pi \times \Theta$. Given an instance $\mathbf{x}$ from a set of instances X, and a model m, we assume a function $Predict : M \times X \rightarrow Y$, where Y is some set of (class) labels. Then, $Predict(\mathbf{x}|m)$, is to be read as "the value of $Predict$ on an instance $\mathbf{x}$ given model m".
- Let $\mathcal{D}$ denote the set of subsets of $X \times Y$. We use the term "model construction" to mean a function $Learn : \mathcal{D} \rightarrow M$. Thus, $m = Learn(d)$ is to be read as "the model m constructed by $Learn$, using data d". Using the function $Transfer : \mathcal{D} \times M \rightarrow M$, by "model transfer from source to a target", we will mean the composition $Transfer \circ Learn$. That is, given *source* data d_s, and *target* data d_t, a model transfer from source to target is the model $Transfer(d_t|Learn(d_s))$. Both $Learn$ and $Transfer$ are conditional on the definition of source-specific and target-specific loss functions. We omit this detail here.
- Testing model performance will require the implementation of $Predict$

Given source-data d_s and target-data d_t, the obvious form of transfer learning is one that is defined by $Transfer(d_t|Learn(d_s))$. Here, a model is first constructed for the source data, and is then transferred to the target data. However, the literature suggests that better models for a dataset d may be obtained by $Transfer(d|Learn(d_g))$, where d_g denotes a large generic dataset (like ImageNet). In the following section, we investigate the performance of model transfer using source-data (d_s) from the brain and target-data (d_t) from the lung, and *vice versa*.

3   Empirical Evaluation of Transferring Lesion Models

3.1   Aim(s)

Given a source-dataset d_s, a target-dataset d_t, and a (large) generic dataset d_g, we distinguish between the following target models: (a) *Baseline*, denoting $Learn(d_t)$; (b) *Lesion-agnostic*, denoting $Transfer(d_t|Learn(d_g))$; and (c)

Lesion-augmented, denoting $Transfer(d_t|Transfer(d_s|Learn(d_g)))$. For reasons of space, we will not consider *lesion-only* models ($Transfer(d_t|Learn(d_s))$).

Our aim is to compare the performance of models (a)–(c), using model transfer from brain-to-lung and lung-to-brain. We clarify our definition of performance in Step 4.

3.2 Materials

The experiments here use the following datasets:

ImageNet data: This constitutes d_g, or the generic data used for model transfer. This dataset is a classical collection of images for visual recognition research [7].

Lung-lesion data: The LIDC-IDRI dataset contains lung CT scans with annotated lesions [6]. The malignancy level for each lesion is annotated in the range 1–5. The tumours with average malignancy values from 1 to 3 were considered as benign, and the rest were considered malignant.

Brain-lesion data: The brain tumour dataset [4] contains 3064 T1-weighted contrast-enhanced images from 233 patients with three kinds of brain tumours - meningioma (708 slices), glioma (1426 slices) and pituitary tumour (930 slices) along with the corresponding lesion masks. We take meningioma and pituitary tumours as benign, and glioma tumours as malignant to form a binary classification problem.

4   Method

To help the model generalise better [2] we process the images by extract the lesions using the segmentation masks and normalise them. We then perform image augmentations such as random horizontal and vertical flips, shifts and rotations using the `ImageDataGenerator` class available in Keras [5].

Our experiments investigate the transfer of lesion models from brain-to-lung, and from lung-to-brain. We examine the effect of: (a) varying target training data size (with a fixed source data size); and (b) varying source data size (with a fixed target data size). For a given source (brain or lung), we adopt the following method:

– Let Te_t denote an independent sample of test instances for assessing the performance of the target model
– **Fixed source data size:** For a given source data set d_s and random samples d_t of target data sizes in $\{High, Medium, Low, VLow\}$:
   1. Construct $Base_{t|s} = Learn(d_t)$
   2. Construct $LesAgn_{t|s} = Transfer(d_t|Learn(d_g))$
   3. Construct $LesAug_{t|s} = Transfer(d_t|Transfer(d_s|Learn(d_g)))$
   4. Compare the performance of $Base_{t|s}$, $LesAgn_{t|s}$, and $LesAug_{t|s}$ on Te_t
– **Fixed target data size:** For a given target data set d_t and random samples d_s of source data sizes in $\{High, Medium, Low, VLow\}$:

1. Construct $Base_{s|t} = Learn(d_t)$
2. Construct $LesAgn_{s|t} = Transfer(d_t|Learn(d_g))$
3. Construct $LesAug_{s|t} = Transfer(d_t|Transfer(d_s|Learn(d_g)))$
4. Compare the performance of $Base_{s|t}$, $LesAgn_{s|t}$, and $LesAug_{s|t}$ on Te_t

The following details are relevant:

- The overall sizes of the brain data and lung data are 3064 and 729 respectively. The sizes of independent test data is 200 images for both brain-to-lung and lung-to-brain transfer. In all cases d_g refers to ImageNet.
- The size of d_s in Step 4 with brain-as-source is 3000 brain images, and d_s with lung-as-source is 700 lung images. In all cases, we use the following target data sizes d_t: 400 (High); 335 (Medium); 250 (Low); and 165 (VLow).
- The size of d_t in Step 4 with brain-as-target and lung-as-target is 400 images. For brain-as-source, the sizes of source data d_s are: 3000 (High); 2000 (Medium); 1000 (Low); and 500 (VLow). For lung-as-source, the size of source data d_s are: 700 (High); 500 (Medium); 300 (Low); and 100 (VLow).
- In all our experiments, we use the model structure of DenseNet-201 [9] to construct the lesion classification models (source and target). The final structure and parameters of models are determined using the Adam optimiser [11] using a binary cross entropy loss function. Early stopping was used while monitoring the validation loss.
- The training hyper-parameters for the algorithm and the models are as follows: the batch size is set to 64, the learning rate is 10^{-4}.
- All the experiments are conducted in Python environment in a machine with 64 GB main memory, 16-core Intel processor and 8 GB NVIDIA P4000 graphics processor.
- We define the performance of a model as the pair (R, P) where R is an unbiased estimate of the recall of the model, and P is an unbiased estimate of the model's precision.[1] The performance of a pair of models will be compared lexicographically. That is, (R_1, P_1) is better than (R_2, P_2) if and only if: $R_1 > R_2$, or if $R_1 = R_2$ and $P_1 > P_2$.

5 Results

The principal findings of our experiments are these: (a) Target models obtained using lesion-augmented transfer perform better than those obtained using lesion-agnostic transfer in most of the cases as seen in Fig. 1; (b) As the lesion-augmented target data d_t decreases, the benefit of lesion-augmented transfer over lesion-agnostic transfer increases (illustrated more clearly with F_2 scores in Fig. 4, we note that the gap in the score is largest in the regime of very low target training data); and (c) As the source data d_s decreases, the benefit

---

[1] We assume that in lesion-identification, the positive class refers to malignant lesions and that false-negatives are costlier than false-positives. That is, recall is more important than precision.

of lesion-augmented transfer over lesion-agnostic transfer decreases as seen in Fig. 2. Surprisingly, we also note that as the source data decreases, sometimes the model ends up performing worse than the baseline models on the test set. We now highlight some additional aspects of our study:

Lower Variance of Lesion-Augmented Models: We find that predictions using lesion-augmented models display a lesser variance than those from lesion-agnostic models. An example of this is shown in Fig. 3, showing the range of estimates of recall, precision obtained when performing a 5-fold cross-validation with $|d_s|$ = 'High', and $|d_t|$ = 'High'. This suggests that these models depend less on a particular training instance. The same trend holds in lung-to-brain transfer, but is less pronounced.

Faster Convergence of Lesion-Augmented models: We note that the gains of using a lesion-augmented model are twofold: (a) The loss starts from a much lower value, and (b) The lesion-augmented model converges in a fraction of the time that either of the other models takes to converge. An example of this is shown in Fig. 5 with the normalised binary cross-entropy loss. This behaviour occurs in lung-to-brain transfer as well, and we haven't shown it here due to space limitations.

	Base		LesAgn		LesAug	
$|d_t|$	R	P	R	P	R	P
High	0.89	0.80	0.93	0.93	**0.96**	**0.94**
Med.	0.69	0.77	0.52	0.83	**0.88**	**0.80**
Low	0.55	0.73	0.14	0.62	**0.71**	**0.77**
VLow	0.02	1.00	0.02	1.00	**0.67**	**0.76**

(a) Brain-to-lung

	Base		LesAgn		LesAug	
$|d_t|$	R	P	R	P	R	P
High	0.71	0.94	0.68	0.88	**0.73**	**0.83**
Med.	0.66	0.82	0.63	0.81	**0.71**	**0.83**
Low	**0.76**	**0.84**	0.66	0.82	0.71	0.83
VLow	0.56	0.82	0.44	0.82	**0.68**	**0.78**

(b) Lung-to-brain

Fig. 1. Recall and Precision scores keeping source data constant. The best performance pair (R, P) in each row is in bold.

	Base		LesAgn		LesAug	
$|d_s|$	R	P	R	P	R	P
High	0.89	0.80	0.93	0.93	**0.96**	**0.94**
Med.	0.89	0.80	**0.93**	**0.93**	0.86	0.70
Low	0.89	0.80	**0.93**	**0.93**	0.81	0.77
VLow	0.89	0.80	**0.93**	**0.93**	0.84	0.79

(a) Brain-to-lung

	Base		LesAgn		LesAug	
$|d_s|$	R	P	R	P	R	P
High	0.71	0.94	0.68	0.88	**0.73**	**0.83**
Med.	**0.71**	**0.94**	0.68	0.88	0.71	0.91
Low	**0.71**	**0.94**	0.68	0.88	0.68	0.88
VLow	**0.71**	**0.94**	0.68	0.88	0.68	0.77

(b) Lung-to-brain

Fig. 2. Recall and Precision scores keeping target data constant. 'Base' and 'LesAgn' models have no access to d_S, therefore R and P scores are the same across $|d_s|$ values.

	Brain-to-lung (Range)		Lung-to-brain (Range)	
	R	P	R	P
LesAgn	0.12–0.82	0.17–1.00	0.67–0.73	0.82–0.87
LesAug	0.94–0.97	0.79–0.99	0.66–0.72	0.84–0.87

Fig. 3. Ranges of recall (R) and precision (P) in a 5-fold cross-validation

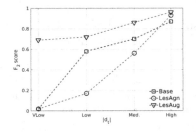

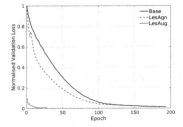

Fig. 4. F_2 score vs d_s for brain-to-lung transfer, varying $|d_t|$

Fig. 5. Loss vs epochs for brain-to-lung transfer, using a "VLow" value of $|d_t|$

6   Conclusion

Transfer learning, especially using deep neural networks, presents one way of dealing with the problems arising from the lack of sufficient data to build good models, that is common in problems involving medical images. This is due to reasons of cost, rarity of disease occurrence, difficulty of annotation, and so on. Although most routine demonstrations of transfer learning with deep networks have involved transfer from general datasets to specific ones, it would seem evident that transfer would be more effective if the source data were in some way related to the target. In this paper we have investigated this for transfer of lesion-models from one organ to another. Although the problem of lack of data persists for lesions for some organs, there are good biological reasons to believe that lesions from a different organ could be useful in constructing a model for the target organ. Our results here show how even small amounts of such lesion-specific source data can make a substantial difference to target models (the augmentation of a dataset of nearly 14 m images, with at most 5000 lesion images for the source organ). Besides better predictive performance, we find that the augmentation results in target models that converge faster and have lower variance.

While our results are largely consistent for both brain-to-lung and lung-to-brain transfer, there is a difference in the gains resulting from the inclusion of lesion-specific data. Additional experiments suggest that this is not due to differences in the quantity or quality of data. There may be some underlying biological reasons for this difference, which needs to be investigated further. We think the experiments could also benefit from the use of data available in [13], which already has segmented lesions throughout the body. If annotations were available for even some small part of this data, they could prove more helpful than using a generic dataset like ImageNet. Additional experiments are also needed with other pairs of organs, to ensure that the observations here can be generalised: for the present, the results show that lesion-knowledge can be transferred usefully from the brain to the lung and *vice versa*.

Acknowledgement. This work is supported by "The DataLab" agreement between BITS Pilani, K.K. Birla Goa Campus and TCS Research, India. AS is a Visiting Professorial Fellow at the School of CSE, UNSW Sydney.

References

1. Alahmer, H., Ahmed, A.: Computer-aided classification of liver lesions from CT images based on multiple ROI. Procedia Comput. Sci. **90**, 80–86 (2016). https://doi.org/10.1016/j.procs.2016.07.027
2. Asano, Y.M., Rupprecht, C., Vedaldi, A.: A critical analysis of self-supervision, or what we can learn from a single image. arXiv preprint arXiv:1904.13132 (2019)
3. Chang, K., et al.: Distributed deep learning networks among institutions for medical imaging. J. Am. Med. Inform. Assoc. JAMIA **25** (2018). https://doi.org/10.1093/jamia/ocy017
4. Cheng, J.: Brain tumor dataset, April 2017. https://figshare.com/articles/dataset/brain_tumor_dataset/1512427
5. Chollet, F., et al.: Keras (2015). https://github.com/fchollet/keras
6. Clark, K., et al.: The cancer imaging archive (TCIA): maintaining and operating a public information repository. J. Digit. Imaging **26**(6), 1045–1057 (2013). https://doi.org/10.1007/s10278-013-9622-7
7. Deng, J., Dong, W., Socher, R., Li, L.J., Li, K., Fei-Fei, L.: ImageNet: a large-scale hierarchical image database. In: CVPR09 (2009)
8. He, K., Girshick, R., Dollár, P.: Rethinking ImageNet pre-training. In: Proceedings of the IEEE International Conference on Computer Vision, pp. 4918–4927 (2019)
9. Huang, G., Liu, Z., Van Der Maaten, L., Weinberger, K.Q.: Densely connected convolutional networks. In: Proceedings of the IEEE Conference on Computer Vision and Pattern Recognition, pp. 4700–4708 (2017)
10. Kaur, B., et al.: Improving pathological structure segmentation via transfer learning across diseases. In: Wang, Q., et al. (eds.) DART/MIL3ID -2019. LNCS, vol. 11795, pp. 90–98. Springer, Cham (2019). https://doi.org/10.1007/978-3-030-33391-1_11
11. Kingma, D.P., Ba, J.: Adam: a method for stochastic optimization. arXiv preprint arXiv:1412.6980 (2014)
12. Raghu, M., Zhang, C., Kleinberg, J., Bengio, S.: Transfusion: understanding transfer learning for medical imaging. In: Advances in Neural Information Processing Systems, pp. 3347–3357 (2019)
13. Yan, K., Wang, X., Lu, L., Summers, R.: DeepLesion: automated mining of large-scale lesion annotations and universal lesion detection with deep learning. J. Med. Imaging **5**, 1 (2018). https://doi.org/10.1117/1.JMI.5.3.036501
14. Yosinski, J., Clune, J., Bengio, Y., Lipson, H.: How transferable are features in deep neural networks? In: Advances in Neural Information Processing Systems, pp. 3320–3328 (2014)

Transfer Learning with Joint Optimization for Label-Efficient Medical Image Anomaly Detection

Xintong Li[1,2], Huijuan Yang[2(✉)], Zhiping Lin[1], and Pavitra Krishnaswamy[2]

[1] School of EEE, Nanyang Technological University, Singapore, Singapore
`I160006@e.ntu.edu.sg, ezplin@ntu.edu.sg`
[2] Institute for Infocomm Research, A*STAR, Singapore, Singapore
`{hjyang,pavitrak}@i2r.a-star.edu.sg`

Abstract. Many medical imaging applications require robust capabilities for automated image anomaly detection. Supervised deep learning approaches can be employed for such tasks, but poses large data collection and annotation burdens. To address this challenge, recent works have proposed advanced unsupervised, semi-supervised or transfer learning based deep learning methods for label-efficient image anomaly detection. However, these methods often require extensive hyperparameter tuning to achieve good performance, and have yet to be demonstrated in data-scarce domain centric applications with nuanced normal-vs-anomaly distinctions. Here, we propose a practical label-efficient anomaly detection method that employs fine-tuning of pre-trained model based on a small target domain dataset. Our approach employs a joint optimization framework to enhance discriminative power for anomaly detection performance. In evaluations on two benchmark medical image datasets, we demonstrate (a) strong performance gains over state-of-the-art baselines and (b) increased label efficiency over standard fine-tuning approaches. Importantly, our approach reduces the need for large annotated datasets, requires minimal hyperparameter tuning, and shows stronger performance boost for more challenging anomalies (Supplement: http://s000.tinyupload.com/?file_id=24916959421870989415).

Keywords: Label-efficient deep learning · Anomaly detection · Transfer learning · Joint optimization

1 Introduction

The ability to automatically detect anomalies in medical images has applications in disease screening, triaging, automated diagnostic systems, and quality control in high-throughput laboratory or clinical settings. Recent studies have

Electronic supplementary material The online version of this chapter (https://doi.org/10.1007/978-3-030-61166-8_16) contains supplementary material, which is available to authorized users.

J. Cardoso et al. (Eds.): iMIMIC 2020/MIL3ID 2020/LABELS 2020, LNCS 12446, pp. 146–154, 2020.
https://doi.org/10.1007/978-3-030-61166-8_16

highlighted the potential of deep learning approaches for screening and triaging applications involving chest X-Rays, retinal images, mammograms, and brain scans. However, these studies frame the anomaly detection task as a supervised classification problem, and hence require sizeable labeled datasets. Creation of these datasets requires laborious and resource-intensive annotation by specialized domain experts, and may result in noisy and biased labels due to the discordant opinions from the experts. Further, it is infeasible to scale collection to large data volumes in many of the above applications. As such, there is a need to advance medical image anomaly detection approaches that can address the annotation burden and data scarcity challenges. Previous works have explored unsupervised [1–5], semi-supervised [6,7] and transfer learning or fine-tuning based [8–11] anomaly detection approaches. The un-/semi-supervised methods commonly employ advanced reconstruction-based frameworks [1,3], generative adversarial networks [3,7] and deep support vectors [12]. These approaches often require hyper-parameter tuning and intensive computational resources, thus hampering use in practice. Furthermore, they do not perform well when applied to the medical images with subtle anomalies [13]. In the realm of transfer learning, the focus is to fine-tune representations learned on a large source domain dataset using a small subset of target domain data. One recent study combined an unsupervised classifier with transfer-learning based feature selection for medical image anomaly detection [13], but did not use target domain data to fine-tune the features or model. Some studies have explored optimal fine-tuning strategies and loss functions in multi- and one-class settings [8–11]. However, these strategies have not been validated in complex real-world tasks involving detection of non-apparent anomalies and performance is insufficient for practical deployment.

In this work, we expand upon [10,13] to propose a practical label-efficient anomaly detection method for challenging medical imaging applications. Specifically, our approach learns the feature representations from a pre-trained ImageNet model by transfer learning, and then fine-tunes these representations based on a small target domain dataset. For the fine-tuning, we combine an adaptive approach with a joint optimization for two concurrent objectives: (a) to learn discriminative features, and (b) to enhance compactness of normal class representation for effective anomaly detection. We evaluate our approach on two medical image benchmark datasets and demonstrate large gains over baselines.

2   Methods

2.1   Network Architecture

The network architecture for the training process in our proposed method is illustrated in Fig. 1. There are two parts: (a) a *feature extractor* denoted as G_f which includes the input layer and all the hidden layers in selected neural network model; and (b) a *classifier*, i.e., a fully connected layer. For G_f, we employ a neural network that is pre-trained on ImageNet for transfer learning. Specifically, we choose the pre-trained ResNet-26 [9,11] as our base model (although our method is applicable across choices of pre-trained models). The output from

G_f informs the calculation of an intra-class variance loss (L_{iv}) [10] to enforce compactness of the cluster representation of normal features. The output from the classifier informs the calculation of a cross-entropy loss (L_{ce}) [14]. These two losses are jointly optimized in the training process.

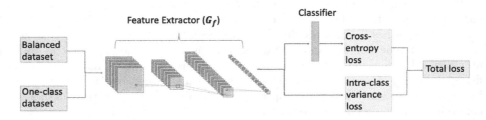

Fig. 1. Network architecture of our proposed method. The data batches are sampled from a balanced dataset and one-class dataset during training. The cross-entropy loss and intra-class variance loss are calculated and back-propagated simultaneously.

2.2   Construction of Datasets for Training

We consider a training dataset comprising normal samples and a small number of abnormal samples. We randomly sample a small subset of this training dataset following the native normal-to-abnormal class ratio in the training data, and name this subset as the small imbalanced (**SI**) dataset. We divide **SI** into two groups, namely, a "balanced dataset (denoted as $\mathbf{D_b}$)" and a "one-class dataset (denoted as $\mathbf{D_o}$)". $\mathbf{D_b}$ comprises all abnormal samples in the **SI** and an equal number of normal samples. It is used to train the feature extractor and classifier, and eventually to inform the cross-entropy loss. $\mathbf{D_o}$ comprises of the remaining normal samples in **SI**, which is used to train the feature extractor and eventually to inform the intra-class variance loss. $\mathbf{D_b}$ helps to enhance the discriminative power of the feature extractor, whereas $\mathbf{D_o}$ helps to enforce compactness of the feature representation for normal class for improved anomaly detection.

2.3   Joint Optimization

In order to concurrently achieve good discriminative power of the feature extractor and enhance compactness of the feature representation for normal class, we employ a joint optimization framework. This framework is based on a dual loss formulation that combines cross-entropy loss (L_{ce}) and intra-class variance loss (L_{iv}). L_{ce} is widely used to characterize the accuracy of a classification model in relation to ground truth labels [14]. Here we use the small balanced dataset to fine-tune the pre-trained model directly and characterize fine-tuning performance with L_{ce}. The compactness of the feature representation for the normal samples can be characterized by distance measures. Here, we employ the

Euclidean distance measure to compute L_{iv}. A low value of L_{iv} indicates that the features extracted from the normal samples are compact and tightly clustered. This means that abnormal samples falling outside the boundary can be easily identified. Formally, consider the data sample x_i with data label y_i. For each sample from a batch of the one-class data, the distance between the given sample and the rest of the samples from the same batch is given by

$$z_i = G_f(x_i) - m_i \tag{1}$$

where x_i and m_i are the given sample, and the mean of the features for the rest of samples, respectively. The intra-class variance loss is the averaged Euclidean distance, which is given by

$$L_{iv} = \frac{1}{n_o} \sum_{i=1}^{n_o} z_i^T z_i \tag{2}$$

where n_o is the number of one-class samples in the batch and each feature is of dimension d. Then, the overall loss function is given by

$$\arg \min_{\theta} \left(\sum_{x_i \in \mathbf{D_b}} \lambda L_{ce}(G_f(x_i), \theta, y_i) + \sum_{x_i \in \mathbf{D_o}} (1 - \lambda) L_{iv}(G_f(x_i), \theta, m_i) \right) \tag{3}$$

where λ is the hyper-parameter used to balance the two losses, and θ denotes the learned parameters.

3   Experimental Procedures

3.1   Datasets

We used two medical imaging datasets (Kaggle Diabetic Retinopathy (DR) challenge [15] and the RSNA Intracranial Hemorrhage Detection (RSNA-IHD) [16]), and one natural image dataset (CIFAR-10) for the experimental evaluations. Figure 2 illustrates examples of the normal and abnormal images in DR and RSNA-IHD. For DR and CIFAR-10, we use the official train and test sets. For RSNA-IHD, we use the official train set and randomly split 10,000 images as the test set. For DR dataset, We consider grade "0" (no diabetic retinopathy) and "4" (proliferative diabetic retinopathy) as normal and abnormal samples, respectively. Results of "0" vs. "2, 3, 4" are presented in Table 4 of supplementary. For RSNA-IHD, we designate slices with any anomaly (epidural, intraparenchymal, intraventricular, subarachnoid and subdural) as abnormal ("1") and those without any anomaly ("0") as normal. For CIFAR-10 dataset, we select one of the six randomly chosen classes as the "normal" class each time, and use data from the other nine classes as the "abnormal" class. We randomly sample 10 abnormal and 150 normal images from the above training datasets as our small imbalanced (**SI**) dataset to simulate the native normal-to-abnormal class ratio. As discussed in Sect. 2.2, with the **SI** dataset, we construct the balanced dataset ($\mathbf{D_b}$) and one-class dataset ($\mathbf{D_o}$). Specifically, we use all 10 abnormal images and randomly sample 10 normal images from **SI** as $\mathbf{D_b}$ to train the classifier. The remaining normal samples from **SI** are formed as $\mathbf{D_o}$.

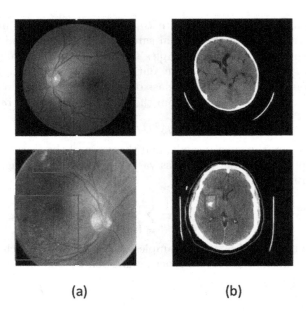

(a)   (b)

Fig. 2. Examples of the normal (upper panel) and abnormal (lower panel) images from: (a) DR, (b) RSNA-IHD datasets. The differences between abnormal vs. normal images are highlighted in rectangle.

Table 1. Number of samples used in training

10/150	FTID, UnSpSm, Ours
150/150	FTBD
234/3520	DAGMM, DSEBM, UnSpSm
0/3520	OCSVM

Note: the abnormal/normal samples are obtained from randomly sampled 5000 images, but to keep the outlier ratio as ours for DAGMM, DSEBM and UnSpSm

3.2   Baselines and Ablation Study

We compare our proposed method against several baselines: a) a shallow support vector machines (SVM)-based model [17]; b) two deep anomaly detection methods based on autoencoders (DAGMM [3], DSEBM [1]); c) an unsupervised anomaly detection approach based on pre-trained ImageNet with features selected using small labeled datasets (UnSpSm) [13]; d) a supervised baseline that fine-tunes a pre-trained ImageNet model based on the balanced dataset (FTBD); e) an **ablation study** which fine-tunes a pre-trained ImageNet based on the relevant imbalanced dataset (FTID). To enable rigorous comparison, the feature selection for UnSpSm employ the same labeled abnormal/normal samples as our method. Further, we note that FTID comprises of only cross-entropy loss and no compactness loss of normal class representation is imposed. Table 1 details the construction of training datasets for the different baselines.

3.3 Training and Testing

During training, the network is initialized with the weights of pre-trained ResNet-26 model. The weights of the first half of the network layers are frozen. Two data batches (with batch size 20) generated from the D_b and D_o are fed to the input of the network simultaneously. The cross-entropy loss and intra-class variance loss are computed and summed to compute the composite loss, which is then back-propagated. The parameters are trained using gradient descent with learning rate of 0.001. Training is stopped when there is no decrease in the composite loss for 20 epochs. During test phase, all the images are fed to the feature extractor to extract features, which are then passed to the classifier (i.e., fully connected layer) to classify the input image as normal or abnormal.

4 Results

We now compare the performance between our proposed method and the baselines. We ran five experiments across random seeds, and characterize the performance in terms of mean and standard deviation of the AUROC and AUPRC across seeds for all the methods, with the results shown in Table 2. First, our method outperforms the baseline unsupervised anomaly detection methods (OCSVM, DAGMM and DSEBM) by a large margin. Our method achieves at least 10% increase in AUROC and far greater increases (over 20%) in AUPRC for both datasets. In many cases, the baselines require large volumes of normal data, while our method does not. Second, the FTID ablation study results demonstrate that the inclusion of compactness loss further increases the AUROC by 2–5%. Third, the method achieves AUROC comparable with the FTBD approach although FTBD imposes far greater annotation burden. To compare the label-efficiency of our method and FTBD, we consider a realistic annotation scenario which involves sampling anomalous images iteratively from the native datasets. Since the proportions of outliers are 6% and 15.1% in DR and RSNA-IHD, we

Table 2. Performance comparison of our method vs. baselines

	Dataset: DR						
	OCSVM	DAGMM	DSEBM	UnSpSm	FTID	FTBD	Ours
AUROC	38.6 ± 1.2	60.8 ± 2.0	45.8 ± 1.4	79.8 ± 0.6	87.5 ± 1.3	**91.2 ± 1.4**	89.3 ± 1.3
AUPRC	2.5 ± 0.1	4.9 ± 0.5	2.6 ± 0.2	13.5 ± 1.0	26.3 ± 2.3	**38.3 ± 3.6**	27.5 ± 2.7
	Dataset: RSNA-IHD						
	OCSVM	DAGMM	DSEBM	UnSpSm	FTID	FTBD	Ours
AUROC	30.1 ± 0.4	45.6 ± 0.6	43.2 ± 3.0	37.0 ± 2.2	71.4 ± 4.6	**78.2 ± 2.5**	76.6 ± 3.1
AUPRC	10.4 ± 1.2	12.3 ± 0.6	12.3 ± 0.6	11.4 ± 0.4	35.4 ± 3.2	**45.3 ± 3.7**	36.7 ± 2.4

need to sample from 167 (proposed) vs. 2500 (FTBD), and 66 (proposed) vs. 993 (FTBD) to obtain the required number of 10 (proposed) and 150 (FTBD) anomalous training samples for DR and RSNA-IHD, respectively. These numbers suggest that the annotation burden of our method is about 15 times less than FTBD (when 10/150 abnormal/normal samples are chosen). We also evaluated the performance on CIFAR-10. The averaged AUROC (%) of proposed method is 88.0 ± 4.0, which is comparable to that achieved with FTBD (94.7 ± 2.6), better than the performance with FTID (85.6 ± 2.7), and far better than unsupervised baseline (UnSpSm) (56.6 ± 2.7). These results demonstrate the general applicability of our proposed method. Further, the performance gains with our method are higher for the medical image datasets than for the natural image CIFAR-10 dataset, suggesting advantages in more challenging applications.

We now evaluate how the proportion of the labelled abnormal data in train data affects the performance. For this purpose, we vary the number of the labelled abnormal samples and evaluate the AUROC for the DR dataset. The results are in Fig. 3. Note that the same number of abnormal/normal samples are used for all the methods. The performance of AUROC of our method is significantly better than the baselines, and performance of our method increases as the proportion of the abnormal sample increases.

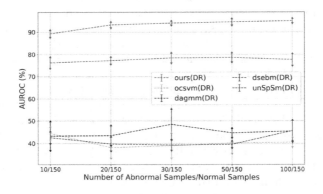

Fig. 3. AUROC for DR dataset as a function of the number of abnormal samples.

5   Conclusions

In this work, we proposed a novel but practical label-efficient anomaly detection method for challenging medical imaging applications. To address the data scarcity problem, our method utilizes a very small number of labelled abnormal and normal samples to fine-tune a pre-trained network, and employs a joint optimization framework to achieve both high discriminative power for classification and compactness of normal class representation for effective anomaly detection.

We showed high performance boost on two medical image datasets with anomalies that are difficult to detect. As such, our work offers a practical benchmark for future work in medical image anomaly detection, and has implications for a range of domain-centric anomaly detection tasks. Future work will focus on expanding our approach for 3D medical images, and performing more extensive evaluations for practical translation.

Acknowledgement. This work was supported by funding and infrastructure support for deep learning and medical imaging research from the Institute for Infocomm Research, A*STAR, Singapore (Grant No. SSF A1718g0045). We also acknowledge insightful discussions with Foo Chuan Sheng and are grateful for the constructive feedback from the anonymous reviewers.

References

1. Zhai, S., Cheng, Y., Lu, W., Zhang, Z.: Deep structured energy based models for anomaly detection. CoRR, abs/1605.07717 (2016)
2. Schlegl, T., Seeböck, P., Waldstein, S.M., Schmidt-Erfurth, U., Langs, G.: Unsupervised anomaly detection with generative adversarial networks to guide marker discovery. In: Niethammer, M., et al. (eds.) IPMI 2017. LNCS, vol. 10265, pp. 146–157. Springer, Cham (2017). https://doi.org/10.1007/978-3-319-59050-9_12
3. Zong, B., et al.: Deep autoencoding Gaussian mixture model for unsupervised anomaly detection. In: 6th International Conference on Learning Representations, ICLR (2018)
4. Zenati, H., Romain, M., Foo, C.-S., Lecouat, B., Chandrasekhar, V.: Adversarially learned anomaly detection. In: 2018 IEEE International Conference on Data Mining (ICDM), pp. 727–736. IEEE (2018)
5. Golan, I., El-Yaniv, R.: Deep anomaly detection using geometric transformations. In: Advances in Neural Information Processing Systems, pp. 9758–9769 (2018)
6. Madani, A., Ong, J.R., Tibrewal, A., Mofrad, M.R.K.: Deep echocardiography: data-efficient supervised and semi-supervised deep learning towards automated diagnosis of cardiac disease. NPJ Digit. Med. **1**(1), 1–11 (2018)
7. Lecouat, B., et al.: Semi-supervised deep learning for abnormality classification in retinal images. arXiv preprint arXiv:1812.07832 (2018)
8. Andrews, J., Tanay, T., Morton, E.J., Griffin, L.D.: Transfer representation-learning for anomaly detection. In: Proceedings of the 33rd International Conference on Machine Learning, pp. 1–5 (2016)
9. Guo, Y., Shi, H., Kumar, A., Grauman, K., Rosing, T., Feris, R.: SpotTune: transfer learning through adaptive fine-tuning. In: Proceedings of IEEE Conference on Computer Vision and Pattern Recognition, pp. 4805–4814 (2019)
10. Perera, P., Patel, V.M.: Learning deep features for one-class classification. IEEE Trans. Image Process. **28**(11), 5450–5463 (2019)
11. Rebuffi, S.-A., Bilen, H., Vedaldi, A.: Learning multiple visual domains with residual adapters. In: Advances in Neural Information Processing Systems, pp. 506–516 (2017)
12. Ruff, L., et al.: Deep one-class classification. In: Proceedings of the 35th International Conference on Machine Learning, pp. 4393–4402 (2018)

13. Ouardini, K., et al.: Towards practical unsupervised anomaly detection on retinal images. In: Wang, Q., et al. (eds.) DART/MIL3ID -2019. LNCS, vol. 11795, pp. 225–234. Springer, Cham (2019). https://doi.org/10.1007/978-3-030-33391-1_26
14. Li, C.H., Lee, C.K.: Minimum cross entropy thresholding. Pattern Recognit. **26**(4), 617–625 (1993)
15. Diabetic retinopathy detection (2015). https://www.kaggle.com/c/diabetic-retinopathy-detection. Accessed 07 July 2020
16. RSNA intracranial hemorrhage detection (2019). https://www.kaggle.com/c/rsna-intracranial-hemorrhage-detection. Accessed 07 July 2020
17. Schölkopf, B., Williamson, R.C., Smola, A.J., Shawe-Taylor, J., Platt, J.C.: Support vector method for novelty detection. In: Advances in Neural Information Processing Systems, pp. 582–588 (2000)

Unsupervised Wasserstein Distance Guided Domain Adaptation for 3D Multi-domain Liver Segmentation

Chenyu You[1(✉)], Junlin Yang[2], Julius Chapiro[3], and James S. Duncan[1,2,3,4]

[1] Department of Electrical Engineering, Yale University, New Haven, CT, USA
`chenyu.you@yale.edu`
[2] Department of Biomedical Engineering, Yale University, New Haven, CT, USA
[3] Department of Radiology and Biomedical Imaging, Yale School of Medicine, New Haven, CT, USA
[4] Department of Statistics and Data Science, Yale University, New Haven, CT, USA

Abstract. Deep neural networks have shown exceptional learning capability and generalizability in the source domain when massive labeled data is provided. However, the well-trained models often fail in the target domain due to the domain shift. Unsupervised domain adaptation aims to improve network performance when applying robust models trained on medical images from source domains to a new target domain. In this work, we present an approach based on the Wasserstein distance guided disentangled representation to achieve 3D multi-domain liver segmentation. Concretely, we embed images onto a shared content space capturing shared feature-level information across domains and domain-specific appearance spaces. The existing mutual information-based representation learning approaches often fail to capture complete representations in multi-domain medical imaging tasks. To mitigate these issues, we utilize Wasserstein distance to learn more complete representation, and introduces a content discriminator to further facilitate the representation disentanglement. Experiments demonstrate that our method outperforms the state-of-the-art on the multi-modality liver segmentation task.

1 Introduction

Accurate and consistent measurements on medical images greatly assist radiologists in making precise and reliable diagnoses and staging the patients. In clinical practices, manual segmentation of anatomical structures from 3D medical images by experienced experts is tedious, time-consuming, and error-prone, which is not suitable for large-scale studies [3]. Besides, different medical imaging modalities, such as Magnetic Resonance Imaging (MRI), Computed Tomography (CT), and Positron Emission Tomography (PET), provide unique views of tissue features at different spatial resolutions with functional information. In particular, CT is

---

C. You and J. Yang—Equal contribution.
This work was supported by NIH Grant 5R01 CA206180.

J. Cardoso et al. (Eds.): iMIMIC 2020/MIL3ID 2020/LABELS 2020, LNCS 12446, pp. 155–163, 2020.
https://doi.org/10.1007/978-3-030-61166-8_17

the most common imaging modality for the diagnosis of hepatocellular carcinoma (HCC), the primary malignant tumor in the human liver. However, the scan is associated with the radiation dosage and provides low soft-tissue contrast, which makes it difficult to visualize tumor boundaries. As a non-invasive technique, MRI offers higher contrast, but has disadvantages in assessment cost, acquisition time, and is more prone to artifacts. In clinical practice, the fusion of multi-modal images allows for capturing more anatomical information and integrating complementary information to minimize redundancy and enhancing the diagnostic potential. Thus, it is a rapidly rising demand to segment cross-modality images for accurate analysis and interpretation.

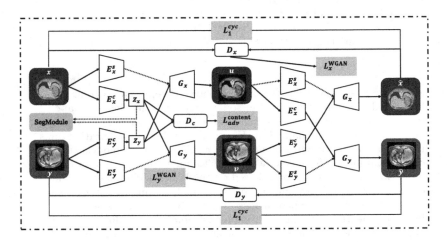

Fig. 1. Overview of the proposed 3D Wasserstein Distance Guided Domain Adaption Model. With the guidance of the following constraints $\mathcal{L}^{\text{WGAN}}$, $\mathcal{L}^{\text{cyc}}$ and $\mathcal{L}^{\text{content}}$, we can learn the cross-domain mapping between unpaired CT and Multi-phasic MRI sequences. The domain discriminators $\{D_{\mathcal{X}}, D_{\mathcal{Y}}\}$, and a content discriminator D^c jointly encourage the model to obtain the well-learned representation disentanglement.

Unsupervised domain adaptation has been widely used for generalizing medical image segmentation models across domains. The major challenge is to mitigate domain gaps between different modalities. Several recent efforts have been made to improve the segmentation performance without label data in the medical imaging community [2,8,9]. For example, Yang *et al.* [9] utilized multi-modal unsupervised image-to-image translation framework (MUNIT) [6] to decompose image into a shared domain-invariant content space and a domain-specific style space. Then, the learned content representations are used to train the segmentation network.

In this paper, we present a novel unsupervised cross-modality domain adaptation method for medical image segmentation. Our proposed method extends upon [9] as follows: Firstly, in order to obtain more complete domain invariant representations, we introduce Wasserstein distance [4,11,12] to reduce the domain discrepancy instead of the negative log-likelihood used in [7]. Secondly, medical imaging data is inherently three-dimensional (3D). However, most of domain adaption methods leverage 2D information. We incorporate 3D volumetric information to improve the image quality of reconstructed images by fully exploiting detailed spatial information along the z dimension. Thirdly, to facilitate the decomposition of domain-invariant shared information and domain-specific features, in our work we propose a content discriminator to distinguish extracted content-level representations between different domains and utilize a cross-cycle consistency loss to enforce many-to-many mappings. We demonstrate that our proposed methods are competitive against other state-of-the-art methods in multi-domain liver segmentation.

2   Method

2.1   Overview

Our goal is to learn a cross-modality mapping between two domains $\mathcal{X}$ and $\mathcal{Y}$ without paired training data. We assume that there exists a potentially many-to-many mapping between two domains. Our approach decomposes images onto a shared content spaces $c \in \mathcal{C}$, and domain-specific space $\mathcal{S}_\mathcal{X}$ and $\mathcal{S}_\mathcal{Y}$ [6,7]. Intuitively, the content encoders are used to map the shared information shared among domains onto $\mathcal{C}$, and the style encoders project the domain-specific information onto $\mathcal{S}_\mathcal{X}$ and $\mathcal{S}_\mathcal{Y}$.

2.2   Model

Let $x \in \mathcal{X}$ and $y \in \mathcal{Y}$ be images from two different domains, or in our task, two different imaging modalities. As shown in Fig. 1, Our method deploys 3D CNN to take advantage of spatial information. Similar to the recent works [6,7], the overall model consists of several networks: jointly trained content encoders $\{E_\mathcal{X}^c,$ $E_\mathcal{Y}^c\}$, style encoders $\{E_\mathcal{X}^s, E_\mathcal{Y}^s\}$, decoders $\{G_\mathcal{X}, G_\mathcal{Y}\}$ and domain discriminators $\{D_\mathcal{X}, D_\mathcal{Y}\}$, and a content discriminator D^c. $i.e.$, given the domain $\mathcal{X}$, the content encoder $E_\mathcal{X}^c$ and the style encoder $E_\mathcal{X}^s$ encode x to a content code z_x^c in a shared, domain-invariant content space $(\mathcal{C})$ and a style code z_x^s in domain-specific style space $(\mathcal{S}_\mathcal{X})$, respectively. The decoders $G_\mathcal{X}$ reconstruct images conditioned on both content and style codes. The discriminator $D_\mathcal{X}$ aims to discriminate between real images and reconstructed images from the domain $\mathcal{X}$. In addition, the content discriminator D^c is trained jointly to distinguish the encoded content features z_x^c and z_y^c between two domain.

Latent Reconstruction. To achieve representation disentanglement and preserve maximal information in the representation of each domain, we use the bidirectional reconstruction loss to encourage the bidirectional mapping which includes the self-reconstruction loss and latent reconstruction loss, *i.e.*,

$$\mathcal{L}_{\mathcal{X}}^{\text{recon}} = \mathbb{E}_x[||G_{\mathcal{X}}(E_{\mathcal{X}}^c(x), E_{\mathcal{X}}^s(x)) - x||_1], \tag{1a}$$

$$\mathcal{L}_{\mathcal{X}^c}^{\text{latent}} = \mathbb{E}_{x,y}[||E_{\mathcal{Y}}^c(G_{\mathcal{Y}}(z_x^c, z_y^s)) - z_x^c||_1], \tag{1b}$$

$$\mathcal{L}_{\mathcal{Y}^s}^{\text{latent}} = \mathbb{E}_{x,y}[||E_{\mathcal{Y}}^s(G_{\mathcal{Y}}(z_x^c, z_y^s)) - z_y^s||_1], \tag{1c}$$

In Domain Reconstruction. In order to facilitate the disentangled content and attribute representations for cyclic reconstruction, we formulate the cross-cycle consistency loss as [7]:

$$\mathcal{L}^{\text{cyc}} = \mathcal{L}_{\mathcal{X}}^{\text{cyc}} + \mathcal{L}_{\mathcal{Y}}^{\text{cyc}} = \mathbb{E}_{x,y}[||G_{\mathcal{X}}(E_{\mathcal{Y}}^c(\hat{x}), E_{\mathcal{X}}^s(\hat{y})) - x||_1 \tag{2}$$
$$+ ||G_{\mathcal{Y}}(E_{\mathcal{X}}^c(\hat{y}), E_{\mathcal{Y}}^s(\hat{x})) - y||_1],$$

where $\hat{y} = G_{\mathcal{X}}(E_{\mathcal{Y}}^c(y), E_{\mathcal{X}}^s(x))$ and $\hat{x} = G_{\mathcal{Y}}(E_{\mathcal{X}}^c(x), E_{\mathcal{Y}}^s(y))$, respectively.

Adversarial Loss. First, we introduce WGAN-GP [4,10] to match the distribution of reconstructed images to the target domain. The generator G and two discriminators $D_{\mathcal{X}}$ and $D_{\mathcal{Y}}$ are trained via alternatively optimizing the corresponding composite loss functions. *i.e.*:

$$\mathcal{L}_{\mathcal{Y}}^{\text{WGAN}} = \mathbb{E}_y[D_{\mathcal{Y}}(y)] - \mathbb{E}_{\hat{y}}[D_{\mathcal{Y}}(\hat{y})] + \alpha \cdot L_{\mathcal{Y}}^{grad}, \tag{3}$$

where $D_{\mathcal{Y}}$ is a discriminator for domain adaption to distinguish between reconstructed images $\hat{y}$ and real images y. The coefficient α is a weighting hyperparameter. The gradient penalty term is $L_{\mathcal{Y}}^{grad} = \mathbb{E}_{\tilde{y}}[(||\nabla_{\tilde{y}} D_{\mathcal{Y}}(\tilde{y}) - 1||_2)^2)]$, where $\tilde{y}$ is uniformly sampled between y and $\hat{y}$. The discriminator $D_{\mathcal{X}}$ and loss $\mathcal{L}_{\mathcal{X}}^{\text{WGAN}}$ are defined similarly. Second, we employ a content discriminator D_C to match the distribution of the encoded content features z_x and z_y of different domains. We formulate the content adversarial loss [7] as:

$$\mathcal{L}^{\text{content}}(E_{\mathcal{X}}^c, E_{\mathcal{Y}}^c, D_C) = \min_G \max_D \mathbb{E}_x[\frac{1}{2}\log D_C(E_{\mathcal{X}}^c(x)) + \frac{1}{2}\log(1 - D_C(E_{\mathcal{X}}^c(x)))] \tag{4}$$
$$+ \mathbb{E}_y[\frac{1}{2}\log D_C(E_{\mathcal{Y}}^c(y)) + \frac{1}{2}\log(1 - D_C(E_{\mathcal{Y}}^c(y)))]$$

Total Loss. We jointly train the encoders, decoders, and discriminators via optimizing the following objective function.

$$\min_{G_{\mathcal{X}},G_{\mathcal{Y}},E_{\mathcal{X}},E_{\mathcal{Y}}} \max_{D_{\mathcal{X}},D_{\mathcal{Y}},D_C} \mathcal{L}(G_{\mathcal{X}}, G_{\mathcal{Y}}, E_{\mathcal{X}}, E_{\mathcal{Y}}, D_{\mathcal{X}}, D_{\mathcal{Y}}, D_C) = \lambda^{\text{WGAN}}\mathcal{L}^{\text{WGAN}} \tag{5}$$
$$+ \lambda^{\text{recon}}\mathcal{L}^{\text{recon}} + \lambda^{\text{cyc}}\mathcal{L}^{\text{cyc}} + \lambda^{\text{latent}}\mathcal{L}^{\text{latent}} + \lambda^{\text{content}}\mathcal{L}^{\text{content}},$$

where $\lambda^{\text{WGAN}}, \lambda^{\text{recon}}, \lambda^{\text{cyc}}, \lambda^{\text{latent}}, \lambda^{\text{content}}$ are weights that control the importance of each term.

SegModule. Once disentangled representation is achieved, the content-only image can be generated given the content code. For both CT and MRI, we assume that their content codes are embedded onto the shared domain-invariant latent space that preserve anatomical information but exclude modality-specific information. We implement DenseNet [5] as the segmentation network. Note that we tailored the network configuration for our task. To address the inherent class imbalance between foreground liver part and the background, we combine the Soft Dice and weighted Cross-Entropy (CE) losses [8] to train the SegModule.

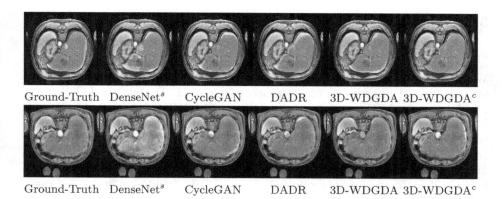

Ground-Truth   DenseNet^s CycleGAN   DADR   3D-WDGDA 3D-WDGDA^c

Ground-Truth   DenseNet^s CycleGAN   DADR   3D-WDGDA 3D-WDGDA^c

Fig. 2. Qualitative results of different methods on segmentation. We list the ground-truth, DenseNet^s, CycleGAN, DADR, 3D-WDGDA, 3D-WDGDA^c (with content discriminater).

Model Implementation. We implement the proposed method in PyTorch, using NVIDA TITAN XP GPUs. For domain adaption tasks, we build our model based on [6,7] with changes as discussed in Sect. 2. The network architecture here includes a VAE with two domain-specific encoders and decoders that is based on [6]. We utilize the Wasserstein distance with gradient penalty instead of the negative log-likelihood. The content discriminator adopts the same architecture as in [7]. We use Adam optimizer with a learning rate of 10^{-4} and set the hyperparameter $\lambda^{\mathrm{WGAN}}, \lambda^{\mathrm{recon}}, \lambda^{\mathrm{cyc}}, \lambda^{\mathrm{latent}}, \lambda^{\mathrm{content}}$, and α as $1.0, 10.0, 0.1, 10$, and 10.0. The content discriminator is updated every 3 iterations. At the rest iterations, other discriminators and generators would be updated jointly, leveraging the advantage of content discriminator to align the content code across different domains.

Table 1. Comparison over domain adaptation.

Method	Dice	Jaccard
DenseNet^s [5]	0.362 ± 0.016	0.325 ± 0.047
CycleGAN [13]	0.753 ± 0.031	0.681 ± 0.083
DADR [9]	0.828 ± 0.072	0.757 ± 0.092
3D-WDGDA	0.837 ± 0.054	0.759 ± 0.065
3D-WDGDA^c	**0.875 ± 0.039**	**0.814 ± 0.027**

3   Experiments

3.1   Datasets and Training Settings

We used two datasets for validation: 1). LiTS - Liver Tumor Segmentation Challenge dataset [1]. It consists of 131 contrast-enhanced 3D abdominal CT scans. 2). Multi-phasic MRI scans of 36 local patients with HCC (note that the CT and MRI scans are **unpaired** and **unmatched**). Considering the clinical practise, we chose CT scans as source domain and MRI scans as target domain. We use 5-fold cross validation on the CT and MRI datasets, and normalized as zero mean and unit variance. In both WDGDA and SegModule part, input size of 3D modules is $256 \times 256 \times 5$, and for 2D modules is 256×256. To avoid over-fitting, we used standard data augmentation methods, including randomly flipping and rotating along the axial plane. We evaluated two variations of the proposed method: our proposed 3D Wasserstein Distance Guided Domain Adaptation model without content discriminator (3D-WDGDA), and 3D-WDGDA with content discriminator (3D-WDGDA^c) (Table 2).

Table 2. Comparison over joint-domain.

	CT		MRI	
	Dice	Jaccard	Dice	Jaccard
DenseNet^s [5]	0.807 ± 0.031	0.793 ± 0.035	0.821 ± 0.017	0.722 ± 0.028
DADR [9]	0.811 ± 0.076	0.780 ± 0.067	0.828 ± 0.022	0.727 ± 0.049
3D-WDGDA	0.885 ± 0.026	0.801 ± 0.058	0.843 ± 0.047	0.735 ± 0.045
3D-WDGDA^c	**0.904 ± 0.012**	**0.831 ± 0.041**	**0.883 ± 0.036**	**0.802 ± 0.038**

In this work, there are three experiment setups: 1). For the domain adaption part, we use 4 folds of CT and 4 folds of pre-contrast MRI for training. Then 4 folds of content-only CT and 1 fold of pre-contrast MRI are used to train and test the SegModule, respectively. 2). We follow the same domain adaption setting in experiment 1, then utilize 4 folds of content-only CT and 4 fold of pre-contrast MRI as network input to train the SegModule, and 1 fold of pre-contrast MRI as test dataset. 3). 4 folds of CT and 4 folds of multi-phasic MRI are used for training. For the segmentation part, we investigate the multi-modal target domain by using 4 folds of CT and 4 folds of multi-phasic MRI. Note that multi-phasic MRIs themselves are multi-modal target domain since they contain several MRI modalities. We evaluate segmentation performance in terms of two metrics: Dice and Jaccard.

3.2 Results

Experiment 1: To demonstrate the domain shift problem, we first evaluate the performance of the unadapted baseline by directly feeding target images to DenseNet^s [5]. We further compare our methods with CycleGAN+DenseNet, DADR [9]. We present two typical results in Fig. 2. The quantitative results are shown in Table 1. Compared with other methods, the proposed 3D-WDGDA^c improves the segmentation performance, and achieves an average Dice of 0.875 and Jaccard of 0.814.

Experiment 2: To show the robustness of our method for joint training, we compare our methods with other state-of-the-art methods. As shown in Table 1, our proposed method 3D-WDGDA^c consistently obtains the highest Dice and Jaccard score over CT and MRI datasets. Visual results of the proposed 3D-WDGDA^c are shown in Fig. 3.

Experiment 3: Multi-phasic MRI are considered as multi-modal target domain with complex statistics. We therefore analyze the effectiveness of the proposed method in multi-modal target domain. The quantitative results are shown in Table 3. *i.e.*, for brevity, CT→MRI denotes that SegModule is trained with content-only CT images and tested by multi-phasic MRI images. We can see that our method clearly remains effective with the multi-modal target domain.

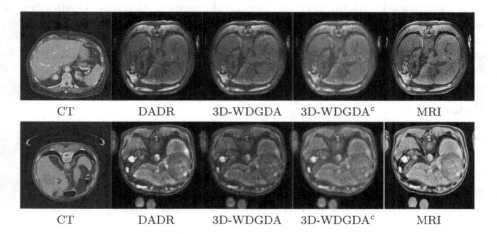

Fig. 3. Visualization of content-only images by different methods. We list the CT images, DADR, 3D-WDGDA, 3D-WDGDA^c, and the reference MRI images.

Table 3. Comparison over multi-modal target domain. For brevity, CT→MRI denotes that SegModule is trained with content-only CT images and tested by multi-phasic MRI images.

	CT→MRI		CT→CT	
	Dice	Jaccard	Dice	Jaccard
DenseNet^s [5]	0.469 ± 0.005	0.289 ± 0.004	0.896 ± 0.048	0.821 ± 0.002
DADR [9]	0.736 ± 0.034	0.619 ± 0.059	0.893 ± 0.038	0.824 ± 0.048
3D-WDGDA	0.776 ± 0.013	0.677 ± 0.053	0.902 ± 0.056	0.832 ± 0.037
3D-WDGDA^c	**0.834 ± 0.029**	**0.707 ± 0.047**	**0.919 ± 0.044**	**0.851 ± 0.053**
	MRI→CT		MRI→MRI	
	Dice	Jaccard	Dice	Jaccard
DenseNet^s [5]	0.766 ± 0.003	0.631 ± 0.038	0.851 ± 0.015	0.725 ± 0.016
DADR [9]	0.782 ± 0.019	0.674 ± 0.015	0.854 ± 0.022	0.739 ± 0.031
3D-WDGDA	0.796 ± 0.016	0.718 ± 0.035	0.869 ± 0.047	0.740 ± 0.064
3D-WDGDA^c	**0.807 ± 0.044**	**0.744 ± 0.057**	**0.881 ± 0.027**	**0.786 ± 0.031**

4   Conclusions and Discussions

We present a novel 3D unsupervised cross-modality Wasserstein distance guided domain adaptation method for medical image segmentation, which would improve clinical decision support systems by leveraging unpaired multi-parametric MRI and CT data. Our method applies Wasserstein distance for the adversarial training, and further takes advantage of 3D CNN to capture spatial information. More importantly, we introduce a content discriminator to encourage content features not to carry modality-specific information, and further preserve feature-level anatomical information for the segmentation task. Qualitative and quantitative results demonstrate the superiority of proposed model over the multi-modal image reconstruction in clinical domains, which is consistent with quantitative evaluations in terms of traditional image segmentation measures. Future work includes improving the efficiency of the proposed methods.

References

1. Christ, P., Ettlinger, F., Grün, F., Lipkova, J., Kaissis, G.: Lits-liver tumor segmentation challenge. ISBI and MICCAI (2017)
2. Dong, N., Kampffmeyer, M., Liang, X., Wang, Z., Dai, W., Xing, E.: Unsupervised domain adaptation for automatic estimation of cardiothoracic ratio. In: Frangi, A.F., Schnabel, J.A., Davatzikos, C., Alberola-López, C., Fichtinger, G. (eds.) MICCAI 2018. LNCS, vol. 11071, pp. 544–552. Springer, Cham (2018). https://doi.org/10.1007/978-3-030-00934-2_61
3. Greenspan, H., Van Ginneken, B., Summers, R.M.: Guest editorial deep learning in medical imaging: overview and future promise of an exciting new technique. IEEE Trans. Med. Imaging **35**(5), 1153–1159 (2016)

4. Gulrajani, I., Ahmed, F., Arjovsky, M., Dumoulin, V., Courville, A.C.: Improved training of Wasserstein Gans. In: Advances in Neural Information Processing Systems, pp. 5767–5777 (2017)
5. Huang, G., Liu, Z., Van Der Maaten, L., Weinberger, K.Q.: Densely connected convolutional networks. In: Proceedings of the IEEE Conference on Computer Vision and Pattern Recognition, pp. 4700–4708 (2017)
6. Huang, X., Liu, M.-Y., Belongie, S., Kautz, J.: Multimodal unsupervised image-to-image translation. In: Ferrari, V., Hebert, M., Sminchisescu, C., Weiss, Y. (eds.) ECCV 2018. LNCS, vol. 11207, pp. 179–196. Springer, Cham (2018). https://doi.org/10.1007/978-3-030-01219-9_11
7. Lee, H.-Y., Tseng, H.-Y., Huang, J.-B., Singh, M., Yang, M.-H.: Diverse image-to-image translation via disentangled representations. In: Ferrari, V., Hebert, M., Sminchisescu, C., Weiss, Y. (eds.) ECCV 2018. LNCS, vol. 11205, pp. 36–52. Springer, Cham (2018). https://doi.org/10.1007/978-3-030-01246-5_3
8. Ouyang, C., Kamnitsas, K., Biffi, C., Duan, J., Rueckert, D.: Data efficient unsupervised domain adaptation for cross-modality image segmentation. In: Shen, D., et al. (eds.) MICCAI 2019. LNCS, vol. 11765, pp. 669–677. Springer, Cham (2019). https://doi.org/10.1007/978-3-030-32245-8_74
9. Yang, J., Dvornek, N.C., Zhang, F., Chapiro, J., Lin, M.D., Duncan, J.S.: Unsupervised domain adaptation via disentangled representations: application to cross-modality liver segmentation. In: Shen, D., et al. (eds.) MICCAI 2019. LNCS, vol. 11765, pp. 255–263. Springer, Cham (2019). https://doi.org/10.1007/978-3-030-32245-8_29
10. You, C., et al.: CT super-resolution GAN constrained by the identical, residual, and cycle learning ensemble (GAN-CIRCLE). IEEE Trans. Med. Imaging **39**(1), 188–203 (2019)
11. You, C., Yang, L., Zhang, Y., Wang, G.: Low-dose CT via deep CNN with skip connection and network-in-network. In: Developments in X-Ray Tomography XII, vol. 11113, p. 111131W. International Society for Optics and Photonics (2019)
12. You, C., et al.: Structurally-sensitive multi-scale deep neural network for low-dose CT denoising. IEEE Access **6**, 41839–41855 (2018)
13. Zhu, J.Y., Park, T., Isola, P., Efros, A.A.: Unpaired image-to-image translation using cycle-consistent adversarial networks. In: 2017 IEEE International Conference on Computer Vision (ICCV) (2017)

HydraMix-Net: A Deep Multi-task Semi-supervised Learning Approach for Cell Detection and Classification

Raja Muhammad Saad Bashir[1(✉)], Talha Qaiser[2], Shan E Ahmed Raza[1], and Nasir M Rajpoot[1,3]

[1] Department of Computer Science, University of Warwick, Coventry, UK
{saad.bashir,shan.raza,n.m.rajpoot}@warwick.ac.uk
[2] Department of Computing, Imperial College London, London, UK
t.qaiser@imperial.ac.uk
[3] The Alan Turing Institute, London, UK

Abstract. Semi-supervised techniques have removed the barriers of large scale labelled set by exploiting unlabelled data to improve the performance of a model. In this paper, we propose a semi-supervised deep multi-task classification and localization approach HydraMix-Net in the field of medical imagining where labelling is time consuming and costly. Firstly, the pseudo labels are generated using the model's prediction on the augmented set of unlabelled image with averaging. The high entropy predictions are further sharpened to reduced the entropy and are then mixed with the labelled set for training. The model is trained in multi-task learning manner with noise tolerant joint loss for classification localization and achieves better performance when given limited data in contrast to a simple deep model. On DLBCL data it achieves 80% accuracy in contrast to simple CNN achieving 70% accuracy when given only 100 labelled examples.

1 Introduction

Deep learning (DL) has revolutionized computer vision in recent years and achieved state-of-the-art performance in various vision-related tasks. The inevitable fact is that most of the DL success is attributed to availability of large scale datasets and compute-power available these days. To achieve state-of-the-art performance, it is incumbent to train models as single-task learning paradigm on large scale datasets with their associated labels. The costs associated with labelling of the datasets is often very high especially for medical imaging data which involves expert knowledge to collect the ground-truth. In contrast, semi-supervised learning (SSL) approaches [1] take advantage of the limited labelled data and leverages readily available unlabelled data to improve

---

Electronic supplementary material The online version of this chapter (https://doi.org/10.1007/978-3-030-61166-8_18) contains supplementary material, which is available to authorized users.

© Springer Nature Switzerland AG 2020
J. Cardoso et al. (Eds.): iMIMIC 2020/MIL3ID 2020/LABELS 2020, LNCS 12446, pp. 164–171, 2020.
https://doi.org/10.1007/978-3-030-61166-8_18

the model performance. This also alleviates the need for time-consuming and laborious task of manual annotations and assist training of more complex models for better performance. Generally, SSL techniques follow a two-step approach a) predict pseudo labels for unlabelled data from the model trained on limited labelled data and b) retrain the model on pseudo labels and limited labelled data to improve the performance. More recently, the trend has been to improve learning ability of SSL by introducing regularization [2,3] and entropy minimization [4] to avoid high-density predictions and train models into an end-to-end manner.

In this work, we propose a multi-task SSL method to alleviate the need of time-consuming and laborious task(s) of manual labelling for histology whole-slide images (WSI). In this regard, we opted to use diffuse large B-cell lymphoma (DLBCL) data because manual annotation of cell type and nuclei localization is very hard due to large number of cells present in WSIs. DLBCL malignancy originates from B-cell lymphocytes and it is the most common high-grade lymphoma among the western population with poor disease prognosis [6]. We propose a novel deep multi-task learning framework, HydraMix-Net, for simultaneous detection and classification of cells, enabling end-to-end learning in a semi-supervised manner. We improve the performance of a semi-supervised approach by enhancing a single loss term with noisy labels for joint training of multi-task problem which to our knowledge has not been performed earlier. Our main contributions are as follows: a) a novel multi-task SSL framework (HydraMix-Net) for cell detection and classification, and b) combating noisy labels using symmetric cross-entropy loss function.

2 Related Work

The purpose of semi-supervised task is to learn from unlabelled data during learning such that it improves the model's performance. To achieve this goal these approaches take advantage of different techniques to mitigate the issues faced during learning e.g., consistency regularization, entropy minimization and noise reduction etc. Decision boundary passing through high-density regions can be minimized using entropy minimization techniques like [4] which minimize entropy with the help of a loss function for the unlabelled data. Consistency regularization can be achieved using standard augmentation such that the network knows if the input was being altered in some ways e.g., rotation, etc. [2,3]. Semi-supervised approaches also suffer from noisy labels as the pseudo labels can introduce noise in the training batches which can be handled using noise reduction methods such as [5]. Using these common approaches there have been semi-supervised methods for classification of natural images e.g., Berthelot et al. [7] used simple data augmentation and mixup [3] for consistency regularization and used sharping [8] for entropy minimization for semi-supervised training. Tarvainen et al. [9] improved the temporal ensembling over labels to use moving average of the weights of student model in teacher model after comparing students prediction with its teacher's prediction, which in turn improves learning of the teacher model. Inspired from all these methods and techniques we propose our novel deep multi-task join training framework for end-to-end classification and detection. Related work regarding fully supervised cell detection and classification is discussed in the Supplementary Material Sect. 1.

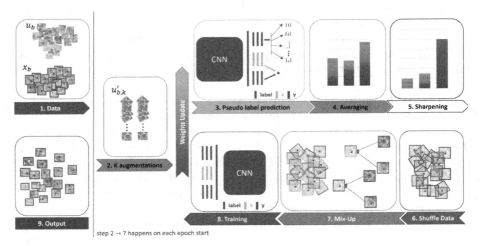

Fig. 1. The schematic diagram of the proposed HydraMix-Net. The unlabelled data u_b is first subjected to k augmentations to generate $u'_{b,k}$ and then process them from the model to generate pseudo labels after which the predictions are averaged and sharpened to minimize entropy in the prediction distribution. Once pseudo labels are assigned, unlabelled set u_b is mixed-up with labelled data x_b to help model iteratively learn more generalized distributions with noise suppression.

3   HydraMix-Net: Cell Detection and Classification

The proposed semi-supervised method HydraMix-Net is a holistic approach consisting of different multi-task and semi-supervised techniques to handle various learning issues e.g., consistency regularization using standard augmentations and mixup techniques [3], entropy minimization with the help of sharpening [8], and handling noisy labels with modified loss terms like symmetric cross entropy (SCE) loss [5]. The proposed HydraMix-Net jointly optimizes the combined loss function for classification and localization of centroids for the cell patches. Our proposed multi-task learning framework consists of a backbone model with three heads responsible for the classification and regression (i.e., localization of cell nuclei). The following sections delineate the data augmentation, pseudo label generation, noise handling and training in the proposed semi-supervised HydraMix-Net model, The schematic diagram of the proposed model can be seen in the Fig. 1.

3.1   Data Augmentation

During training the model takes an input batch of labelled x_b images from $X = \{x_b\}_{b=1}^{B}$ and unlabelled u_b images from $U = \{u_b\}_{b=1}^{B}$, where B was the total number of batches, with known one-hot encoded labels l_c and l_x, l_y representing nuclei centroid. To generate the pseudo labels and their centroids l_{uc}, l_{ux}, l_{uy} using the model, k augmentations like horizontal flip, vertical flip, random rotate, etc., were applied to u_b to yield an augmented batch u'_b as $u'_{b,k} = augment(k, u_b), k \in (1, .., K)$. x_b is also subject to single augmentation per image such that it generates x'_b as $x'_b = augment(k, x_b), k = 1$.

3.2 Pseudo Label Generation

To generate pseudo labels l_{uc} for the batch u_b, predictions from the models φ for k augmented images u'_b were averaged out on class distributions. While for pseudo centroids, prediction on only the original image from the model was used. This is due to the fact that after various augmentations, the centroids are not in the same place because of transformations and hence averaging the centroids of augmentations will lead of incorrect centroids as in Eq. (1).

$$l_{uc}, l_x, l_y = \begin{cases} \frac{1}{k} \sum_{k=1}^{N} \varphi(y'|u_{b,k}; \theta), & \text{if} \quad c = 1 \\ \varphi(y'|u_b; \theta), & \text{otherwise} \end{cases} \tag{1}$$

where φ is the model and θ are the corresponding weights yielding the prediction y' which was split into patch label $l_u c$ when $c = 1$, otherwise centroids l_{ux} and l_{uy}.

Pseudo Label Sharpening. The generated pseudo labels l_{uc} tend to have large entropy in the prediction as a result of averaging of different distributions. Therefore, sharpening [8] was used to reduce or minimize entropy of predictions by adjusting temperature of the categorical distribution as in Eq. (2).

$$sharpening(l_{uc}, T)_i := l_i^{\frac{1}{T}} \Big/ \sum_{j}^{L} l_j^{\frac{1}{T}} \tag{2}$$

where l_{uc} is the categorical distribution of predictions averaged over k augmentations and T temperature is the hyper-parameter which controls the output distribution. When T approaches to 0 it will produce the one-hot encoded output meaning lowering the temperature will yield in low entropy output distributions.

3.3 Mixup

To bridge the gap between unseen examples and remove over-fitting and achieve generalization in semi-supervised approaches mix-up [3] technique was used. Given a pair of images and their labels as (x_1, l_1) and (x_2, l_2). Images were mixed along with their one-hot encoded labels in an appropriate proportion γ. However, the centroids were not mixed due to their numeric nature and transformations. Therefore, centroids from x_1 were used after fusion as shown in (3). In our method, we have used the modified mix-up [7] technique where γ was extracted from beta distribution and then max between γ and $1 - \gamma$ was taken as γ, this ensures that maximum of the original image was preserved and output was closer to x_1.

$$
\begin{aligned}
\gamma &= \max(Beta(\alpha, \beta), 1 - Beta(\alpha, \beta)) \\
x_m &= \gamma x_1 + (1 - \gamma)x_2 \\
l_m &= \gamma l_1 + (1 - \gamma)l_2 \\
l_{mx}, l_{my} &= l_{x_1}, l_{y_1}
\end{aligned}
\tag{3}
$$

In order to apply this technique here x_b' and u_b' were concatenated and shuffled into W and were used for the mix-up. Afterwards, x_b' was mixed-up with $W_{0...|x_b'|}$ and u_b' was mixed-up with $W_{|x_b'|....N}$ where $|x_b'|$ is the length of the augmented mixed-up set x_b' and N is the total number of samples in W.

3.4 Noise Reduction

To handle noise, symmetric cross entropy (SCE) loss [5] was used for both labelled and unlabelled loss instead of just relying on categorical cross-entropy for labelled loss and mean squared loss for the guessed labels. SCE handles the noisy labels by incorporating cross-entropy term for labelled loss as well reverse cross-entropy for predictions loss. This provides a way to learn from model predictions as well instead of just relying on given labels as in Eq. (4). As with iterative progressive learning, the model gets more confident in it's learning and predictions, which is why for unlabelled loss more weight is assigned to predictions and in labelled loss more weight is assigned to labels.

$$
l_{sl} = \delta(-\sum_{c=1}^{C} q(c|x_m) \log p(c|x_m)) + \rho(-\sum_{c=1}^{C} p(c|x_m) \log q(c|x_m))
\tag{4}
$$

where δ and ρ controls the effect of input labels and models predictions.

3.5 Model Training

The learning mechanism of the HydraMix-Net jointly optimizes the combined loss function for classification and regression to predict label and location tuple for labelled and unlabelled batches as in Eq. (5).

$$
l_{total} = \mu(l_{c-sce} + l_{uc-sce}) + (1 - \mu)(l_{rx} + l_{ryx} + l_{ruy} + l_{ry})
\tag{5}
$$

where l_{c-sce} represents the symmetric cross-entropy loss for the labelled part where l_{uc-sce} represents the symmetric cross-entropy loss for the unlabelled part, both coupled together in weight μ which weights the classification head more to provide more accurate labels. While the l_{rx} and l_{ry} are the mean squared error loss terms for the labelled data whereas the l_{rux} and l_{ruy} are the mean squared error loss terms for the unlabelled data for the regression head being weighted by the $(1 - \mu)$. While calculating loss for regression heads the predictions of the classification head were multiplied by regression heads in order to avoid the loss incorporated by background patches which is why the classification head was given more weight in the loss term.

4   Results

The data set used for the study is a private data for DLBCL [10]. Patches of size 41×41 were extracted from 10 manually annotated WSI's resulting in 12553 patches and after offline augmentations, 24000 patches were used for this study. 3 WSI's were selected for the test purposes while 7 WSI's were used for the training purposes, splitting on 70–30 basis which resulted in 18000 training patches and 6000 test patches. **See Supplementary Material section for the detailed description of the data set, implementation details, comparative and ablation study.**

4.1   Experimental Settings

The experimental settings used to test the effectiveness of the proposed approach were i) fully supervised ii) partial data iii) semi-supervised, In the first one all of the available data was used to train a simple CNN i.e., WideRes-Net [11], while in partial setting WideRes-Net was trained on partially labelled data. Lastly, HydraMix-Net used semi-supervised approach for training where both labelled and unlabelled data were used in a way discussed earlier in the Sect. 3. Further, for labelled and unlabelled data we tested different configurations from 50 labelled images to 100, 200, 300, 500, 700 and so on.

4.2   Quantitative Results

Table 1 shows the accuracy achieved by the HydraMix-Net in contrast to the simple CNN on partially labelled data e.g., when provided with the random 50 labelled examples the simple CNN model under-performed by achieving 62% accuracy where the proposed approach leveraged the unlabelled data and achieved superior performance with 66% accuracy. Similarly, when increased the data from 50 labelled examples to 100 and 300 the HydraMix-Net achieved higher performance and reached up to 81% accuracy while simple CNN model trained on only these labelled examples only gave the best performance of 76% accuracy which shows higher efficiency of the proposed approach in scarcity of the labelled examples. Confusion matrix for 100 labelled examples is shown along with the cell centroid detection in the Fig. 2. Figure 3 shows the actual predictions for the proposed approach for the 100 labelled training set. When trained with all the data the highest accuracy achieved is 90% where this threshold is reached by approx. 3000 labelled data by both the techniques.

Table 1. Test accuracy of the HydraMix-Net and partial data approaches with various amount of labelled data provided.

Labelled data	50	100	300	500	700	1000	3000
Simple CNN	0.62	0.70	0.76	0.83	0.85	0.84	**0.90**
HydraMix-Net w/o SCE	0.66	0.70	0.70	0.35	0.35	0.35	.–
HydraMix-Net	**0.66**	**0.80**	**0.81**	**0.85**	**0.85**	**0.85**	0.88

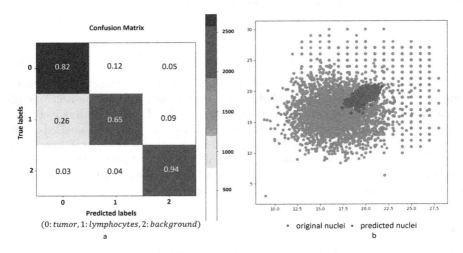

Fig. 2. (a) Represents the confusion matrix for the HydraMix-Net while (b) Represents the prediction and distribution of the centroid in the HydraMix-Net trained on 100 labelled instances where the output size is 32×32.

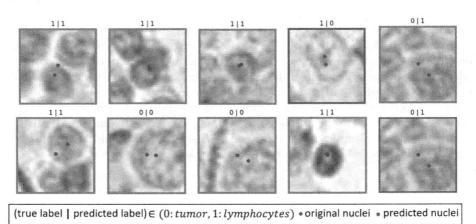

Fig. 3. The prediction of labels and distribution of the centroid on a example set where the HydraMix-Net was trained on 100 labelled examples

5 Conclusion

In this study, we proposed a novel end-to-end holistic multi-task SSL approach for simultaneous classification and localization of nuclei in DLBCL. Further, we plan to extend this work by improving the technique with the help of strong augmentations and validating the performance of our HydraMix-Net on larger cohorts from multiple tumour indications. The cell detection and classification may also be help in performing follow-up analysis like survival prediction and understanding the spatial arrangement of malignant cells within tumour micro-environment to predict other clinical outcomes.

References

1. Chapelle, O., Scholkopf, B., Zien, A.: Semi-supervised learning (chapelle, o. et al., eds.; 2006) [book reviews]. IEEE Trans. Neural Netw. **20**(3), 542 (2009)
2. Cireşan, D.C., et al.: Deep, big, simple neural nets for handwritten digit recognition. Neural Comput. **22**(12), 3207–3220 (2010)
3. Zhang, H., et al.: mixup: beyond empirical risk minimization. arXiv preprint arXiv:1710.09412 (2017)
4. Grandvalet, Y., Bengio, Y.: Semi-supervised learning by entropy minimization. In: Advances in Neural Information Processing Systems (2005)
5. Wang, Y., et al.: Symmetric cross entropy for robust learning with noisy labels. In: Proceedings of the IEEE International Conference on Computer Vision (2019)
6. Coiffier, B., et al.: CHOP chemotherapy plus rituximab compared with CHOP alone in elderly patients with diffuse large-B-cell lymphoma. N. Engl. J. Med. **346**(4), 235–242 (2002)
7. Berthelot, D., et al.: MixMatch: a holistic approach to semi-supervised learning. In: Advances in Neural Information Processing Systems (2019)
8. Goodfellow, I., Bengio, Y., Courville, A.: Deep Learning. MIT Press, Cambridge (2016)
9. Tarvainen, A., Valpola, H.: Mean teachers are better role models: weight-averaged consistency targets improve semi-supervised deep learning results. In: Advances in Neural Information Processing Systems (2017)
10. Qaiser, T., Pugh, M., Margielewska, S., Hollows, R., Murray, P., Rajpoot, N.: Digital tumor-collagen proximity signature predicts survival in diffuse large B-cell lymphoma. In: Reyes-Aldasoro, C.C., Janowczyk, A., Veta, M., Bankhead, P., Sirinukunwattana, K. (eds.) ECDP 2019. LNCS, vol. 11435, pp. 163–171. Springer, Cham (2019). https://doi.org/10.1007/978-3-030-23937-4_19
11. Zagoruyko, S., Komodakis, N.: Wide residual networks. arXiv preprint arXiv:1605.07146 (2016)

Semi-supervised Classification of Chest Radiographs

Eduardo H. P. Pooch[1(✉)], Pedro Ballester[2], and Rodrigo C. Barros[1]

[1] Machine Learning Theory and Applications Research Group (MALTA),
School of Technology, Pontifícia Universidade Católica do Rio Grande do Sul,
Porto Alegre, Rio Grande do Sul, Brazil
eduardo.pooch@edu.pucrs.br
[2] Neuroscience Graduate Program, McMaster University, Hamilton, ON, Canada

Abstract. To train deep learning models in a supervised fashion, we need a significant amount of training data, but in most medical imaging scenarios, there is a lack of annotated data available. In this paper, we compare state-of-the-art semi-supervised classification methods in a medical imaging scenario. We evaluate the performance of different approaches in a chest radiograph classification task using the ChestX-ray14 dataset. We adapted methods based on pseudo-labeling and consistency regularization to perform multi-label classification and to use a state-of-the-art model architecture in chest radiograph classification. Our proposed approaches resulted in average AUCs up to 0.6691 with only 25 labeled samples per class, and an average AUC of 0.7182 when using only 2% of the labeled data, achieving results superior to previous approaches on semi-supervised chest radiograph classification.

Keywords: Medical imaging · Semi-supervised learning · Deep learning

1 Introduction

With the digitization of radiology, computer-aided diagnosis systems can be integrated into the radiological practice workflow, giving support via automated diagnosis tools. The development of automated diagnosis methods involves knowledge from software development, digital image processing, and machine learning. Automated diagnosis tools might deal with classic computer vision problems, such as image classification, object detection, and segmentation, which are usually solved by image feature extraction and classification algorithms. Some of the methods typically used for medical image classification are decision trees, linear classifiers, and artificial neural networks [1].

Convolutional neural networks and other deep learning methods are becoming the method of choice for most medical imaging applications in recent years [9], mostly due to its high performance in image classification when a large amount of data is available for training. Convolutional neural networks

© Springer Nature Switzerland AG 2020
J. Cardoso et al. (Eds.): iMIMIC 2020/MIL3ID 2020/LABELS 2020, LNCS 12446, pp. 172–179, 2020.
https://doi.org/10.1007/978-3-030-61166-8_19

and other deep learning methods advanced the state-of-the-art in many data processing tasks [7], and achieved radiologist-level performance on some medical imaging tasks, such as detecting pneumonia [12] or hip fractures [5].

To train deep learning models in a supervised fashion, we need a significant amount of training data. One of the medical imaging tasks with large datasets available is chest radiography classification. Public datasets provide over 100,000 chest radiographs labeled with the most common findings [15]. These datasets have automatically extracted labels obtained via natural language processing algorithms on radiological reports and have been used to build radiologist-level models [12]. However, in most medical imaging scenarios, there is a lack of annotated data available [9], since, for most tasks, the samples need to be manually annotated by an expert, which is an expensive and time-consuming task.

Recently, research in semi-supervised learning for image classification had some considerable progress [11]. These methods use labeled and unlabeled data to build a machine learning model. Methods based on consistency regularization such as Mean Teacher [14], Unsupervised Data Augmentation [16], MixMatch [3], and FixMatch [13] achieved results comparable to supervised training but with only a fraction of the training samples. For instance, training a model on the SVHN dataset in a supervised fashion using all training data (73,257 labeled samples) results in an error rate of 2.59%, whereas training the same model with the MixMatch approach and only 250 labeled samples achieves an error of 3.78% [3]. However, these recent methods were still not thoroughly validated and compared in a medical imaging scenario.

Our objective in this paper is to compare state-of-the-art semi-supervised classification methods in a medical imaging scenario. We chose chest radiograph classification since it is a common examination, has a lot of available data, and, therefore, a strong baseline to compare. We adapt the semi-supervised classification methods to a multi-label scenario and compare them to a strong supervised baseline in chest radiograph classification, the CheXNet architecture [12].

2   Background

2.1   Supervised Learning

In the supervised learning approach, the model learns based on labeled examples. As the system is presented to input and output variables from the training set, it seeks to create a model that represents this data distribution. Then, this model is extrapolated to infer the output variable of an unseen input sample. Formally, the training data comprises samples $\{x_1, x_2 \ldots, x_n\}, x_i \in \mathcal{X}$ along with their corresponding labels $\{y_1, y_2 \ldots, y_n\}, y_i \in \mathcal{Y}$. We use the training set in order to model the function $f(x) : \mathcal{X} \rightarrow \mathcal{Y}$, where $\mathcal{X}$ is the s-dimensional feature space and $\mathcal{Y}$ is the c-dimensional label space. We can use the final model m to predict the labels of previously unseen samples.

2.2   Semi-supervised Learning

Semi-supervised learning is a learning paradigm intersecting supervised and unsupervised learning. In this scenario, besides $\mathbb{L} = \{(x_1, y_1), (x_2, y_2) \ldots, (x_n, y_n)\}$ we also have unlabeled samples $\mathbb{U} = \{u_1, u_2 \ldots, u_n\}$ that are also within the feature space $\mathcal{X}$ but whose corresponding labels within label space $\mathcal{Y}$ are unknown. We can use $\mathbb{U}$ in the training set alongside $\mathbb{L}$ in order to improve the modeling of the function $f(x)$. Intuitively, the unlabeled samples provide important clues on the data distribution based on sample similarity, and they help to add robustness to the model by exploring this distribution [11].

3   Related Work

The work of Rivero et al. [2] aims at reducing the need for annotated data in medical imaging. They propose GraphX^{NET}, a graph-based semi-supervised learning approach for X-ray data classification. It is a graph model that contains all the training samples with only a limited amount of them are labeled. They tested the approach in the ChestX-ray14 dataset. When using only 20% of the data, they achieve results close to a fully-supervised model. However, under extreme minimal supervision (2% labeled data) the model does not perform well, having an average AUC of 0.53.

Tanan et al. [10] perform semi-supervised classification in skin lesion classification and thoracic image analysis. The proposed method is called SRC-MT. It is a semi-supervised classifier based on Mean Teacher [14] and introduces a sample relation consistency term to the optimization function. This enforces the consistency based on the relationship information among different samples instead of individual predictions. They achieve using similar results to GraphX-NET when using 20% of ChestX-ray14, but when using only 2% of labeled data, they achieve an average AUC of 0.67.

4   Materials and Methods

4.1   Dataset

ChestX-ray14 [15] from the National Institute of Health contains 112,120 frontal-view chest radiographs from 32,717 different patients labeled with 14 radiological findings. In this work we use the official split which contains 78.468 training samples, 11.219 validation samples and 22.433 test samples.

4.2   Semi-supervised Learning Methods

Pseudo-labeling is a simple semi-supervised approach and used as a semi-supervised baseline. In this approach, the model is trained with a regular cross-entropy loss on labeled data, and we also take the top prediction made in unlabeled data and use it as a pseudo-label, compute the unsupervised loss and add it to the combined loss. Our approach was based on the work of Lee [8], since our task is a multi-label scenario, we use a binary cross-entropy (BCE) loss and also tested a soft label approach, in which the pseudo-label is the classes score prediction, and a hard label approach, in which the pseudo-label is a one-hot vector with the top prediction as one and the rest as zero.

Mean Teacher. [14] consists of using two models with identical architecture, which are called *student* and *teacher*. At every training iteration, both models are fed the same inputs with different augmentation policies, then, a consistency loss is computed based on the distance between both models predictions. Finally, the student weights Θ^s are updated via loss optimization, and the teacher weights Θ^t are updated via an exponential moving average (EMA) of the student weights after each training step e. A hyperparameter ρ controls the EMA decay rate to update the teacher's weights, as in $\Theta_e^t = \rho\Theta_{e-1}^t + (1 - \rho)\Theta_e^s$. A combined loss function $\mathcal{L}_{comb}$ is used to update the student's weights. This loss is the sum of the task loss $\mathcal{L}_{task}$ with the consistency loss $\mathcal{L}_{cons}$ controlled by a consistency weight hyperparameter γ as in $\mathcal{L}_{comb} = \mathcal{L}_{task} + \gamma\mathcal{L}_{cons}$. The task loss $\mathcal{L}_{task}$ is a regular cross-entropy loss between the ground-truth labels y and the predictions of the student model $m_s(x)$, which is only computed on labeled instances. Since our problem is multi-label we replaced the original cross-entropy loss with a binary cross-entropy (BCE) loss over all labels as in $\mathcal{L}_{task} = BCE(m_s(x), y)$. The consistency loss is a mean-squared error of the predictions from the student and the teacher on unlabeled data u when submitted to two different augmentation policies ϕ_s and ϕ_t therefore $\mathcal{L}_{cons} = ||m_s(\phi_s(u)) - m_t(\phi_t(u))||^2$.

Unsupervised Data Augmentation (UDA). [16] uses advanced data augmentation techniques to input noise on the training data and compute consistency between non-augmented samples. For image classification tasks, the authors propose using RandAugment [4] as the data augmentation technique. RandAugment randomly selects transformations for each sample from a collection of transformations. Global parameters m and n control the distortions' magnitude and the number of augmentations applied in each image. UDA uses only one model $m(\cdot)$, which is updated by a combined loss similar to Mean Teacher's, except that the consistency loss $\mathcal{L}_{cons}$ is a KL divergence between the predictions for strongly augmented ($\Phi(u)$) and non-augmented (u) unlabeled data. Since there is usually a limited amount of labeled data, the authors propose a technique called training signal annealing (TSA) to prevent overfitting the labeled data and underfitting the unlabeled data. It consists of defining a confidence threshold for the model's predictions to use the training signals of the labeled sample, gradually increasing the threshold T from $\frac{1}{c}$ to 1 (where c is the number of classes) according to a schedule. This technique prevents over-training on easy samples and focuses the initial stage of the training on complex samples. In our approach, we also replaced the supervised loss with a BCE loss.

MixMatch. [3] is an algorithm that combines techniques from different semi-supervised learning regularization approaches. It starts by sampling and augmenting labeled and unlabeled samples. Each unlabeled sample is augmented Q

times, and the model computes predictions for each augmented sample. These predictions are averaged and sharpened to become pseudo-labels $\hat{y}$, in order to force predictions to be closer to a one-hot distribution. Then, the augmented labeled and unlabeled data form a batch with their respective labels and pseudo-labels. This batch is shuffled and regularized using the MixUp regularizer [17], which interpolates data points using values sampled from a beta distribution to create a smoother training manifold, and the model is trained using the interpolated points $\tilde{x}$ and their interpolated labels $\tilde{y}$ and $\tilde{\hat{y}}$. The loss function combines the losses on labeled and unlabeled data controlled by a hyperparameter γ like the previous approaches. The loss for labeled data is a cross-entropy as the one in UDA, except that it uses the data points $\tilde{x}$ and labels $\tilde{y}$ generated by MixUp. The loss for unlabeled data is a mean-squared error between generated pseudo-labels $\tilde{\hat{y}}$ and the predictions on mixed-up unlabeled inputs $\tilde{u}$. In our experiments, we replaced the sharpen done on the softmax function to one made in the sigmoid function as in $\sigma(m(.))^{\frac{1}{\tau}}$, and also used a BCE loss.

FixMatch. [13] is a simple yet effective approach which holds the state-of-the-art in many datasets. This approach combines consistency regularization similar to UDA with pseudo-labeling. It leverages strong and weak augmentation policies. At first, an input sample u_i is weakly augmented with a policy ϕ and fed to a model $m(\cdot)$. Its output becomes a pseudo-label for u_i using $\hat{y}_i = argmax(m(\phi(u_i)))$. Then, the input u_i is strongly augmented with a policy Φ, and the model is trained with a regular cross-entropy loss using the previously generated pseudo-label $\hat{y}_i$. FixMatch optimizes a combined loss of labeled and unlabeled data controlled by a hyperparameter γ like previous methods. The task loss $\mathcal{L}_{task}$ is a cross-entropy between predictions for weakly augmented inputs $\phi(x_i)$ and their ground-truth labels y_i. The consistency loss $\mathcal{L}_{cons}$ is also a cross-entropy, but between the strongly-augmented $\Phi(u)$ and the pseudo-labels $\hat{y}$ that have a confidence score $max(m(\phi(x_i)))$ higher than a threshold T. In our approach, we adapted the method to compute a BCE loss with a one-hot vector containing the top prediction higher than T.

4.3 Experiment Settings

We employ a multi-label classification approach reproducing the CheXNet model [12], a popular approach that achieved state-of-the-art results in classifying multiple pathologies using a DenseNet121 convolutional neural network architecture [6]. We use it as our supervised baseline and also as the model architecture for the semi-supervised methods. The model is pre-trained on the ImageNet dataset, and the images are resized to 224×224 pixels and normalized using the ImageNet mean and standard deviation. We use a learning rate of 0.01, a cosine learning rate schedule, and a Stochastic Gradient Descent optimizer with 0.9 momentum and a weight decay of 0.001 and a mini-batch size of 16. In semi-supervised methods, we use 8 labeled and 8 unlabeled samples for each batch. The weak augmentations are the same ones performed in the supervised baseline [12], the strong augmentations are done by RandAugment [4] with $n = 2$ and $m = 10$.

Every method is trained for 20 epochs, as we empirically observed that a longer training does not show improvement. We use the same model hyperparameters for supervised training in all methods, varying only the hyperparameters referring to the semi-supervised training. We use subsets containing 25, 100, and 400 labeled samples per class for each method and leave the rest of the training set as unlabeled samples, which is a common setup for semi-supervised evaluation. We have three different subsets with different samples used as labeled for each labeled amount, and we report the mean and standard deviation of the top performance on the three experiments. We evaluate the models' performance computing the area under the receiver operating characteristic curve (AUC) for each label.

To select better hyperparameters for our objective task, we performed a random hyperparameter search in each method using a 25 labels subset. We trained the model with different hyperparameters for 20 epochs and selected the ones that achieved a higher AUC on the validation set. In all methods, we searched for a consistency weight between 0.5 and 100. For Pseudo-labeling, we selected 1 as the unsupervised weight and also searched for two different pseudo-labeling strategies using soft and hard pseudo-labels and hard pseudo-labels had the best performance. For the Mean Teacher, we selected a consistency weight of 100, and also searched for an EMA decay rate ρ for the teacher model between 0.8 and 0.99 and selected 0.99. In UDA, we selected a consistency weight of 2 and searched for a TSA schedule using linear, exponential, and logarithmic. The logarithmic schedule showed the best results, but using none was still better, so we did not use TSA in our experiments. For MixMatch, we selected a consistency weight of 10 and also searched for the α of the $\beta(\alpha, \alpha)$ distribution between 0.1 and 50, we selected 0.1. For FixMatch, we selected a consistency weight of 1, and also searched for a threshold between 0.7 and 0.95 and selected 0.8. Since in the original paper, the authors reported that a larger ratio of unlabeled samples increased the model performance, we also searched for a ratio of 2,3, and 4, but the ratio of 1 still showed the best results.

5   Results and Discussion

We summarize the results for each label subset in Table 1. Our strongest baseline is the fully-supervised CheXNet [12], which achieves an average AUC of 0.8414. The results of all the semi-supervised approaches are very similar, with the most gain being obtained by Mean Teacher using 400 labels, achieving an average AUC 9% higher than the one obtained by a supervised training. With 25 labels, the highest average result was obtained from UDA, improving supervised training in 5%, and using 100 labels, the best performance was with Pseudo-label, improving the baseline in 6%. Comparing to the original baseline results presented by Wang et al. [15] in ChestX-ray14's release, our approaches were able to outperform their fully supervised model using only 400 labeled samples per class and achieved similar results when using 100.

Table 1. AUCs of our proposed approaches and baselines using different amounts of labeled samples on ChestX-ray14.

	25 labels						100 labels					
	Supervised	Pseudo-label	Mean Teacher	MixMatch	FixMatch	UDA	Supervised	Pseudo-label	Mean Teacher	MixMatch	FixMatch	UDA
Atelectasis	59.87	66.37	66.09	66.65	**67.75**	67.36	64.64	**71.70**	71.12	68.33	69.66	71.21
Cardiomegaly	56.82	64.21	62.83	62.70	66.46	**66.71**	64.89	**79.86**	76.22	77.91	76.90	77.73
Consolidation	67.53	68.16	68.82	67.71	67.97	**69.45**	71.17	72.92	72.30	72.22	73.15	**73.60**
Edema	77.48	80.82	79.53	**80.86**	79.95	80.43	81.30	83.66	**83.78**	83.50	82.36	82.89
Effusion	65.87	**76.35**	75.91	74.64	75.45	74.82	72.25	80.55	**80.68**	78.70	79.89	80.32
Emphysema	54.47	65.79	**68.64**	67.61	66.00	65.35	63.81	76.28	77.17	76.63	75.15	**78.50**
Fibrosis	65.94	65.86	**66.87**	64.74	63.36	64.78	66.03	68.29	**70.35**	67.77	69.41	69.76
Hernia	69.92	80.05	**81.92**	78.51	78.11	81.41	78.34	**89.45**	88.44	89.23	86.92	87.43
Infiltration	**60.99**	60.28	60.31	60.42	60.80	59.43	62.68	62.98	**63.93**	63.14	63.09	62.55
Mass	52.43	53.62	54.95	51.48	55.90	**56.47**	56.13	62.21	**62.83**	61.92	62.38	62.35
Nodule	55.12	57.52	56.40	**58.26**	57.13	55.37	56.48	58.81	**58.80**	58.28	57.33	58.14
Pleural Thickening	56.50	62.31	**63.53**	62.48	60.71	62.33	57.97	65.15	**65.44**	62.58	62.73	64.66
Pneumonia	60.88	**63.11**	61.44	61.80	62.67	62.61	63.56	**66.57**	64.95	65.77	66.16	65.52
Pneumothorax	57.99	69.97	69.15	**70.18**	68.70	70.35	65.61	76.09	76.61	75.12	74.57	**77.80**
No Finding	59.53	66.81	65.21	65.98	65.53	**66.83**	64.50	70.30	70.89	69.70	69.68	**71.29**
Average	61.42 ± 02.91	66.75 ± 01.55	66.77 ± 01.55	66.27 ± 01.95	66.43 ± 01.86	**66.91 ± 01.76**	65.96 ± 03.00	**72.32 ± 00.14**	72.23 ± 01.02	71.39 ± 00.45	71.29 ± 01.10	72.25 ± 01.25

	400 labels						Wang et al. [15]	Fully-supervised CheXNet [12]
	Supervised	Pseudo-label	Mean Teacher	MixMatch	FixMatch	UDA	Wang et al. [15]	CheXNet [12]
Atelectasis	67.39	74.48	**74.89**	74.04	74.17	73.43	71.6	80.94
Cardiomegaly	68.84	85.16	**86.42**	86.39	85.47	85.01	80.7	92.48
Consolidation	72.20	73.67	73.19	74.30	**76.13**	74.30	70.8	79.01
Edema	81.96	85.36	**86.74**	86.26	86.13	85.63	83.5	88.78
Effusion	72.82	82.86	**83.65**	83.49	83.17	82.61	78.4	86.38
Emphysema	67.61	84.15	**87.06**	84.96	83.82	85.64	81.5	93.71
Fibrosis	68.73	72.49	**75.75**	75.55	73.85	75.03	76.9	80.47
Hernia	81.25	89.34	89.08	88.41	**90.89**	88.63	76.7	91.64
Infiltration	63.45	65.06	65.62	64.38	**65.80**	64.10	60.9	73.45
Mass	53.10	69.09	**72.26**	70.56	69.87	70.40	70.6	86.76
Nodule	58.93	63.28	**65.29**	64.39	65.05	64.63	67.1	78.02
Pleural Thickening	60.73	68.59	69.56	67.72	70.42	**70.60**	70.8	80.62
Pneumonia	63.73	66.51	**70.71**	67.21	69.19	66.60	63.3	76.8
Pneumothorax	68.89	81.13	**81.78**	81.20	77.67	81.77	80.6	88.87
No Finding	65.18	73.53	**74.18**	72.98	73.42	73.38	-	-
Average	68.05 ± 06.75	75.65 ± 00.50	**77.08 ± 00.13**	76.12 ± 00.48	76.34 ± 00.29	76.12 ± 00.65	73.8	84.14

The other two works that performed semi-supervised classification on ChestX-ray14 [2,10] evaluated their methods using percentages of the training data as labeled samples. To compare our results with theirs, we trained our best approach, the Mean Teacher based, in subsets of 2% and 5% of labeled data. Table 2 shows the results. Our approach shows almost 5% of improvement over the previous state-of-the art when using 2% labels.

Table 2. Comparison of average AUC of our best approach on different amounts of labeled data with two previous approaches of semi-supervised classification on ChestX-ray14. Results as reported on the original papers.

	2%	5%
GraphX^{NET} [2]	53	58
SRC-MT [10]	66.95	72.29
Mean Teacher	**71.82**	**74.82**

6   Conclusion

In this work, we evaluated different semi-supervised learning methods performing multi-label classification in a medical imaging scenario and achieved state-of-the-art results on semi-supervised classification on ChestX-ray14. Most of the trained methods showed similar results, with Mean Teacher having a slightly better gain in performance when compared to a supervised baseline. The improvement over a supervised baseline is not as high as the ones reported by the original methods in common computer vision datasets like CIFAR-10, highlighting the need for semi-supervised approaches specifically designed for medical imaging.

References

1. Amir, G.J., Lehmann, H.P.: After detection: the improved accuracy of lung cancer assessment using radiologic computer-aided diagnosis. Acad. Radiol. **23**(2), 186–191 (2016)
2. Aviles-Rivero, A.I., et al.: GraphXNET chest X-ray classification under extreme minimal supervision. In: International Conference on Medical Image Computing and Computer-Assisted Intervention, pp. 504–512. Springer (2019)
3. Berthelot, D., Carlini, N., Goodfellow, I., Papernot, N., Oliver, A., Raffel, C.: MixMatch: a holistic approach to semi-supervised learning, May 2019
4. Cubuk, E.D., Zoph, B., Shlens, J., Le, Q.V.: Randaugment: practical data augmentation with no separate search. arXiv preprint arXiv:1909.13719 (2019)
5. Gale, W., Oakden-Rayner, L., Carneiro, G., Bradley, A.P., Palmer, L.J.: Detecting hip fractures with radiologist-level performance using deep neural networks. arXiv preprint arXiv:1711.06504 (2017)
6. Huang, G., Liu, Z., Van Der Maaten, L., Weinberger, K.Q.: Densely connected convolutional networks. In: Proceedings of the IEEE Conference on Computer Vision and Pattern Recognition, pp. 4700–4708 (2017)
7. LeCun, Y., Bengio, Y., Hinton, G.: Deep learning. Nature **521**(7553), 436 (2015)
8. Lee, D.H.: Pseudo-label: the simple and efficient semi-supervised learning method for deep neural networks. In: Workshop on Challenges in Representation Learning, ICML (2013)
9. Litjens, G., et al.: A survey on deep learning in medical image analysis. Med. Image Anal. **42**, 60–88 (2017)
10. Liu, Q., Yu, L., Luo, L., Dou, Q., Heng, P.A.: Semi-supervised medical image classification with relation-driven self-ensembling model. IEEE Trans. Med. Imaging (2020)
11. Qi, G.J., Luo, J.: Small data challenges in big data era: a survey of recent progress on unsupervised and semi-supervised methods. arXiv preprint (2019)
12. Rajpurkar, P., et al.: ChexNet: radiologist-level pneumonia detection on chest X-rays with deep learning. arXiv preprint arXiv:1711.05225 (2017)
13. Sohn, K., et al.: FixMatch: simplifying semi-supervised learning with consistency and confidence (2020)
14. Tarvainen, A., Valpola, H.: Mean teachers are better role models: weight-averaged consistency targets improve semi-supervised deep learning results, March 2017
15. Wang, X., Peng, Y., Lu, L., Lu, Z., Bagheri, M., Summers, R.M.: ChestX-Ray8: hospital-scale chest X-ray database and benchmarks on weakly-supervised classification and localization of common thorax diseases. CoRR (2017)
16. Xie, Q., Dai, Z., Hovy, E., Luong, M.T., Le, Q.V.: Unsupervised data augmentation. arXiv preprint arXiv:1904.12848 (2019)
17. Zhang, H., Cisse, M., Dauphin, Y.N., Lopez-Paz, D.: Mixup: beyond empirical risk minimization. arXiv preprint arXiv:1710.09412 (2017)

LABELS 2020

Risk of Training Diagnostic Algorithms on Data with Demographic Bias

Samaneh Abbasi-Sureshjani[1], Ralf Raumanns[1,2], Britt E. J. Michels[1],
Gerard Schouten[1,2], and Veronika Cheplygina[1(✉)]

[1] Eindhoven University of Technology, Eindhoven, The Netherlands
{s.abbasi,r.raumanns,v.cheplygina}@tue.nl
[2] Fontys University of Applied Science, Eindhoven, The Netherlands
g.schouten@fontys.nl

Abstract. One of the critical challenges in machine learning applications is to have fair predictions. There are numerous recent examples in various domains that convincingly show that algorithms trained with biased datasets can easily lead to erroneous or discriminatory conclusions. This is even more crucial in clinical applications where predictive algorithms are designed mainly based on a given set of medical images, and demographic variables such as age, sex and race are not taken into account. In this work, we conduct a survey of the MICCAI 2018 proceedings to investigate the common practice in medical image analysis applications. Surprisingly, we found that papers focusing on diagnosis rarely describe the demographics of the datasets used, and the diagnosis is purely based on images. In order to highlight the importance of considering the demographics in diagnosis tasks, we used a publicly available dataset of skin lesions. We then demonstrate that a classifier with an overall area under the curve (AUC) of 0.83 has variable performance between 0.76 and 0.91 on subgroups based on age and sex, even though the training set was relatively balanced. Moreover, we show that it is possible to learn unbiased features by explicitly using demographic variables in an adversarial training setup, which leads to balanced scores per subgroups. Finally, we discuss the implications of these results and provide recommendations for further research.

Keywords: Computer-aided diagnosis · Demographic bias · Classification parity

1 Introduction

In medical image analysis, machine learning algorithms can be on par with or even exceed the performance of experts. However, for reliable generalization, large datasets are needed, that are representative of the population on which they are ultimately applied. In the medical domain, this is often not the case [6,15]. A further requirement is that the properties of the training data are similar to the test data, which is sometimes overlooked. For example, some patient

© Springer Nature Switzerland AG 2020
J. Cardoso et al. (Eds.): iMIMIC 2020/MIL3ID 2020/LABELS 2020, LNCS 12446, pp. 183–192, 2020.
https://doi.org/10.1007/978-3-030-61166-8_20

groups (based on age, sex, ethnicity among others) can be overrepresented in the data, biasing the model. Besides the notorious discriminatory face recognition example [5], detrimental effects of such bias have been demonstrated in various domains, varying from predictions of recidivism, to job offers or loan decisions. For medical imaging, the problem seems relatively unexplored, despite the potentially harmful consequences.

We aim to quantify whether and how bias is addressed in medical imaging papers focusing on the diagnosis. We first survey proceedings from a recent conference. For selected papers, we report the sample size, whether any demographic measures are available, whether these are used by the algorithm and whether demographics/bias are discussed in the paper. Using a dataset of skin lesions, we then demonstrate that a classifier trained on a relatively balanced dataset in terms of age and sex already shows biased results on the held-out test set. In terms of [10], we apply the principle of *classification parity*, meaning that we aim at making the predictive performance (AUC in our case) equal across subgroups defined by so-called protected features (in our case sex and age). As is also explained by [10], this mitigation strategy is not necessarily synonymous with fair machine learning, since equitable and fair decisions are very much context-dependent. Finally, we provide some guidelines for evaluating algorithms concerning this important topic.

1.1   Related Work

One form of dataset bias refers to a distribution shift between datasets, such that models trained on one dataset, show a drop in performance on the other. This idea has been studied in computer vision [18,31]. In medical imaging, such drops in performance can be experienced in datasets collected at different centers [2,25,32]. Such differences are often addressed with transfer learning [6] techniques, which either align the data distributions or learn dataset-independent representations.

A more specific case of dataset bias is when the bias is based on the demographics of the training subjects including differences in ages, sexes, diets, habits, genetics and so on. Collected data often inadvertently encode human preferences. As an example, it has been demonstrated that face recognition algorithms can discriminate based on e.g. skin color and perform poorly on under-represented groups [5]. In medical imaging, similar factors might influence the data, thus have an impact on the incidence of disease too, as shown in some studies. For instance, [8,21] show that signs of brain aging as a biomarker of aging can be predicted from brain and retinal images; or the work by [9] demonstrates the relation between the human immunodeficiency virus (HIV) and the aging process of the brain. A paper published after the first version of ours, shows gender bias in classification of lung diseases from chest X-ray images [20]. Thus it is essential to include or at least take into account the demographics in the data analysis.

Various algorithms to mitigate this type of bias have been proposed. The first set of approaches focuses on preventing this bias in the first place i.e., creating a balanced set in the data preparation step [28]. However, this is not always an option especially for medical data which is rare and where new acquisitions are often costly. Therefore, recent studies have focused mainly on learning representations that are not only predictive of the actual outputs but also invariant to the extraneous factors [1,11,26,34]. In most cases, by including the additional available demographic information during training, their predictive power is mitigated by an adversarial loss and the features become invariant to them.

Due to the rise of machine learning diagnostic applications in the medical image analysis domain, we conduct a survey of the published techniques in MICCAI 2018 [12] to investigate the inclusion of demographics in addition to the medical images. Our results show that even though the demographics might impact the outcome of the models, it is not a widely discussed topic in medical imaging. Most of the datasets do not include the demographic information, and the proposed techniques rarely propose to correct for potential biases in their models. Additionally, we use a relatively balanced dataset of skin lesions [7] and highlight the importance of correction for age and sex biases in this dataset. The closest study to this analysis is [19], where they show that skin lesion datasets over-represent lighter skin, but do not find large differences in performance for different skin types.

2   Methods

2.1   Paper Analysis

We screened the MICCAI 2018 proceedings [12] for papers on diagnosis using macroscopic images. We, therefore, focused on the chapters "Machine Learning in Medical Imaging", "Optical and Histology Applications", "Cardiac, Chest and Abdominal Applications" and "Neuroimaging and Brain Segmentation Methods: Neuroimaging". Papers were included if they focused on the diagnosis or detection of abnormalities. For each selected paper, one of the authors quantified the following: number of public or private datasets used, number of subjects, whether demographic information was given, and whether demographics were discussed.

2.2   Classifier Analysis

To understand potential differences in the performance of a classifier for different demographic groups, we set up a baseline binary classification experiment. We used the ISIC 2017 skin lesion dataset [7] for the diagnosis of melanoma skin cancer since the age and sex were available for over 75% of the subjects. We included only the subjects for which both variables were available in our analysis. Age was provided to the nearest 5 years. To create large enough subgroups for evaluation, we split the subjects by calculating the median age in the training set (equal to 60) and using that as a threshold. The numbers of subjects in each group are provided in Table 1.

Table 1. Demographics of the used datasets.

ISIC subset	Total	Included	Male	Female	<60	≥60
Train	2000	1744	886	858	1087	657
Validation	150	149	90	59	87	62
Test	600	553	283	270	302	251

Baseline Network. We trained an Inception-v4 [30] network as our baseline model using the training procedure from [24], which has outperformed the top result (0.874) from the ISIC 2017 challenge. The network uses data augmentation based on adjusting the color (saturation, contrast, brightness, hue) and geometry (affine transformations, flips, random crops) of the image. It is initialized with ImageNet weights, and then further trained on randomly augmented training images resized to 299 × 299. Training is then done with stochastic gradient descent with a momentum factor of 0.9, batch size of 40, and learning rate of 1e−3 which is reduced to 1e−4 after the 10th epoch. Early stopping is used if the validation area under the curve (AUC) does not improve after 8 epochs. At test time, an image is randomly augmented 32 times, and the predictions are averaged. All parameters are used as defined by [24] and not specifically optimized for the subset of data that we used. We evaluated the classifiers with AUC for the following groups: all subjects, male, female, young (<60) and old (≥60).

Bias-Aware Network. To evaluate whether the learned representation has any relations to the available demographics, we use the method proposed by [1]. Thus we employ an ensemble network with a shared feature encoder (the same as the baseline model) and two classifier heads. One classifier is in charge of classifying the skin cancer and it consists of a fully connected layer followed by average pooling and softmax layers (similar to the baseline model). The other head is supposed to predict the confounding parameter and it consists of a fully connected layer followed by an average pooling layer. Parameters of the encoder, cancer classifier and bias predictor are denoted by θ_e, θ_c, θ_{bp} respectively. Three losses are used for training the network. For training the skin cancer classifier head and encoder a cross-entropy loss (L_c) is used. While for optimizing the bias predictor head, a bias prediction loss (L_{bp}) is defined as the negative-squared Pearson correlation coefficient ($-Corr^2$). By minimizing $-Corr^2$, the correlation between the predicted and true confounding parameter should increase. Since sex is a binary parameter, in some experiments we define L_{bp} as a binary cross-entropy loss (BCE). The third (bias resilient) loss is defined as $L_{br} = -\lambda L_{bp}$ and is used to optimize the encoder adversarially to reduce the predictive power of the encoded features for the confounding parameter. λ determines how much the encoder is penalized for leading to correct predictions of the target demographic parameter.

The ensemble network is trained iteratively with three main steps: (a) updating θ_e and θ_c based on the L_c loss; (b) updating only the θ_{bp} parameters based on L_{bp} loss; (c) and finally updating θ_e adversarially based on L_{br} loss. Note that the encoder weights are not updated in the second step, and the bias predictor weights are not updated in the third step. The updates are done one-by-one iteratively. The learning rates and optimizers of the three update steps are the same as the baseline model. It is worth mentioning that for the steps involving the bias prediction, we only use the control data to make sure that the confounding parameters are reliably estimated from healthy subjects. Multiple experiments are performed to see whether it is possible to weaken the potential relationship between the encoded features from images and the confounding parameters, in our case age or sex.

3 Results

3.1 Paper Analysis

A total of 65 papers fit our inclusion criteria. Several statistics of the datasets used, and the inclusion of demographic information by the papers are shown in Fig. 1. In total there were 52 papers using 1 dataset, 11 papers using 2 datasets, and 2 papers using 3 or more datasets. Nearly half (32 papers) did not use any public datasets. The sizes of the datasets varied between 10 subjects and 112K subjects, with 217 subjects as the median size.

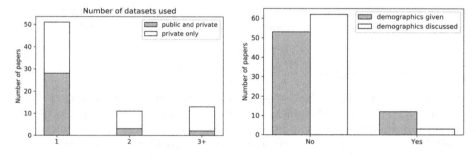

Fig. 1. Number of papers using a certain number of private/public datasets (left), and including demographic information (right).

In this set of 65 papers, 12 papers described at least age or sex. Notably, 10 of these were neuroimaging papers. Of the 12 papers, only 3 also evaluate or discuss their results with respect to the demographics. [23] test whether their glaucoma risk index differs significantly between the healthy and patient groups, while also checking whether these groups have statistically different age and sex distributions. [14] stratify their results of detecting brain malformations by age group (children vs adults). Finally [16] corrects their Alzheimer's score estimation for brain images, with a factor based on linear regression of cognitively normal subjects.

Table 2. An overview of the AUCs obtained in each experiment. The most balanced performances after correction for the bias are bolded.

Experiment	Confounder	λ	L_{bp}	All	Young	Old	Male	Female
1. Baseline	N/A	N/A	N/A	0.83	0.76	0.84	0.76	0.90
2. Ensemble	Age	0	$-Corr^2$	0.83	0.78	0.83	0.77	0.90
3. Ensemble	Age	5	$-Corr^2$	0.80	**0.77**	**0.80**	0.73	0.90
4. Ensemble	Sex	0	$-Corr^2$	0.83	0.77	0.84	0.76	0.91
5. Ensemble	Sex	5	$-Corr^2$	0.72	0.64	0.75	**0.73**	**0.78**
6. Ensemble	Sex	0	BCE	0.84	0.77	0.85	0.78	0.90
7. Ensemble	Sex	0.5	BCE	0.83	0.78	0.84	0.77	0.91

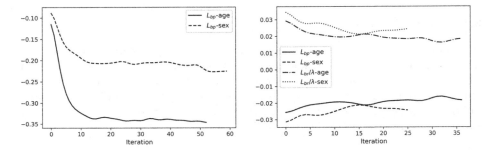

Fig. 2. Left: The L_{bp} loss of the ensemble network with $\lambda = 0$ (experiments 2 and 4 in Table 2); right: The L_{bp} and L_{br}/λ losses when $\lambda > 0$ (experiments 3 and 5).

3.2   Classifier Analysis

The AUC performances on the test set across different subgroups and all the subjects are shown in Table 2. For each experiment, we have specified the L_{bp}, demographic, and λ parameters used during training. The baseline model achieves an overall AUC of 0.83 that is slightly lower than the AUC of 0.88 reported in [24] because we only use the subset of subjects with known demographics and we do only half of the test time augmentations used by [24]. Moreover, the ensemble networks achieve the same performances as the baseline model when λ is set to 0 (experiments 2, 4 and 6) because there is no back-propagation from the bias predictor head to the encoder i.e., $L_{br} = 0$. In these experiments, both subgroup splits show large differences between them: depending on age, the AUC varies between 0.76 and 0.85 (9%), and depending on sex, between 0.76 and 0.91 (15%). The obtained L_{bp} values for experiment 2 and 4 ($\lambda = 0$) are also shown in Fig. 2 (left). As seen in this figure, there exist some correlations between the predicted and true confounding parameter when the encoder is only optimized for classifying skin cancer. This correlation is stronger for age than sex.

When we use the ensemble network to predict the age of the control subjects in an adversarial setting (experiment 3), we see that the differences between the performances of young and old subgroups decrease (only 3%), while that is not the case for male/female subgroups. Similarly, when the sex is used as the confounding parameter (experiment 5), the AUC of male/female subgroups get closer (5% difference), while the difference between young/old subgroups remains large (11%). The values of L_{bp} and L_{br}/λ for experiments 3 and 5 are visualized in Fig. 2 (right). Since the training is a min-max optimization problem, compared to the left figure, the correlation does not increase during training i.e., L_{bp} does not decrease. Additionally, the training stops much earlier resulting in a drop in the overall AUC of the skin cancer classifier.

Since sex is a binary parameter unlike age, the BCE loss is used in experiments 6 and 7. As depicted by the results, the BCE loss is not as effective as the $-Corr^2$ and the AUCs are almost the same as the baseline model. Note that λ is determined heuristically based on the ratio between L_c and L_{bp} loss in order to have an effective penalty in updating the encoder weights.

4   Discussion and Conclusions

Our paper analysis showed that demographics are rarely discussed and used in diagnostic algorithms. This is surprising, given the importance of demographics in diagnosis. For example, men and women have different distributions of melanoma subtypes [3], which can affect the final diagnosis.

Our classifier analysis results showed large differences in performance between male and female subjects, and between different groups of age for the baseline model. The male/female difference is somewhat surprising, given that the training data was relatively balanced. This suggests that these factors might influence how difficult a skin lesion is to diagnose.

Additionally, we demonstrated the possibility to correct for the potential bias in predictions to some extent by using an adversarial training setting. The same method was used in [1] to investigate whether diagnosis of HIV based on brain MRIs is dependent on subjects' age instead of true HIV markers or not. Their results suggest that predictions from the baseline may be biased, whereas the bias-aware network results in a space with no apparent bias to age.

Our results indicate that age, sex and possibly other characteristics might bias the results differently. There might be some correlations between different confounders, or a case of Simpson's paradox [33]. Moreover, there might be additional unknown factors (for instance the skin color or the hairs on the skin) that need to be identified and treated appropriately. In general, correction is more effective, when all confounding parameters are known and used simultaneously along with training for the main target task. Additionally, we treated the age as a continuous parameter, but the evaluation was done for two subgroups (young and old). The fairness of this evaluation strategy needs to be investigated in future works.

A possible way to address the bias problem would be to standardize what information about the data or model needs to be included in a research paper.

This could be inspired by datasheets for datasets [13], which describes the dataset design and collection procedure, and relatedly model cards [22], which describe in detail the choices made to train and optimize the model. Enforcing such standards would require large-scale collaboration between journals and conferences, but researchers could already include such information to increase awareness in the community as a whole. Although this type of measure does not remove bias, it can show that a bias potentially might exist. The exact sources of this bias could then be quantified, for example following the framework proposed by [29].

A recent interesting approach is the development of tools for assessing and mitigating the amount of bias and fairness in machine learning models and datasets. For example, Aequitas [27] is an open source toolkit developed at Center for Data Science and Public Policy University of Chicago in order to enable users to test models for several bias and fairness metrics in relation to multiple population sub-groups. Another example is the AI Fairness 360 (AIF360) [4] open source toolkit for checking unwanted bias and moreover mitigating it. This toolkit developed at IBM focuses on industrial usability and software engineering. These tools can be helpful for data scientists, researchers, policy makers and software engineers.

Another important direction is building bias-aware algorithms and removing the bias in the final predictions. Even though that might be at the expense of overall model accuracy [17,36]. Thus dataset and model interventions are both necessary [35]. Once an algorithm is designed to be sensitive to bias, we need to evaluate whether it is successful at this. Therefore, we need ways to quantify what performance gap is evidence of bias or not.

In conclusion, we highlighted the importance of bias in medical datasets and diagnostic algorithms, since ignoring it could affect the generalization across different demographic subgroups. We believe that this is an important point of attention for researchers working in medical image analysis community.

References

1. Adeli, E., et al.: Representation learning with statistical independence to mitigate bias (2019)
2. Ashraf, A., Khan, S., Bhagwat, N., Chakravarty, M., Taati, B.: Learning to unlearn: building immunity to dataset bias in medical imaging studies. arXiv preprint arXiv:1812.01716 (2018)
3. Beddingfield III, F.: The melanoma epidemic: res ipsa loquitur. Oncologist **8**(5), 459 (2003)
4. Bellamy, R.K.E., et al.: AI fairness 360: an extensible toolkit for detecting and mitigating algorithmic bias. IBM J. Res. Dev. **63**(4/5), 4:1–4:15 (2019)
5. Buolamwini, J., Gebru, T.: Gender shades: intersectional accuracy disparities in commercial gender classification. In: Conference on Fairness, Accountability and Transparency, pp. 77–91 (2018)
6. Cheplygina, V., de Bruijne, M., Pluim, J.P.: Not-so-supervised: a survey of semi-supervised, multi-instance, and transfer learning in medical image analysis. Med. Image Anal. **54**, 280–296 (2019)

7. Codella, N.C., et al.: Skin lesion analysis toward melanoma detection: a challenge at the 2017 International Symposium on Biomedical Imaging (ISBI), hosted by the International Skin Imaging Collaboration (ISIC). arXiv preprint arXiv:1710.05006 (2017)

8. Cole, J.H., et al.: Predicting brain age with deep learning from raw imaging data results in a reliable and heritable biomarker. NeuroImage **163**, 115–124 (2017)

9. Cole, J.H., Underwood, J., et al.: Increased brain-predicted aging in treated HIV disease. Neurology **88**(14), 1349–1357 (2017)

10. Corbett-Davies, S., Goel, S.: The measure and mismeasure of fairness: a critical review of fair machine learning. arXiv preprint arXiv:1808.00023 (2018)

11. Creager, E., et al.: Flexibly fair representation learning by disentanglement. In: Chaudhuri, K., Salakhutdinov, R. (eds.) International Conference on Machine Learning. Proceedings of Machine Learning Research, vol. 97, pp. 1436–1445. PMLR, Long Beach, California, USA, 09–15 June 2019

12. Frangi, A.F., Schnabel, J.A., Davatzikos, C., Alberola-López, C., Fichtinger, G. (eds.): MICCAI 2018. LNCS, vol. 11073. Springer, Cham (2018). https://doi.org/10.1007/978-3-030-00937-3

13. Gebru, T., et al.: Datasheets for datasets. CoRR abs/1803.09010 (2018)

14. Gill, R.S., et al.: Deep convolutional networks for automated detection of epileptogenic brain malformations. In: Frangi, A.F., Schnabel, J.A., Davatzikos, C., Alberola-López, C., Fichtinger, G. (eds.) MICCAI 2018. LNCS, vol. 11072, pp. 490–497. Springer, Cham (2018). https://doi.org/10.1007/978-3-030-00931-1_56

15. Greenspan, H., Van Ginneken, B., Summers, R.M.: Guest editorial deep learning in medical imaging: overview and future promise of an exciting new technique. IEEE Trans. Med. Imaging **35**(5), 1153–1159 (2016)

16. Hett, K., Ta, V.-T., Manjón, J.V., Coupé, P.: Graph of brain structures grading for early detection of Alzheimer's disease. In: Frangi, A.F., Schnabel, J.A., Davatzikos, C., Alberola-López, C., Fichtinger, G. (eds.) MICCAI 2018. LNCS, vol. 11072, pp. 429–436. Springer, Cham (2018). https://doi.org/10.1007/978-3-030-00931-1_49

17. Kamishima, T., Akaho, S., Sakuma, J.: Fairness-aware learning through regularization approach. In: International Conference on Data Mining Workshops, pp. 643–650 (2011)

18. Khosla, A., Zhou, T., Malisiewicz, T., Efros, A.A., Torralba, A.: Undoing the damage of dataset bias. In: Fitzgibbon, A., Lazebnik, S., Perona, P., Sato, Y., Schmid, C. (eds.) ECCV 2012. LNCS, vol. 7572, pp. 158–171. Springer, Heidelberg (2012). https://doi.org/10.1007/978-3-642-33718-5_12

19. Kinyanjui, N.M., et al.: Estimating skin tone and effects on classification performance in dermatology datasets. arXiv preprint arXiv:1910.13268 (2019)

20. Larrazabal, A.J., Nieto, N., Peterson, V., Milone, D.H., Ferrante, E.: Gender imbalance in medical imaging datasets produces biased classifiers for computer-aided diagnosis. In: Proceedings of the National Academy of Sciences (2020)

21. Liu, C., et al.: Biological age estimated from retinal imaging: a novel biomarker of aging. In: Shen, D., et al. (eds.) MICCAI 2019. LNCS, vol. 11764, pp. 138–146. Springer, Cham (2019). https://doi.org/10.1007/978-3-030-32239-7_16

22. Mitchell, M., et al.: Model cards for model reporting. In: Fairness, Accountability, and Transparency (FAccT), pp. 220–229. ACM (2019)

23. Orlando, J.I., Barbosa Breda, J., van Keer, K., Blaschko, M.B., Blanco, P.J., Bulant, C.A.: Towards a glaucoma risk index based on simulated hemodynamics from fundus images. In: Frangi, A.F., Schnabel, J.A., Davatzikos, C., Alberola-López, C., Fichtinger, G. (eds.) MICCAI 2018. LNCS, vol. 11071, pp. 65–73. Springer, Cham (2018). https://doi.org/10.1007/978-3-030-00934-2_8

24. Perez, F., Vasconcelos, C., Avila, S., Valle, E.: Data augmentation for skin lesion analysis. In: Stoyanov, D., et al. (eds.) CARE/CLIP/OR 2.0/ISIC -2018. LNCS, vol. 11041, pp. 303–311. Springer, Cham (2018). https://doi.org/10.1007/978-3-030-01201-4_33
25. Pooch, E.H., Ballester, P.L., Barros, R.C.: Can we trust deep learning models diagnosis? The impact of domain shift in chest radiograph classification. arXiv preprint arXiv:1909.01940 (2019)
26. Roy, P.C., Boddeti, V.N.: Mitigating information leakage in image representations: a maximum entropy approach. In: Computer Vision and Pattern Recognition (CVPR), pp. 2581–2589, June 2019
27. Saleiro, P., et al.: Aequitas: a bias and fairness audit toolkit. arXiv preprint arXiv:1811.05577 (2018)
28. Salimi, B., Rodriguez, L., Howe, B., Suciu, D.: Interventional fairness: causal database repair for algorithmic fairness. In: International Conference on Management of Data, pp. 793–810. Association for Computing Machinery (2019)
29. Suresh, H., Guttag, J.V.: A framework for understanding unintended consequences of machine learning. arXiv preprint arXiv:1901.10002 (2019)
30. Szegedy, C., Ioffe, S., Vanhoucke, V., Alemi, A.A.: Inception-v4, inception-ResNet and the impact of residual connections on learning. In: AAAI Conference on Artificial Intelligence (2017)
31. Torralba, A., Efros, A.A.: Unbiased look at dataset bias. In: Computer Vision and Pattern Recognition (CVPR), pp. 1521–1528 (2011)
32. Wachinger, C., Becker, B.G., Rieckmann, A.: Detect, quantify, and incorporate dataset bias: a neuroimaging analysis on 12,207 individuals. arXiv preprint arXiv:1804.10764 (2018)
33. Wagner, C.H.: Simpson's paradox in real life. Am. Stat. **36**(1), 46–48 (1982)
34. Wang, T., Zhao, J., Yatskar, M., Chang, K.W., Ordonez, V.: Balanced datasets are not enough: estimating and mitigating gender bias in deep image representations. In: International Conference on Computer Vision (2019)
35. Yang, K., Qinami, K., Fei-Fei, L., Deng, J., Russakovsky, O.: Towards fairer datasets: filtering and balancing the distribution of the people subtree in the ImageNet hierarchy. In: Fairness, Accountability, and Transparency (FAccT), FAT* 2020, pp. 547–558 (2020)
36. Zemel, R., Wu, Y., Swersky, K., Pitassi, T., Dwork, C.: Learning fair representations. In: Dasgupta, S., McAllester, D. (eds.) International Conference on Machine Learning. Proceedings of Machine Learning Research, vol. 28, no. 3, pp. 325–333. PMLR, Atlanta, Georgia, USA, 17–19 June 2013

Semi-weakly Supervised Learning for Prostate Cancer Image Classification with Teacher-Student Deep Convolutional Networks

Sebastian Otálora[1,2(✉)], Niccolò Marini[1,2], Henning Müller[1,3], and Manfredo Atzori[1]

[1] Institute of Information Systems, HES-SO (University of Applied Sciences and Arts Western Switzerland), Sierre, Switzerland
juan.otaloramontenegro@hevs.ch
[2] Centre Universitaire d'Informatique, University of Geneva, 1227 Carouge, Switzerland
[3] Medical Faculty, University of Geneva, 1211 Geneva, Switzerland

Abstract. Deep Convolutional Neural Networks (CNN) are at the backbone of the state–of–the art methods to automatically analyze Whole Slide Images (WSIs) of digital tissue slides. One challenge to train fully-supervised CNN models with WSIs is providing the required amount of costly, manually annotated data. This paper presents a semi-weakly supervised model for classifying prostate cancer tissue. The approach follows a teacher-student learning paradigm that allows combining a small amount of annotated data (tissue microarrays with regions of interest traced by pathologists) with a large amount of weakly-annotated data (whole slide images with labels extracted from the diagnostic reports). The task of the teacher model is to annotate the weakly-annotated images. The student is trained with the pseudo-labeled images annotated by the teacher and fine-tuned with the small amount of strongly annotated data. The evaluation of the methods is in the task of classification of four Gleason patterns and the Gleason score in prostate cancer images. Results show that the teacher-student approach improves significatively the performance of the fully-supervised CNN, both at the Gleason pattern level in tissue microarrays (respectively $\kappa = 0.594 \pm 0.022$ and $\kappa = 0.559 \pm 0.034$) and at the Gleason score level in WSIs (respectively $\kappa = 0.403 \pm 0.046$ and $\kappa = 0.273 \pm 0.12$). Our approach opens the possibility of transforming large weakly–annotated (and unlabeled) datasets into valuable sources of supervision for training robust CNN models in computational pathology.

S. Otálora and N. Marini—Equal contribution.

Electronic supplementary material The online version of this chapter (https://doi.org/10.1007/978-3-030-61166-8_21) contains supplementary material, which is available to authorized users.

© Springer Nature Switzerland AG 2020
J. Cardoso et al. (Eds.): iMIMIC 2020/MIL3ID 2020/LABELS 2020, LNCS 12446, pp. 193–203, 2020.
https://doi.org/10.1007/978-3-030-61166-8_21

Keywords: Computational pathology · Deep learning · Semi-weakly supervision · Prostate cancer · Knowledge distillation

1   Introduction

Prostate cancer (PCa) is the fourth most common cancer worldwide, with 1.2 million new cases in 2018, and it has the second-highest incidence of all cancers in men. The gold standard for the diagnosis of PCa is the visual inspection of needle biopsies or tissue samples such as prostatectomies. Currently, the Gleason score (GS) is the standard grading system used to determine the aggressiveness of PCa. The GS system is based on the architectural patterns shown in prostate tissue samples that describe tumor appearance and the presence of alterations in the glands. The Gleason score results from the sum of the two patterns (Gleason patterns from 1 to 5) most present in the tissue slide producing a final grade in the range of 2 to 10. Typical scores range from 6 to 10, where cases with higher values are more likely to grow and spread faster. The Gleason score system has been revised in 2016 [5] to propose a simpler system by having a smaller number of grades (five-groups) with the most significant prognostic differences, Nevertheless, GS is still commonly used in pathology reports, in conjunction with the new five-groups classes. Thanks to the recent improvements in digital microscopy, the diagnosis is increasingly made through the visual inspection of high-resolution scans of a tissue sample or a Whole-Slide Image (WSI).

One of the current challenges in medical imaging and particularly in computational pathology (CP), is the lack of datasets with copious region annotations for training robust supervised deep convolutional neural networks (CNN) [4]. For example, to train the deep learning models in Nagpal et al. [9], the authors collected 112 million image patches derived from 912 slides, which required approximately 900 pathologist hours to annotate. Such efforts raise the question of investigating models that minimize this costly labeling effort and reuse publicly available data to train CNN-based models.

While there is an increasing amount of available raw data, it is well known that finding reliable annotations accompanying the WSI, which are made of up to 100000^2 pixels, is a problem in this field. Examples of valuable, publicly available datasets are the Camelyon dataset for breast cancer [8] and The Cancer Genome Atlas datasets, containing up to 500 Whole slide images for individual organs, including the prostate (TCGA-PRAD)[1]. The main drawback of the TCGA datasets is that the repository does not provide region annotations for the images. The lack of strong labels poses a challenge to use the dataset to train state–of–the–art supervised CNN models for CP tasks such as the classification and segmentation of tissue subtypes of PCa. The available strongly annotated datasets in CP usually contain few images annotated or small regions of larger images [2], since the annotation of such large slides is a costly process that takes a considerable amount of time from highly-specialized personnel. In machine

---

[1] https://portal.gdc.cancer.gov/projects/TCGA-PRAD Retrieved 1st of July, 2020.

learning and computer vision, the use of semi-supervised and semi–weakly supervised learning has recently shown the potential of leveraging on large unlabeled and weakly–labeled datasets, reaching better performance than state–of–the–art supervised models in the classification of the ImageNet dataset [14]. Also, combining few strongly labeled and many weakly labeled images has been proposed in [11], achieving competitive results on natural image datasets, while requiring significantly less annotation effort.

Recently in CP, deep CNN approaches using weakly supervision have reached good performance for automatic Gleason scoring in WSI [10]. Obtaining *pseudo-labeled* data that is automatically annotated and that can improve the robustness against dataset heterogeneity and performance of CNN models is highly valuable, given a large amount of unlabeled (and weakly annotated) datasets that are publicly available and the improvement that it can bring to the results.

In this paper, the simple, yet effective, teacher-student approach of fine-tuning very large pre-trained models to generate pseudo-labeled examples is explored for the first time in the task of classifying prostate cancer tissue. Our approach employs a high-capacity (22 million parameters) ResNext-based model as a teacher. The teacher is pre-trained with a dataset of nearly one billion natural images retrieved from Instagram and its hashtags, and fine-tuned with both, weakly–annotated images from TCGA-PRAD, and annotated tissue microarrays. The smaller student model, a DenseNet-BC-121 with 7 million parameters, is then trained with the TCGA-PRAD pseudo-labeled regions annotated by the teacher and fine-tuned with the tissue microarray strong pixel-wise labels. Experimental results show that the teacher-student approach improves with statistical significance the performance of the fully-supervised CNN, both at the Gleason pattern level in tissue microarrays (respectively $\kappa = 0.594 \pm 0.022$ and $\kappa = 0.559 \pm 0.034$) and at the Gleason score level in WSI (respectively $\kappa = 0.403 \pm 0.046$ and $\kappa = 0.273 \pm 0.12$).

2   Experimental Setup

The overall workflow of the proposed semi-weakly supervised approach for classifying PCa images is summarized in Fig. 1. The details of each step involved in the training of the models are further explained in Sect. 2.2. The cardinality and characteristics of the datasets used in the article are described in Sect. 2.1.

2.1   Datasets

The two datasets of prostate images are gathered from two different sources. The TCGA-PRAD WSI repository and Tissue Microarrays (TMA). TCGA-PRAD includes WSIs from 19 different medical centers. It implies visual heterogeneity between dataset content, even though the tissues are stained in both datasets with the same reagents: hematoxylin and eosin (H&E). The dataset is comprised of pairs of WSIs, up to 100000^2 pixels, scanned at 40x resolution and the corresponding weak labels (one label per WSI) from the diagnostic report of prostate cancer cases with Gleason scores between 6 and 10.

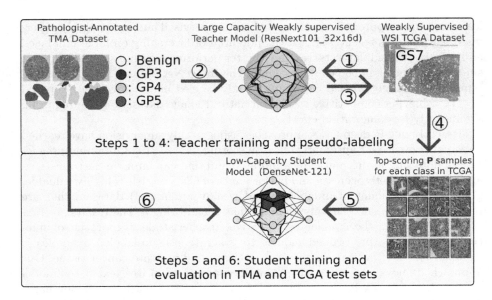

Fig. 1. The teacher-student approach: The teacher model is involved in the steps 1 to 4 (yellow background, top) and the student model is steps 5 and 6 (green background at bottom). The teacher model is first fine-tuned (from the trained model of [14]) to predict the weak labels of the TCGA-PRAD patches (primary GP) and then fine-tuned with the strongly-annotated patches from the TMA dataset. The teacher then pseudo-annotate the TCGA-PRAD patches, and the student is pre-trained using the top-ρ ranked patches. Finally, the student is fine-tuned with the strongly annotated patches from the TMA dataset. (Color figure online)

The WSIs are available from The Cancer Genome Atlas (TCGA), which is an extensive publicly available collection of data including digital pathology images that contains 500 cases of prostate adenocarcinoma (TCGA-PRAD). The used WSIs are a subset of the data containing only images used for diagnostic purposes (no frozen sections). The division of the dataset is the same as in baseline sets for cross-validation: 171 cases for training, 84 for validation, and 46 for testing. Each WSI is paired with its global Gleason score. For the task of Gleason pattern prediction at the patch level, the reported primary Gleason pattern of the WSI is used as a weak label. The patches are densely extracted only from tissue-regions of the WSI. For this, the HistoQC tool [7] is used first to generate tissue masks of the WSIs. Then, the blue-ratio mapping described in Chang et al. [3] is used to prevent selecting areas without nuclei such as those containing fat, connective tissue, or background.

The TMA dataset includes pixel-wise annotations, made by pathologists, of 886 prostate TMA cores. Each core is 3100^2 pixels, scanned at 40x resolution (0.23 microns per pixel). The training, validation and test sets as well as the patches are the same as in the study of Arvaniti et al. [2]. The total number of microarrays, WSIs and patches extracted from them is shown in Table 1.

Table 1. Left: Number of patches for each Gleason pattern class in the TMA dataset and for the weakly-annotated patches from TCGA-PRAD, after the semicolon. Right: Number of microarrays in the TMA dataset and WSI (after the semicolon) for TCGA-PRAD.

Class	Train	Val	Test
Benign	1830;1710	1260;840	127;460
GP3	5992;28919	1352;15443	1602;4000
GP4	4472;48398	831;22500	2121;13633
GP5	2766;8000	457;4000	387;3000
Total	15060;87027	3900;42783	4237;23093

Class	Train	Val	Test
Benign	61;–	42;–	12;–
GS6	158;13	35;20	79;5
GS7: 3 + 4	47;42	14;10	28;6
GS7: 4 + 3	18;30	11;14	23;11
GS8	119;37	15;12	84;13
GS9 & GS10	105;49	16;28	19;11
Total	508;171	133;84	245;46

2.2 Weakly Semi-supervised Teacher-Student Approach

The hypothesis in the semi-supervised setting is that if one has a dataset with labeled data and another without, it is possible to train a model that can use both sources, of which the performance is higher than the one obtained using only used the labeled samples [15].

The teacher-student paradigm is a semi-supervised strategy where the teacher's role is to transform the labels from the relevant examples of the weakly–annotated (or unlabeled) data. The teacher model output is pseudo–labels for the unlabeled data (resembling the strong labels) for training the student model with both sources of supervision, the strong annotations, and the pseudo-annotated dataset. Formally, if we denote the loss of a model M trained with a dataset X by $\mathcal{L}_{\mathcal{M}}(\mathcal{X})$, then ideally, $\mathcal{L}_M(S) > \mathcal{L}_M(S \cup T(U))$, where S stand for the strongly-annotated, and $T(U)$ for a pseudo-labeled set transformed using a mapping T of the unlabeled (or weakly labeled) dataset U.

The six-steps setup presented bellow resembles the best-performing configuration from the weakly-supervised teacher-student setup originally presented by Yalniz et al. [14]. In the weakly-supervised setup, the authors exploit the weak labels and characteristics of the datasets resembling the characteristics in our application to computational pathology, where it is feasible to use image-level labels as a weak form of supervision. Our main methodological novelties are the use of very high resolution and highly heterogeneous images with weak labels and the student variants, which are specifically designed for the prostate cancer image classification problem and not presented in the baseline paper [14]. While our approach might resemble commonly used bootstrapping techniques, our method differs from them because there is no random sampling involved since the teacher makes a non-trivial selection of unlabeled samples, and the models do not use subsets of the same training set to estimate the performance measures.

1) Weakly Supervised Teacher Fine-tuning: In this first step, the model is fine-tuned with the TCGA-PRAD dataset to predict the primary Gleason pattern label extracted from the reports. The teacher model weights are initialized from the trained model of [14]. The pre-trained model from Instagram, a ResNext-50 is a high-capacity model with 22 million parameters, that better fits with noisy labels [6]. TCGA-PRAD can be considered a noisy dataset since only a subset of patches actually contains the relevant pattern reported as primary Gleason pattern. In this step, the model is trained for ten epochs with a categorical-cross entropy loss to predict the primary Gleason pattern and stopped if convergence is reached early.

2) Fine-tuning of the Teacher with Strong Annotations: In this step, the weights of the model are refined to classify the TMA patches with ground-truth data. In this case the teacher is also presented with samples from the benign class. Ten models (with different initialization) are trained for 15 epochs, as the TMA dataset is not as large as TCGA-PRAD. Then, the model with the best average performance in the validation TMA partition and validation TCGA partition is kept to pseudo-annotate the patches in the next step. The performance of the teacher up to this step is reported in the results section.

3) & 4) Pseudo-labeling and Patch Selection of TCGA-PRAD: In this step, the previously selected teacher model is used to infer the class-wise probabilities of all the TCGA-PRAD patches. For each class, the ρ highest-ranked patches per class are selected according to the softmax probability of the output of the last fully connected layer. The trade-off between performance and ρ is shown in the results section.

5) Pre-training of the Student Model with Pseudo-labeled Data: The student model is trained in a supervised fashion using the pseudo-labeled images annotated by the teacher. The distillation procedure aims at training the student in such a way that it best reproduces the output of the teacher. This strategy was shown to be successful for several image recognition tasks [14]. The student has a smaller architecture than the teacher model because it is more efficient for evaluation: the student model is the one for which the hyper-parameter selection and test set evaluations are made. Therefore, it is better to have a smaller, faster inference architecture. In the fifth step, the student model is pre-trained with the ρ patches per Gleason pattern that are pseudo-labeled by the teacher. Ten models are trained in this step for 15 epochs. The best student model is then selected (i.e., the one that has the best performance in κ-score in the TMA validation partition).

6) Training of the Student and Variants: In the last step, the best student is trained with the strongly-annotated TMA patches. Ten model runs are trained for 15 epochs, selecting the best (the best average run) and reporting the final performance in the κ-score, both in the TMA and in the TCGA-PRAD test sets.

Four training variants of the student are evaluated. A) Fully-supervised training: here, only the TMA annotated patches are used for training the student; the training scheme is similar to the one described in [2]. B) Using only the pseudo-labeled images: in this case, the student never sees any patch with ground-truth data from the pathologist annotations, just the pseudo-labeled patches from the teacher model. C) Pre-training with pseudo-labeled samples and then fine-tuning with the strong annotations. D) Combining the pseudo-labeled and strongly annotated patches in one single training set: this variant is similar to C), with the difference that all the TMA and TCGA-PRAD patches are mixed at training, instead of having two training stages. These three ablation experiments results for the student model, are reported in the results sections Table 2.

2.3 Implementation, Architectures and Hyperparameter Selection

The implementation of all models was done in PyTorch, initialized with the Instagram/ImageNet pre–trained weights for the teacher and student models, respectively. Batch sizes of 128 samples were used for the first weakly supervised pre-training of the teacher (step 1), and the fine-tuning of the teacher was done with a batch size of 32 TMA patches (step 2). Several CNN models, namely, DenseNet121, DenseNet161, MobileNet, MobileNetV2, were tested for the student. Among these, the one that showed the best performance in the validation TMA set was DenseNet121. Therefore, this architecture was chosen to train the four variants of the student. The choice of a pre–trained network is done for speeding up the convergence of the model, as described for the teacher model. The CNN parameters were selected using a grid search over the validation sets of both TCGA and TMA. The best values found on the validation set are the ones used for training the ten repetitions. Specifically, the values explored for the learning rate are in the set $\{10^{-5}, 10^{-4}, 10^{-3}, 10^{-2}\}$. In each of the student training variants, the Adam optimizer is used with a learning rate of 0.001 and a decay rate of 10^{-6}.

3   Results and Analysis

Table 2. Performance measures for the semi–weakly supervised approaches, as evaluated with κ–score. For the TMA test set, the reported measure is at the patch-level Gleason pattern, while for TCGA-PRAD is at the WSI level. The '*' indicates statistically significant differences with a p-value < 0.05 from the baseline fully supervised CNN, using a Wilcoxon signed-rank test.

Variant	TMA	TCGA-PRAD
A) Fully supervised [2]	0.5590 ± 0.0346	0.2732 ± 0.1207
B) Pseudo-labeled	0.5197 ± 0.0407*	0.3648 ± 0.0571
C) Pre-training → fine-tuning	0.5928 ± 0.0178*	0.3748 ± 0.0438
D) Pseudo-labeled ∪ TMA	**0.5941 ± 0.0225***	**0.4029 ± 0.0450***
Teacher performance	0.5601 ± 0.0440	0.1910 ± 0.1102*

There are two evaluation criteria: patch-level Gleason pattern classification and image-level GS classification. For the GS classification, the models are evaluated using the revised Gleason score as defined by the International Society of

Urological Pathology. All model performances are measured as the inter-rater agreement and pathologist ground-truth. A performance measure that is often used [1,13] is Cohen's kappa, that is defined as $\kappa = 1 - \frac{\sum_{i,j} w_{i,j} O_{i,j}}{\sum_{i,j} w_{i,j} E_{i,j}}$, $w_{i,j} = \frac{(i-j)^2}{(N-1)^2}$ Where i, j are the ordered scores, $N = 5$ is the total number of Gleason scores (or $N = 4$ Gleason pattern classes). $O_{i,j}$, is the number of images that were classified with a score of i by the first rater and j by the second. $E_{i,j}$ denotes the expected number of images receiving rating i by the first expert and rating j by the second. The quadratic term $w_{i,j}$ penalizes the ratings that are not close. When the predicted Gleason score is far from the ground-truth class, $w_{i,j}$ gets closer to 1. For obtaining the GS using the patch probabilities, all the predicted probabilities are combined and a majority voting decides the GS, as in [1].

In Table 2 the test set performance for the four variants of the student models is shown. The best model is variant four, where both TMA and pseudo-labeled patches from TCGA-PRAD are mixed in one single training set. The teacher-student approach improves the performance of the fully-supervised CNN, both at the Gleason pattern level in tissue microarrays (respectively $\kappa = 0.594 \pm 0.022$ and $\kappa = 0.559 \pm 0.034$) as well as in the Gleason score level performance in WSI (respectively $\kappa = 0.403 \pm 0.046$ and $\kappa = 0.273 \pm 0.12$). The results entries with '*' also show that the only student variant performs significantly better than the baseline in both test sets is the combination of pseudo-labeled and strongly-annotated samples, despite the other variants showing relative improvements.

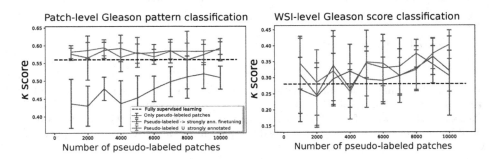

Fig. 2. Performance of the student model, depending on the number ρ of pseudo-labeled images presented. The three strategies are displayed, the two of semi-weakly are better than the fully supervised one.

4   Discussion

An analysis of the optimal ρ for the number of examples presented to the student is shown in Figure 2. The performance of two of the student variants for Gleason pattern classification remains flat with respect to the number of pseudo-labeled patches, likely because the student saturates with few pseudo-labeled patches. Similar behavior was shown in the baseline method of Yalniz et al. [11] where the student reaches a maximum performance with ~10% of the pseudo-labeled data and then starts decreasing, probably due to the introduction of many noisy samples.

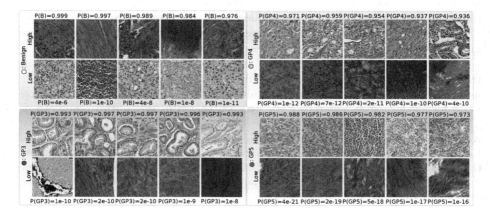

Fig. 3. Example of TCGA-PRAD patches pseudo-labeled by the teacher model: each class-box has five uniformly sampled patches from the top hundred ranked samples by the teacher and in the second row five from the hundred lowest ranked for that class. The probability of each patch belonging to the class is shown on top (first row) and in the bottom (second row). The Xe-Y is shorthand for $X \times 10^{-Y}$.

In Figure 3, a set of pseudo-labeled patches from the teacher are shown. Most of the top-ranked patches match the tissue morphology from the strongly-annotated data. There are a few noisy patches at the lowest probabilities, suggesting that the model is also lowering the relevance of artifacts and other sources of noise. The top-ranked patches for GP3, GP4, and GP5 are similar and typical for the class morphology.

The code and datasets generated during the current study are available from the corresponding author on request. Also, a supplemental document accompanying this paper, details the training of the teacher and each of the three student variants.

Concurrently to the publication of this work, Shaw et al. [12] extended the teacher-student model by generating a chain of student models for the application of classifying colon cancer regions. The results obtained by the authors showed that with the chain of students, using only 0.5% of the original labeled data, is possible to obtain the same performance as using 100% of the annotations, showing the potential for use of this approach in other computational pathology tasks.

5   Conclusion

We present a simple yet effective approach for increasing the training dataset size by obtaining pseudo–labeled regions in the task of prostate cancer classification. The evaluation of the proposed semi-weakly supervised teacher-student approach yielded better quantitative results than a fully supervised approach in two highly heterogeneous datasets of PCa. A qualitative assessment also shows how the annotated images by the teacher follow the same gland morphology patterns of the strongly annotated data. The assessment of the trade-off between performance and the amount of pseudo-labeled data shows that increasing the number of patches can deteriorate the student performance by introducing noise in training. We are now working on the semi-supervised approach only, i.e., without using any weak label, as well as the evaluation of the approach in classification tasks for other tissues, validating the pseudo-labeled images with pathologists.

Acknowledgements.. This project has received funding from the European Union's Horizon 2020 research and innovation programme under grant agreement No. 825292 (ExaMode, https://www.examode.eu). Infrastructure from the SURFsara HPC center was used to train the CNN models in parallel. Otálora thanks Minciencias through the call 756 for Ph.D. studies.

References

1. Arvaniti, E., Claassen, M.: Coupling weak and strong supervision for classification of prostate cancer histopathology images. In: Medical Imaging Meets NIPS Workshop, NIPS 2018 (2018)
2. Arvaniti, E., et al.: Automated Gleason grading of prostate cancer tissue microarrays via deep learning. Sci. Rep. **8**, 1–11 (2018)
3. Chang, H., Loss, L.A., Parvin, B.: Nuclear segmentation in H&E sections via multi-reference graph cut (MRGC). In: International Symposium Biomedical Imaging (2012)
4. Cheplygina, V., de Bruijne, M., Pluim, J.P.: Not-so-supervised: a survey of semi-supervised, multi-instance, and transfer learning in medical image analysis. Med. Image Anal. **54**, 280–296 (2019)
5. Epstein, J.I., et al.: A contemporary prostate cancer grading system: a validated alternative to the Gleason score. Eur. Urol. **69**(3), 428–435 (2016)
6. Han, B., et al.: Co-teaching: robust training of deep neural networks with extremely noisy labels. In: Advances in Neural Information Processing Systems, pp. 8527–8537 (2018)
7. Janowczyk, A., Zuo, R., Gilmore, H., Feldman, M., Madabhushi, A.: HistoQC:: an open-source quality control tool for digital pathology slides. JCO Clin. Cancer Inf. **3**, 1–7 (2019)
8. Litjens, G., et al.: 1399 H&E-stained sentinel lymph node sections of breast cancer patients: the Camelyon dataset. GigaScience **7**(6), giy065 (2018)
9. Luo, F., Nagesh, A., Sharp, R., Surdeanu, M.: Semi-supervised teacher-student architecture for relation extraction. In: Proceedings of the Third Workshop on Structured Prediction for NLP, pp. 29–37. Association for Computational Linguistics, Minneapolis, Jun 2019. https://doi.org/10.18653/v1/W19-1505, https://www.aclweb.org/anthology/W19-1505

10. Otálora, S., Atzori, M., Khan, A., Jimenez-del Toro, O., Andrearczyk, V., Müller, H.: A systematic comparison of deep learning strategies for weakly supervised Gleason grading. In: Medical Imaging 2020: Digital Pathology, vol. 11320, p. 113200L. International Society for Optics and Photonics (2020)
11. Papandreou, G., Chen, L.C., Murphy, K.P., Yuille, A.L.: Weakly-and semi-supervised learning of a deep convolutional network for semantic image segmentation. In: Proceedings of the IEEE International Conference on Computer Vision, pp. 1742–1750 (2015)
12. Shaw, S., Pajak, M., Lisowska, A., Tsaftaris, S.A., O'Neil, A.Q.: Teacher-student chain for efficient semi-supervised histology image classification. arXiv preprint arXiv:1911.04252 (2020)
13. Ström, P., et al.: Artificial intelligence for diagnosis and grading of prostate cancer in biopsies: a population-based, diagnostic study. Lancet Oncol. **21**(2), 222–232 (2020). https://doi.org/10.1016/S1470-2045(19)30738-7
14. Yalniz, I.Z., Jégou, H., Chen, K., Paluri, M., Mahajan, D.: Billion-scale semi-supervised learning for image classification. arXiv preprint arXiv:1905.00546 (2019)
15. Zhu, X., Goldberg, A.B.: Introduction to Semi-supervised Learning. Synthesis Lectures on Artificial Intelligence and Machine Learning, pp. 1–130. Morgan Kaufmann Publishers, San Francisco (2009)

Are Pathologist-Defined Labels Reproducible? Comparison of the TUPAC16 Mitotic Figure Dataset with an Alternative Set of Labels

Christof A. Bertram[1], Mitko Veta[2], Christian Marzahl[3], Nikolas Stathonikos[4], Andreas Maier[3], Robert Klopfleisch[1], and Marc Aubreville[3(✉)]

[1] Institute of Veterinary Pathology, Freie Universität Berlin, Berlin, Germany
[2] Medical Image Analysis Group, Eindhoven University of Technology, Eindhoven, Netherlands
[3] Pattern Recognition Lab, Computer Science, Friedrich-Alexander-Universität Erlangen-Nürnberg, Erlangen, Germany
marc.aubreville@fau.de
[4] Department of Pathology, University Medical Center Utrecht, Utrecht, Netherlands

Abstract. Pathologist-defined labels are the gold standard for histopathological data sets, regardless of well-known limitations in consistency for some tasks. To date, some datasets on mitotic figures are available and were used for development of promising deep learning-based algorithms. In order to assess robustness of those algorithms and reproducibility of their methods it is necessary to test on several independent datasets. The influence of different labeling methods of these available datasets is currently unknown. To tackle this, we present an alternative set of labels for the images of the auxiliary mitosis dataset of the TUPAC16 challenge. Additional to manual mitotic figure screening, we used a novel, algorithm-aided labeling process, that allowed to minimize the risk of missing rare mitotic figures in the images. All potential mitotic figures were independently assessed by two pathologists. The novel, publicly available set of labels contains 1,999 mitotic figures (+28.80%) and additionally includes 10,483 labels of cells with high similarities to mitotic figures (hard examples). We found significant difference comparing F_1 scores between the original label set (0.549) and the new alternative label set (0.735) using a standard deep learning object detection architecture. The models trained on the alternative set showed higher overall confidence values, suggesting a higher overall label consistency. Findings of the present study show that pathologists-defined labels may vary significantly resulting in notable difference in the model performance. Comparison of deep learning-based algorithms between independent datasets with different labeling methods should be done with caution.

Keywords: Breast cancer · Mitotic figures · Computer-aided annotation · Deep learning

© Springer Nature Switzerland AG 2020
J. Cardoso et al. (Eds.): iMIMIC 2020/MIL3ID 2020/LABELS 2020, LNCS 12446, pp. 204–213, 2020.
https://doi.org/10.1007/978-3-030-61166-8_22

1   Introduction

Deep learning-based methods have shown to be powerful in the development of automated image analysis software in digital pathology. This innovative field of research has been fostered by creation of publicly available data sets of specific histological structures. One of the most extensively researched cell structures in current literature are mitotic figures (microscopic appearance of a cell undergoing cell division) in neoplastic tissue. Quantification of the highest density of mitotic figures is one of the most important histological criteria for assessment of biological tumor behavior and this pattern has therefore drawn much research attention for computerized methods.

Manual enumeration of mitotic figures by pathology experts has some limitations including high inter-rater inconsistency of pathologists in classifying individual cells as mitotic figures as they exhibit a high degree of morphological variability and similarity to some non-mitotic structures. In previous studies, disagreement of classification occurred in 6.4–35.3% [8], and 68.2% [13] of labels. This calls for algorithm-assisted approaches in order to increase reproducibility as it has been proven that algorithms can have substantial agreement with pathologists on the object level [15]. Poor consistency of expert classification is, however, also a potential bias for deep learning-based methods, as pathologists are the current gold standard for assessment of morphological patterns, including mitotic figures, and creation of histological ground truth datasets. Due to the high inter-observer discordance of pathologists, we suspect some variability in assigned labels if images are annotated a second time. The usage of pathologist-defined labels for machine learning methods are thus somewhat a paradox as algorithmic methods, which are trained with and tested on these partially noisy ground truths, aim to overcome cognitive and visual limitations of pathologists.

In order to assess the robustness of algorithms and the reproducibility of newly developed deep learning-based methods it is necessary to test on several independent ground truth datasets. For these aspects, images should be independent but the ground truth should ideally be consistent throughout the datasets. To date, several open access datasets are available with labels for mitotic figures in digitalized microscopy images of human breast cancer [10,11,14] and canine cutaneous mast cell tumors [5], which have been developed by three research groups with somewhat variable labeling methods. As several publications have compared their algorithmic approaches between these publicly available datasets (for example [1,4,6]), a strong difference in test performance is known for these datasets. However, the influence of variability in the ground truth labels on training and test performance is currently unknown. In the present work, we have developed an alternative ground truth dataset for the largest of those publicly available images sets and assessed the difference to the original dataset. This was done using a new labeling methodology, targeted towards improved identification of mitotic figure events, and supported by the use of deep learning.

2   Related Work

Most publicly available data sets with annotations for mitotic figures are from human breast cancer, due to the high prevalence and high prognostic importance of the mitotic count for this tumor type. Roux *et al.* were the first to present a data set, consisting of five cases scanned by two whole slide scanners (and one multi-spectral scanner) and annotated by a single pathologist (ICPR MITOS 2012, [11]). A year later, the MICCAI AMIDA 13 challenge introduced a new data set, covering in total 23 cases, which were evenly spread between training and test set [13]. They were the first to acknowledge potential bias (inter-rater variance) by a single pathologist and thus perform the task by two pathologists independently, with a panel of two additional pathologists judging discordant annotations (see Fig. 1). The following year, the group behind the MITOS 2012 data set introduced an extended data set at ICPR (ICPR MITOS 2014, [10]), consisting of 16 cases (11 for training and 5 for test), again scanned using two scanners, but this time including annotations from two pathologists. In case the pathologists disagreed, a third pathologist decided for the particular cell. The data sets includes also an expert confidence score for each mitotic figure as well as for cells probably not mitotic figures (hard negative cells). The most recent mitotic figure dataset was part of the TUPAC16 challenge [14], incorporating all 23 AMIDA13 cases in the training set in addition to 50 new training cases and 34 new test cases. This dataset comprises the currently the highest number of mitotic figure labels in human breast cancer.

Data about the agreement of experts in the MITOS 2014 data set can be extracted from the labels given by the challenge. Out of all 1,014 cells that were flagged by at least one pathologist as *mitosis* or *probably mitosis*, only 317 (31.26%) were agreed by all pathologists to be mitotic figures, but for 749 (73.87%) the expert consensus was *mitosis*. For the MICCAI AMIDA 13 data set, Veta *et al.* reported an agreement in 649 out of 2038 (31.84%) annotated cells by the two initial readers, and the consensus found 1157 (56.77%) to be actual mitotic figures [13]. The fact that for both data sets the final consensus significantly exceeds the initial agreement highlights that spotting of rare mitotic figure events is a difficult component in the labeling process which might lead to data set inconsistency.

For data sets, inclusion of real-life variance of stain and tissue quality is an advantage, as the data is much more representative of a realistic use case. Current datasets on mitotic figures exhibit some differences in staining and other characteristics causing a certain domain shift [4] and somewhat limiting dataset transferability/robustness. Of the aforementioned datasets, the TUPAC16 dataset likely includes the highest variability due to inclusion of currently highest number of cases that were retrieved from three laboratories and scanned with two different scanners [14]. The consequence of the higher variability is an increased difficulty for the pattern recognition task of automatic mitotic figure detection, as also reflected by lower recognition scores achieved on the data set compared to the other data sets. However, this variability represents a more realistic use-case, and is highly beneficial for the development of algorithms to be used in heterogeneous clinical environments.

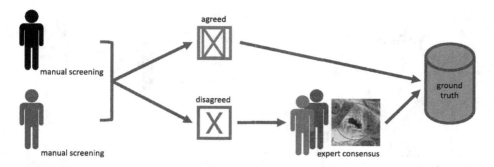

Fig. 1. Annotation workflow in the original AMIDA13/TUPAC16 data sets [13,14]. The images were independently screened by two pathologists. All agreed mitotic figures were directly accepted as ground truth, while disagreed cases were submitted to a panel of two additional experts.

3 Materials and Methods

3.1 Development of an Alternative Set of Labels

Due to the relevance of the TUPAC16 dataset (see above), we have decided to use these images in the present study for assessment of reproducibility of pathologist-defined labels. Available images from the TUPAC16 test and training set (N = 107 cases [14]) were retrieved from the TUPAC challenge website. Cases from the AMIDA13 challenge were available as several separate, but often flanking image sections, which we stitched to single images by utilizing correlation at the image borders, wherever possible. The alternative dataset was developed in a similar way as published by Bertram *et al.* [5]: First, one pathology expert screened all images twice (see Fig. 2) with an open source software solution with a guided screening mode [3]. Mitotic figures (MF) and similar structures (hard negatives, HN) were labeled. The dataset from the first screening of the training set included 5,833 labels (2,188 MF; 3,645 HN), and from the second screening 7,220 labels (2,218 MF; 5,002 HN).

The dataset was given to a second pathologist, who assigned a second label (MF or HN) in a blinded manner (label class obscured) supported by the annotation software through automatic presentation of image section with unclassified objects. The second pathologist assigned the MF label in 2,272 cases and the HN label in 4,978 cases. Initial agreement for the class MF was found for 1,713 cells (61.69%), the pathologists disagreed on 1,064 cells (14.74% of all cells). All disagreed cells were re-evaluated by both experts, and the consensus of the manual dataset contained 1,898 MF and 5,340 HN.

Subsequently, labels from the first expert were used for an algorithmic-aided pipeline for detection of missed objects with high sensitivity (low number of false negatives) and low specificity (very high number of false positives), like described in [5]. The pipeline extracted image patches around additionally detected mitotic

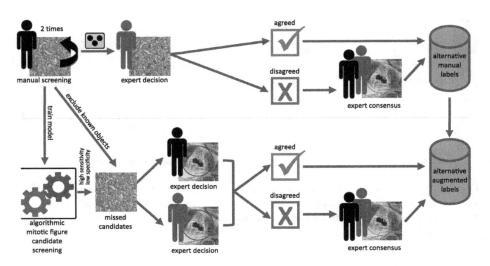

Fig. 2. Labeling approach used for the creation of the alternative set of labels consisting of two steps. First, an expert screened the slides twice, and another expert performed a blind evaluation of all cells. Second, an algorithmic pipeline was used to detect cells potentially missed by the manual screening. For both steps, disagreed labels were re-evaluated by both experts in order to find a common consensus.

figure candidates, sorted according to their model score. The algorithm-based screening additionally found 5,824 objects (mitotic figure candidates), which were then extracted as 128×128 PX image patches centered around the detection. Two experts assessed (MF or HN) these patches independently and agreed on all but 142 patches. All agreed objects were assigned to the dataset immediately and disagreed objects were re-evaluated by joint assessment for consensus. The final augmented data set contains 1,999 MF (+5.32% by augmentation) and 10,483 HN (+96.31% HN by augmentation). Please note that all numbers are given only for the training part of the set to not reveal information about the test set for further usage.

3.2 Automatic Mitosis Detection Methods

We evaluated the alternative labels using a standard, state-of-the-art object detection approach: We customized RetinaNet [7] based on a pre-trained ResNet-18 stem with an input size of 512×512 px to have the object detection head only attached at the 32×32 resolution of the feature pyramid network. We chose four different sizes (scales) of bounding boxes to enable augmentation by zooming and rotation, but only used an 1:1 aspect ratio, since the bounding boxes were defined to be squares. We randomly chose 10 tumor cases to be our validation set, which was used for model selection based on the mAP metric. After model selection, we determined the optimum detection threshold on the concatenated training and validation set. Models were trained using the same pipeline and optimization criteria on both, the original TUPAC16 label set and the novel, alternative set, and evaluated on the respective test sets using F_1 as metric.

Additionally, we calculated the model scores for individual cells of the data sets to assess model confidence. Since the test set of both label sets are not available publicly, we used a three-fold cross-validation on the training set. For this, we disabled the threshold normally used within the non-maximum-suppression of the model post-processing, which enabled us to derive model scores from the classification head of the model for all cells of our data set. We matched annotations in both data sets under the assumption that all annotations within a distance of 25 pixels refer to the same object.

The complete training set and all code that was used for the evaluation is made available online[1]. We encourage other research groups to use this alternative dataset for training their algorithms and we will provide evaluation of the performance of detection results on the augmented test set upon a reasonable request to the corresponding author.

4   Results

Comparing the original and the new, alternative training label sets, we find that they agree for 1,239 MF annotations (53.59%), while the two expert groups disagreed on 1,073 cells (46.41%). As depicted in Fig. 3, 246 of MF identified in the original TUPAC16 label set were assigned to be hard examples in the alternative set, while 67 were not annotated at all. Our experts revisited these 67 labels and classified 11 as MF and 56 as HN by consensus. The alternative set assigned 760 further cells with the label MF, that were not annotated in the original label set.

Looking at the concatenated model scores from the cross-validation experiment, we can state that the model trained on the alternative set shows an overall higher confidence for agreed mitotic figures. In contrast, MF labels only present in the original TUPAC16 dataset had an overall lower model score with a tendency towards higher values in the models trained on the original set (median values are 0.326 and 0.284). The group of labels newly assigned in the alternative set shows higher scores for the model trained on the alternative set (median value of 0.503 vs. 0.266), while the group of hard negatives has a very similar distribution with low model scores for both training label sets.

The higher model confidence for mitotic figures on the alternative dataset in Fig. 3 coincides with a generally higher F_1 score in model performance on the test set (see Table 1). We see a small increment for using the data set using the machine-learning-aided detections for potentially missed cells, related to a notable increase in precision.

---

[1] https://github.com/DeepPathology/TUPAC16_alternativeLabels.

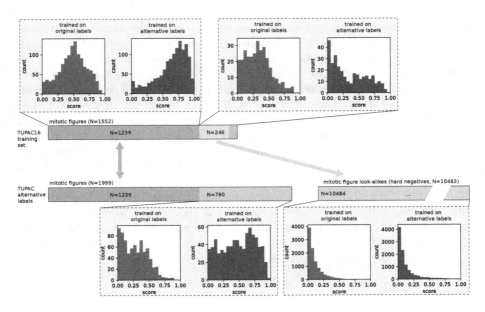

Fig. 3. Comparison of the original TUPAC16 and alternative label sets (training part only). The two expert teams agreed upon 1,239 mitotic figures, while the new set contains 760 additional labels for mitotic figures and 246 out 309 disagreed cells were labeled as hard negatives. The plot shows also the concatenated model scores given by a RetinaNet-approach trained in three-fold cross-validation on the original (blue) and alternative (green) label set. (Color figure online)

5   Discussion

Labeled data is the foundation for training and testing of deep learning-based algorithms. Although a vast diversity of labeling methods have been applied for mitotic figure dataset development [5,9–11,13,14], the effects of these methods on algorithmic performance are currently not fully understood. However, with recent improvements of deep learning methods, the demand for high-quality data is increasing. The currently highest reported F_1 score on the original TUPAC16 dataset is 0.669 [6], which is significantly higher than the value achieved by our standard RetinaNet approach on the same labels (F_1 score: 0.549). Considering the difference between the present and the state-of-the-art results by Li *et al.* [6] on the original TUPAC16 dataset, it seems likely that also the results on the alternative datasets may be further improved by more advanced methods, which we encourage as we have made the alternative datasets publicly available. However, instead of aiming to achieve highest possible performance, we wanted to assess effects of using different ground truth datasets of the same images with the same deep learning method and model optimization. The major finding of the present study was that pathologists-defined labels are not necessarily reproducible even when using annotation protocols that take the consensus of several

Table 1. Comparison of F_1 score, precision and recall achieved on the different label sets with a customized RetinaNet [7] approach.

Metric	Training	Test		
		TUPAC original	Alternative (manual)	Alternative (augmented)
F_1 score	TUPAC original	0.549	0.587	0.563
	Alternative (augmented)	0.555	0.719	0.735
Precision	TUPAC original	0.540	0.682	0.699
	Alternative (augmented)	0.477	0.713	0.772
Recall	TUPAC original	0.559	0.515	0.471
	Alternative (augmented)	0.665	0.725	0.701

experts as the ground truth, and differences may lead to notable variation in performance. In this case, the model trained and tested on the alternative dataset yielded an higher F_1 score (+18.6% points) compared to the same model architecture trained and tested on the original label set. Both, 1) testing the same algorithm on different dataset variants and 2) training algorithms with different dataset variants had notable influence on performance. The present results indicate that comparing model performance between two different datasets should be done with caution.

We believe that higher label consistency between the training and test set and decreased numbers of false negatives contributed to higher performance of the alternative dataset. While the alternative dataset was labeled within a short period of time by the same experts, different parts of the original dataset were created for different challenges (AMIDA 13 and TUPAC16) by different experts, which potentially has contributed to a somewhat higher degree of label inconsistency. Also, the alternative datasets contains 28.80 % more mitotic figure labels in the training set. Some of these additional mitotic figures have a relatively low model score, which could question the unambiguous nature of the labels regardless of the overall higher F_1 score. However, the increased model scores for the algorithms trained on the alternative data, in comparison to the original data, indicates a overall higher consistency. Regardless, both datasets include numerous labels with low model score, which could potentially be explained by the high morphological variability of mitotic figures and availability of very few patches of some morphological variants for training. Large-scale datasets with even higher numbers of mitotic figure labels might potentially overcome this limitation. Additionally, different degrees of inconsistency have been described between pathologists [8,10,14] and pathologist-defined labels represent a somewhat noisy reference standard regardless agreement or consensus by several pathologists.

Besides the difficulties in classification of mitotic figures, differences of expert-defined labels may arise from lack of identifying rare events [15]. The higher number of mitotic figure labels with presumably high label consistency in the alternative datasets (see above), suggests that fewer mitotic figures were overlooked. To avoid bias introduced by the method chosen for generating the augmented missed cells, the labeling method of the alternative dataset follows the paradigm of Viola and Jones [16], of having an initially highly sensitive detection, followed by a secondary classification with high specificity achieved through dual expert consensus. High sensitivity in detecting potential mitotic figure labels was achieved by repeated manual screening of the images and an additional algorithmic augmentation. As the algorithmic detection of missed objects may potentially introduce a confirmation bias, image patches were reviewed by two pathologists independently. Final agreement on being mitotic figures was only obtained for 2.4% of the machine learning-proposed candidate cells, illustrating the desired high sensitivity/low specificity of this approach for algorithmic mitotic figure detection. Of note, adding this relatively low number of labels to the ground truth had a notable effect on performance of up to 1.6% points (difference of test performance between manual and augmented dataset), consistent with previous findings [5]. As the network architecture for the augmented labeling method was based on the initial dataset by the first expert, a certain percentage of true mitotic figure candidates may still be missed. Considering that the original dataset only contains 15 additional MF, while the alternative datasets contains 760 additional MF, we assume that amount of false negative labels in the alternative dataset are negligible.

Algorithmic approaches for dataset development have become more popular in recent years due to increasing demand on datasets that are difficult to accomplish with solely manual approaches. As described above, algorithmically supported identification of missed candidates may improve dataset quality and requires algorithms with high sensitivity [5]. In contrast, enlargement of datasets (higher quantity) may be facilitated through algorithmic detections with high specificity in order to ensured that mainly true positives and only few false positive labels are generated. This approach can be used for the creation of datasets with reduced expenditure of expert labor (crowd sourcing [2] or expert-algorithm-collaboration [9]), or fully automated generation of additional data without pathologists-defined labels (pseudo-labels) [1]. Tellez et al. [12] recently investigated another approach, that used an specific staining for mitotic figures (immunohistochemistry with antibodies against phosphohistone H3) with computerized detection of reference labels and subsequent registration to images of the same tissue section with standard, non-specific hematoxylin and eosin stain. Besides requiring minimal manual annotation effort, this methods may eliminate expert-related inconsistency and inaccuracy.

In conclusion, this study shows considerable variability in pathologists-defined labels. A subsequent effect was evident on training the models (variation of model scores) and performance testing (variation of F_1 score). This needs to be considered when robustness of algorithms or reproducibility of developed deep learning methods are to be tested on independent ground truth datasets with different labeling methods. Therefore, scores should be interpreted in relation to reference results on that specific datset. Further studies on reduction of expert-related inconsistency and inaccuracy are encouraged.

Acknowledgement. CAB gratefully acknowledges financial support received from the Dres. Jutta & Georg Bruns-Stiftung für innovative Veterinärmedizin.

References

1. Akram, S.U., Qaiser, T., Graham, S., Kannala, J., Heikkilä, J., Rajpoot, N.: Leveraging unlabeled whole-slide-images for mitosis detection. In: Stoyanov, D., et al. (eds.) OMIA/COMPAY -2018. LNCS, vol. 11039, pp. 69–77. Springer, Cham (2018). https://doi.org/10.1007/978-3-030-00949-6_9
2. Albarqouni, S., Baur, C., Achilles, F., Belagiannis, V., Demirci, S., Navab, N.: AggNet: deep learning from crowds for mitosis detection in breast cancer histology images. IEEE Trans. Med. Imaging **35**(5), 1313–1321 (2016)
3. Aubreville, M., Bertram, C., Klopfleisch, R., Maier, A.: Sliderunner. Bildverarbeitung für die Medizin 2018, pp. 309–314. Springer, Heidelberg (2018). https://doi.org/10.1007/978-3-662-56537-7_81
4. Aubreville, M., Bertram, C.A., Jabari, S., Marzahl, C., Klopfleisch, R., Maier, A.: Learning new tricks from old dogs-inter-species, inter-tissue domain adaptation for mitotic figure assessment. Bildverarbeitung für die Medizin **2020**, 1–7 (2020). https://doi.org/10.1007/978-3-658-29267-6_1
5. Bertram, C.A., Aubreville, M., Marzahl, C., Maier, A., Klopfleisch, R.: A large-scale dataset for mitotic figure assessment on whole slide images of canine cutaneous mast cell tumor. Sci. Data **6**(274), 1–9 (2019)
6. Li, C., Wang, X., Liu, W., Latecki, L.J., Wang, B., Huang, J.: Weakly supervised mitosis detection in breast histopathology images using concentric loss. Med. Imaging Anal. **53**, 165–178 (2019)
7. Lin, T.Y., Goyal, P., Girshick, R., He, K., Dollár, P.: Focal loss for dense object detection. In: Proceedings of the IEEE International Conference on Computer Vision, pp. 2980–2988 (2017)
8. Malon, C., et al.: Mitotic figure recognition: agreement among pathologists and computerized detector. Anal. Cell Pathol. **35**(2), 97–100 (2012)
9. Marzahl, C., et al.: Are fast labeling methods reliable? A case study of computer-aided expert annotations on microscopy slides. Preprint, arXiv:2004.05838 (2020)
10. Roux, L., et al.: MITOS & ATYPIA - detection of mitosis and evaluation of nuclear atypia score in breast cancer histological images. Agency Sci., Technol. Res. Inst. Infocom Res., Singapore, Image Pervasive Access Lab IPAL (2014)
11. Roux, L., et al.: Mitosis detection in breast cancer histological images An ICPR 2012 contest. J. Pathol. Inf. **4**(1), 8 (2013)
12. Tellez, D., et al.: Whole-slide mitosis detection in H&E breast histology using PHH3 as a reference to train distilled stain-invariant convolutional networks. IEEE Trans. Med. Imaging **37**(9), 2126–2136 (2018)
13. Veta, M., et al.: Assessment of algorithms for mitosis detection in breast cancer-histopathology images. Med. Imaging Anal. **20**(1), 237–248 (2015)
14. Veta, M., et al.: Predicting breast tumor proliferation from whole-slide images - the TUPAC16 challenge. Med. Imaging Anal. **54**, 111–121 (2019)
15. Veta, M., Van Diest, P.J., Jiwa, M., Al-Janabi, S., Pluim, J.P.: Mitosis counting in breast cancer: object-level interobserver agreement and comparison to an automatic method. PloS One **11**(8), e0161286 (2016)
16. Viola, P., Jones, M.: Rapid object detection using a boosted cascade of simple features. In: Proceedings of the 2001 IEEE Computer Society Conference on Computer Vision and Pattern Recognition, vol. 1, pp. I. IEEE (2001)

EasierPath: An Open-Source Tool for Human-in-the-Loop Deep Learning of Renal Pathology

Zheyu Zhu[1], Yuzhe Lu[1], Ruining Deng[1], Haichun Yang[2], Agnes B. Fogo[2], and Yuankai Huo[1(✉)]

[1] Department of Electrical Engineering and Computer Science,
Vanderbilt University, Nashville, TN 37235, USA
yuankai.huo@vanderbilt.edu
[2] Department of Pathology, Microbiology and Immunology,
Vanderbilt University Medical Center, Nashville, TN 37232, USA

Abstract. Considerable morphological phenotyping studies in nephrology have emerged in the past few years, aiming to discover hidden regularities between clinical and imaging phenotypes. Such studies have been largely enabled by deep learning based image analysis to extract sparsely located targeting objects (e.g., glomeruli) on high-resolution whole slide images (WSI). However, such methods need to be trained using labor-intensive high-quality annotations, ideally labeled by pathologists. Inspired by the recent "human-in-the-loop" strategy, we developed EasierPath, an open-source tool to integrate human physicians and deep learning algorithms for efficient large-scale pathological image quantification as a loop. Using EasierPath, physicians are able to (1) optimize the recall and precision of deep learning object detection outcomes adaptively, (2) seamlessly support deep learning outcomes refining using either our EasierPath or prevalent ImageScope software without changing physician's user habit, and (3) manage and phenotype each object with user-defined classes. As a user case of EasierPath, we present the procedure of curating large-scale glomeruli in an efficient human-in-the-loop fashion (with two loops). From the experiments, the EasierPath saved 57% of the annotation efforts to curate 8,833 glomeruli during the second loop. Meanwhile, the average precision of glomerular detection was leveraged from 0.504 to 0.620. The EasierPath software has been released as open-source to enable the large-scale glomerular prototyping. The code can be found in https://github.com/yuankaihuo/EasierPath.

Keywords: Open-source · Human-in-the-loop · Renal pathology · Glomerular detection

1 Introduction

In the past decades, the digital image processing algorithms have been widely applied to digital renal pathology images, especially with advanced deep learning

© Springer Nature Switzerland AG 2020
J. Cardoso et al. (Eds.): iMIMIC 2020/MIL3ID 2020/LABELS 2020, LNCS 12446, pp. 214–222, 2020.
https://doi.org/10.1007/978-3-030-61166-8_23

algorithms [1–10,14]. However, one major challenge of employing deep learning algorithms is the requirement of massive training data with manual annotation. Moreover, the annotation on pathological images requires extensive professional knowledge and resources, which is resource-intensive and tedious for pathologists. To alleviate the human efforts, the human-in-the-loop deep learning strategy has become a promising direction [11,15], whose aim is to integrate human and machine intelligence for efficient deep model deployment.

In this paper, we propose an open-source tool EasierPath, which integrates human physicians and deep learning algorithms, for efficient large-scale object detection of renal pathology (Fig. 1). Briefly, CircleNet [16] was employed to perform automatic glomerular detection. Then, automatic object detection results are globally curated by adjusting the optimal detection threshold for maximizing the true positive and minimizing the false positive. Next, pathologists perform manual quality assurance (QA) and correction upon detection results. The manual QA and correction are not only enabled by using EasierPath, but also seamlessly compatible with the most prevalent commercial software ImageScope[12]. Last, glomeruli were extracted, managed, and documented by EasierPath with customized classification.

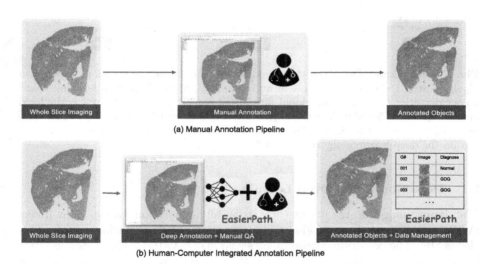

(a) Manual Annotation Pipeline

(b) Human-Computer Integrated Annotation Pipeline

Fig. 1. This figure illustrates two different strategies for glomerular annotation. The upper panel represents the traditional method that the entire dataset is annotated directly by human doctors. The lower panel represents the proposed human-in-the-loop framework, enabled by our EasierPath software. Such framework integrates computers and humans for more efficient data annotation and curation ("GDG" means Global Disappearing Glomerulosclerosis and "GOG" means Global Obsolescent Glomerulosclerosis).

2   Methods

Figure 2 shows the entire workflow, which consists of deep learning based detection, filtering, manual QA, and data management.

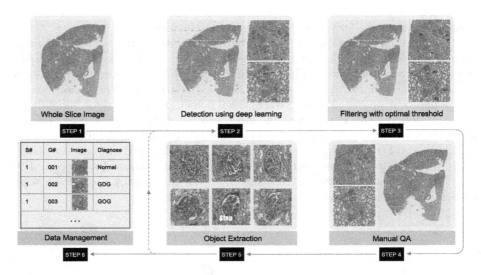

Fig. 2. This figure illustrates the workflow of the human-in-the-loop annotation pipeline. Step 1 shows an example of an input WSI. Step 2 is a deep learning based object detection. Step 3 is the global filtering of detection results. Step 4 is the local manual QA. Step 5 and 6 show the object extraction as well as the corresponding class annotation results. The dashed line indicates that the curated data could be used to retrain the deep learning algorithm as a "loop".

2.1   Detection Using Deep Learning

The input image of the entire flow is a high-resolution whole slide image (WSI). Then, the automatic glomerular detection results are achieved from CircleNet [16]. The detection outcomes are saved in one XML file that contains circle location, type, and the detection score for each detected object. The detection score is a score within 0 to 1, where a larger score indicates the stronger confidence to believe the detected object is a glomerulus. The XML format file can be loaded with the corresponding intensity image in EasierPath, for further QA and data curation.

2.2   Filtering with Optimal Threshold

After deep learning based detection, we can visualize the results by importing the XML file into the EasierPath software. Using EasierPath, we can alter the threshold of detection scores to decide which glomeruli should be kept.

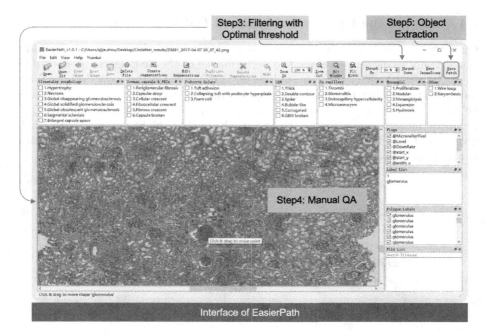

Fig. 3. This figure presents the interface of EasierPath. Toolbox and where each step is operated. Step 3 provides buttons to control the global false positive and false negative by filtering the detection score with different thresholds. Then, the local image QA and curation (step 4) can be performed using our annotation tool. The click on "Save Patch" in step 5 will automatically extract all the detected objects and save each object as an individual image.

The threshold is adjusted by clicking the button "Thresh Up" or "Thresh Down" (Fig. 3). All the circles with detection scores lower than that threshold will disappear for each threshold, leaving behind circles with detection scores higher than or equal to that threshold. The threshold adjustment allows the global balance of false positive and false negative to minimize the manual efforts for the following QA.

2.3 Manual Quality Assurance

After selecting an optimal threshold, doctors can perform manual QA using EasierPath software. As the most labor intensive step in the pipeline, to leverage the annotation efficiency for clinicians who would like to use ImageScope (www. leicabiosystems.com) without changing their user habits, we provide the alternative option to export the detection results after thresholding as an ImageScope compatible format(by simply clicking "Save ImageScope" button). It enables the seamless annotation format conversion between EasierPath and ImageScope. Using EasierPath or ImageScope, clinicians are able delete false positive, annotate false negative, and correct detection results.

2.4   Improve the Deep Learning Based Detection as a Loop

Once we get the manually curated dataset, we can use those datasets as additional training data to leverage the performance of the deep learning algorithm as a "loop", which is the crucial idea in the human-in-the-loop design. With more training data, the performance of the deep learning network will typically be improved for the next loop. That will further help the following manual QA, upon the more accurate automatic results. Note that the Step 2 and Step 3 will not be performed during the first loop, since we don't have any annotated data to train the detection model.

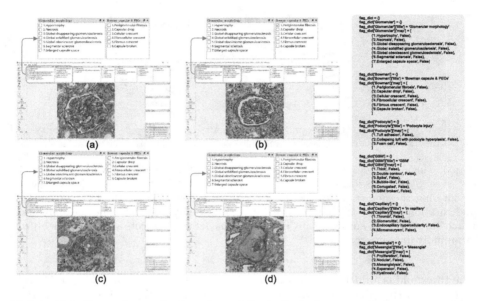

Fig. 4. This figure shows the interface of EasierPath for manual class annotation. (a)–(d) show examples of how to use EasierPath to perform glomerular classification. The definition of the classes can be easily changed in the configuration file (right panel) by clinicians without programming skills.

2.5   Object Extraction and Management

After completing the manual QA, all targeting objects have been annotated. As the histology images are typically high resolution Gigabytes images, while the objects of interest typically only exhibit small portions of the entire image. Therefore, we could only extract, manage, and save those meaningful pixels, to accelerate the following secondary analysis, data retrieval, and model training.

First, all the image patches that contain the targeting objects can be saved as individual images by a single click (Fig. 4). Then, the cropped patch samples can be loaded and labeled efficiently using the same EasierPath software. The physicians are able to define the categories of labels by editing a configuration file conveniently without programming skills. Once the configuration is confirmed, all the image patches can be loaded to the EasierPath and be annotated efficiently (Fig. 4). After clicking, the annotations will be saved into a json format file. The json files and the patch files will be saved as a database for future utilization. For instance, we can efficiently extract all glomeruli from the database with "global glomerulosclerosis" in the future, we would like to investigate such phenotype or train a machine learning based classification method.

3 Data

WSI from renal biopsies and human kidney nephrectomy tissues were utilized for performing the glomerular quantification at the first and second loop respectively. In the first loop, the renal biopsies were quantified, whose kidney biopsy tissues were routinely processed, paraffin-embedded, and $2\,\mu$m thickness sections cut and stained with hematoxylin and eosin (HE), periodic acid-Schiff (PAS) or Jones. In the second loop, the human nephrectomy tissues were quantified, whose tissues were routinely processed, paraffin-embedded, and $3\,\mu$m thickness sections cut and stained with PAS. The data were de-identified, and studies were approved by the Institutional Review Board (IRB). After the first loop, the CircleNet was trained by 704 glomeruli from 42 biopsy samples to perform initial detection for the second loop. In the second loop, 7,449 glomeruli from 18 human nephrectomy images were curated by using both automatic detection and manual QA in EasierPath framework. After the second loop, all curated data are used as training and validation data for retraining the CircleNet. We manually annotate 1384 glomeruli from five untouched human nephrectomy images as testing data to evaluate the performance of detection after the first and second loops.

Table 1. Detection results with CircleNet

	Loop	AP	AP_{50}	AP_{75}	AP_S	AP_M	AP_L
CircleNet	Loop = 1	0.504	0.729	0.511	0.363	0.721	0.625
CircleNet	Loop = 2	0.620	0.915	0.602	0.531	0.756	0.668

4 Experiments and Results

4.1 Labor Cost Analysis

In the second loop, we randomly chose one complete human nephrectomy image to evaluate the labor cost between the two strategies in Fig. 1. Using pure manual annotation, it took about 7 seconds per glomeruli by a renal pathologist with more than 20 years of experience. It took about 3 seconds per glomeruli for the same pathologists using the proposed EasierPath pipeline with the initial CircleNet (after the first loop). The temporal gap between the pure manual annotation and using EasierPath pipeline was more than two weeks to avoid the annotator remember the same human nephrectomy image. From the test, 57.1% of the manual efforts are reduced using the proposed framework.

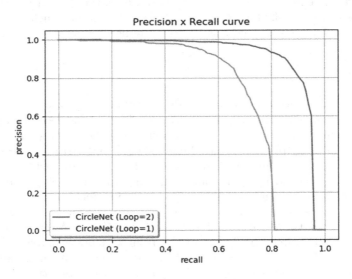

Fig. 5. This figure shows the precision-recall curves from the CircleNet detection after the first and second loop respectively.

4.2 Detection Performance

1384 glomeruli from five untouched human nephrectomy images were manually annotated by another independent annotator as testing data to evaluate the performance of deep learning detection after the first and second loops. We report average precision (AP) related canonical detection metrics overall Intersection over Union (IOU) thresholds (Table 1), which shows the result of CircleNet after the first loop (Loop = 1) and the second loop (Loop = 2).

According to Table 1, CircleNet after the second loop achieved a better accuracy for all types of APs. Especially when area = small, average precision for CircleNet after the second loop was approximately 46.28% higher than the performance after the first loop. When IOU = 0.5, the average precision for CircleNet achieved approximately 25.51% improvements. Better accuracy indicated that the performance of CircleNet was leveraged after when performing more loops.

To visualize the performance, precision-recall curves for CircleNet after the first and second loops were shown in Fig. 5. Precision, the ratio of TP/(TP +

FP) [13], was higher for CircleNet with manual annotation for any recall value, which indicated the performance of CircleNet was leveraged from more training data with more loops. As recall value represented the ratio of FN/(TN + FN), a higher recall curve was obtained after the second loop. When the precision rate was about 0.8, the recall rate for CircleNet (Loop = 1) was approximately 0.68, while the recall rate for CircleNet (Loop = 2) was about 0.9, which equaled to about 32.35% improvements. Using human in the loop for only two loops, CircleNet with manual annotation has provided decent performance on detection of glomeruli.

5   Conclusion

In this paper, we introduced EasierPath, an efficient large-scale pathological image quantification tool with human-in-the-loop deep learning of renal pathology. The proposed method reduced the 57% labor cost for curating large-scale target objects in high-resolution pathological WSI. Meanwhile, the performance of deep learning detection was leveraged after performing each loop. Last, the EasierPath tool provided a easy-to-adapt function to curate, extract, and manage each detected object for future usage.

Acknowledgments. This work was supported by NIH NIDDK DK56942(ABF).

References

1. Gadermayr, M., Dombrowski, A.K., Klinkhammer, B.M., Boor, P., Merhof, D.: CNN cascades for segmenting sparse objects in gigapixel whole slide images. Comput. Med. Imaging Graph. **71**, 40–48 (2019)
2. Gadermayr, M., Klinkhammer, B.M., Boor, P., Merhof, D.: Do we need large annotated training data for detection applications in biomedical imaging? A case study in renal glomeruli detection. In: Wang, L., Adeli, E., Wang, Q., Shi, Y., Suk, H.-I. (eds.) MLMI 2016. LNCS, vol. 10019, pp. 18–26. Springer, Cham (2016). https://doi.org/10.1007/978-3-319-47157-0_3
3. Ginley, B., Tomaszewski, J.E., Yacoub, R., Chen, F., Sarder, P.: Unsupervised labeling of glomerular boundaries using Gabor filters and statistical testing in renal histology. J. Med. Imaging **4**(2), 021102 (2017)
4. Ginley, B.G., Tomaszewski, J.E., Jen, K.Y., Fogo, A., Jain, S., Sarder, P.: Computational analysis of the structural progression of human glomeruli in diabetic nephropathy. In: Medical Imaging 2018: Digital Pathology, vol. 10581, p. 105810A. International Society for Optics and Photonics (2018)
5. Grimm, P.C., et al.: Computerized image analysis of Sirius red-stained renal allograft biopsies as a surrogate marker to predict long-term allograft function. J. Am. Soc. Nephrol. **14**(6), 1662–1668 (2003)
6. Hermsen, M.: Deep learning-based histopathologic assessment of kidney tissue. J. Am. Soc. Nephrol. **30**(10), 1968–1979 (2019)
7. Kato, T., et al.: Segmental hog: new descriptor for glomerulus detection in kidney microscopy image. BMC Bioinform. **16**(1), 316 (2015)

8. Klapczynski, M., Gagne, G.D., Morgan, S.J., Larson, K.J., LeRoy, B.E., Blomme, E.A., et al.: Computer-assisted imaging algorithms facilitate histomorphometric quantification of kidney damage in rodent renal failure models. J. Pathol. Inform. **3**, 20 (2012)
9. Litjens, G., et al.: A survey on deep learning in medical image analysis. Med. Image Anal. **42**, 60–88 (2017)
10. Litjens, G., et al.: Deep learning as a tool for increased accuracy and efficiency of histopathological diagnosis. Sci. Rep. **6**, 26286 (2016)
11. Lutnick, B., et al.: An integrated iterative annotation technique for easing neural network training in medical image analysis. Nature Mach. Intell. **1**(2), 112–119 (2019)
12. Murray, M.E., Graff-Radford, N.R., Ross, O.A., Petersen, R.C., Duara, R., Dickson, D.W.: Neuropathologically defined subtypes of Alzheimer's disease with distinct clinical characteristics: a retrospective study. Lancet Neurol. **10**(9), 785–796 (2011)
13. Ozenne, B., Subtil, F., Maucort-Boulch, D.: The precision-recall curve overcame the optimism of the receiver operating characteristic curve in rare diseases. J. Clin. Epidemiol. **68**(8), 855–859 (2015)
14. Servais, A., et al.: Interstitial fibrosis evolution on early sequential screening renal allograft biopsies using quantitative image analysis. Am. J. Transplant. **11**(7), 1456–1463 (2011)
15. Wang, Y., et al.: Weakly supervised universal fracture detection in pelvic x-rays. In: Shen, D., et al. (eds.) MICCAI 2019. LNCS, vol. 11769, pp. 459–467. Springer, Cham (2019). https://doi.org/10.1007/978-3-030-32226-7_51
16. Yang, H., et al.: CircleNet: anchor-free detection with circle representation. arXiv preprint arXiv:2006.02474 (2020)

Imbalance-Effective Active Learning in Nucleus, Lymphocyte and Plasma Cell Detection

Chao-Ting Li[1], Hung-Wen Tsai[2], Tseng-Lung Yang[3], Jung-Chi Lin[3],
Nan-Haw Chow[4], Yu Hen Hu[5], Kuo-Sheng Cheng[1],
and Pau-Choo Chung[1(✉)]

[1] National Cheng Kung University, Tainan, Taiwan
pcchung@ee.ncku.edu.tw
[2] Department of Pathology, National Cheng Kung University Hospital,
College of Medicine, National Cheng Kung University, Tainan, Taiwan
[3] Kaohsiung Veterans General Hospital, Kaohsiung City, Taiwan
[4] College of Medicine, NCKU, Tainan, Taiwan
[5] University of Wisconsin at Madison, Madison, USA

Abstract. An Imbalance-Effective Active Learning (IEAL) based deep neural network algorithm is proposed for the automatic detection of nucleus, lymphocyte and plasma cells in hepatitis diagnosis. The active sampling approach reduces the training sample annotation cost and mitigates extreme imbalances among the nucleus, lymphocytes and plasma samples. A Bayesian U-net model is developed by incorporating IEAL with basic U-Net. The testing results obtained using an in-house dataset consisting of 43 whole slide images (300 256 * 256 images) show that the proposed method achieves an equal or better performance compared than a basic U-net classifier using less than half the number of annotated samples.

Keywords: Active learning · Class imbalance · Lymphocyte detection · Histopathological image · Convolutional neural networks

1 Introduction

Lymphocytes are the main type of immune/inflammatory cells. Infiltrating immune cells are a host presence in immune reaction and serve as an important indicator in acute inflammatory response diagnosis and tumor prognosis treatment. Plasma cells are one of the main lymphocyte subpopulations, and often occur together with lymphocytes. The presence of plasma cell-rich mononuclear infiltration may be associated with an increased risk of autoimmune hepatitis (AIH). Thus, the numbers and distribution of lymphocytes and plasma cells with respect to liver cells (i.e., nucleus) play an important role in disease diagnosis and prognosis.

However, in performing the automatic detection of nucleus, lymphocyte and plasma cells in digital histopathological images, the lack of sufficient annotated samples is a major concern. Typically, the insufficiency of samples stems from the fact that the annotation process requires specialty-oriented knowledge and experience, which

© Springer Nature Switzerland AG 2020
J. Cardoso et al. (Eds.): iMIMIC 2020/MIL3ID 2020/LABELS 2020, LNCS 12446, pp. 223–232, 2020.
https://doi.org/10.1007/978-3-030-61166-8_24

makes crowd leveraging difficult. Moreover, labeling a digital histopathological image requires extensive manual effort and time since the images usually have a size of several gigabytes (approximately $50,000 \times 50,000$ pixels at 20x) and contain a huge number of object instances (e.g. cells) to be labeled. Furthermore, some cells only exist in certain disease conditions, and hence finding these cells in such a large size image may take a lot of effort and time.

Extreme class imbalance among the nucleus, lymphocytes and plasma cells is also a major problem. For example, compared to liver cells and lymphocytes, plasma cells are always only present with an extremely small quantity. Similarly, plasma cell-rich infiltration, which is one of the main indicators of AIH, only exists in a few cases. Hence, the situation of extreme class imbalance makes it difficult to find minority samples; especially, these samples are vital indicators in assisting disease diagnosis.

Active learning (AL) is a promising approach for reducing the labeling workload through an intelligent sample query strategy. Some of the most widely used criteria for measuring uncertainty in AL are reported in [1–3]. However, recently, the issue of model uncertainty (epistemic) has received significant attention. Exactly, Bayesian network adopts model uncertainty for estimating uncertainty. However, it is not easily implemented. Thus, Gal et al. [4] showed that dropout at test time can be cast as a Monte Carlo dropout (i.e. MC Dropout) for getting the Bayesian approximation to represent the model uncertainty.

The authors in [5, 6] found that AL is innately a good choice for overcoming class imbalance problems. However, standard AL approaches following the query strategy of uncertainty sampling disregards the classes. Such an approach cannot solve the extreme "one-vs-all" class imbalance problem which occurs when the probability of the plasma cells in the query is virtually zero. That is, the approach cannot ensure that the plasma cells are adequately queried. Other publications [7, 8] used asymmetric query for online imbalanced data of a two-class positive and negative samples. [9] measured the uncertainty through the dropout on fast R-CNN and lowered the classification threshold of the minority class. These approaches still did not consider the extremely imbalance in the active learning. Therefore, the present study proposes an Imbalance-Effective Active Learning (IEAL) algorithm to query a more balanced training dataset to solve the extreme "one-vs-all" class imbalance problem in plasma cell detection [10–17].

In IEAL, both uncertainty measurement and class type estimation are performed on the data pool. That is, the selection strategy in IEAL selects the most informative samples for annotation based not only on their uncertainty values, but also their predicted class types, and combines under-sampling majority and over sampling minority across the predicted nucleus, lymphocyte and plasma cells. The over sampling minority intentionally further adds samples which are predicted to contain minority instances. IEAL also included minority data augmentation, which flips and hue transforms the minority data to increase the number of minority samples. In this way, IEAL is more likely to select samples that truly contain minority instances to resolve the extreme data imbalance problem.

The IEAL algorithm is embedded into U-Net for the detection of nucleus, lymphocytes and plasma cells. To further reduce the data imbalance problem, a modified dice loss function is used to increase the significance of the minority data. To evaluate the effectiveness of the IEAL algorithm, the detection results are compared with those obtained from U-Net without IEAL. The experimental results show that with less than 50% of the number of annotated samples, U-Net embedded with IEAL achieves an equal or higher performance than U-Net without IEAL.

The main contributions of this paper are as follows. An improved Imbalance-Effective Active Learning (IEAL) algorithm is proposed which achieves an effective annotation of the nucleus, lymphocytes and plasma cells, while simultaneously addressing the extreme class imbalance problem. Notably, the proposed IEAL algorithm can be used with any deep neural network model to mitigate the extreme data imbalance problem while still achieving effective annotation.

2 Method

This section describes the proposed IEAL model, which is used to simultaneously address both annotation loading and extreme data imbalance problem in the detection and classification of nucleus, lymphocyte, and plasma cells in liver pathology images. In this paper, the U-Net is adopted as a multi-class semantic segmentation model to detect and classify the nucleus, lymphocyte and plasma cells. The IEAL algorithm is combined with U-Net and performs two main operations, namely Make Balanced and Minority Sampling, in order to reduce the data imbalance problem. The "Make Balanced" operation applies model uncertainty to query informative samples from each class, thereby achieving the effect of under-sampling the majority samples. The "Minority Sampling" operation then further selects samples which are predicted to be minority-class in order to achieve oversampling of the minority samples. In addition, data augmentation of the minority-class samples is also conducted. Finally, a minority-sensitive loss function is used in conjunction with U-Net to increase the contribution of the minority samples to the final loss. The details of the proposed IEAL algorithm are described in the following sections.

2.1 Imbalance-Effective Active Learning

In the most widely-used active learning scenario, it is assumed that there exists a small set of training data $L = \{(X_l, Y_l)\}_{l=1}^{m}$ that consists of m labeled image patches and a large pool of unlabeled patches $UL = \{X_l\}_{l=m+1}^{n}$. In each consecutive iteration, the trained model is run on the unlabeled dataset UL to select b number of patches for human labeling. Then, the new labeled patches are moved from UL to L for training. The process is repeated iteratively in this way until the performance cannot be improved any further or the update process reaches a preset maximum number of iterations. In contrast to such standard active learning methods, the selection process in IEAL is performed within each predicted class. By so designed, selection of patches are conducted following class pre-estimation. Therefore, boosting on the selection of the minority class is possible and the imbalance will be alleviated. The details of the proposed method are described in the following.

2.2   Make Balanced: Active Candidate Selection

In any AL method, the query strategy is the key to selecting "informative" samples for further training. In the present study, two query strategies are adopted to reduce the total labeling cost and solve the extreme class imbalanced problem. Firstly, *model uncertainty* measured from each class is used to query the training data; thereby achieving a better class balanced. When the query is performed in each individual class, the process can be regarded as under-sampling the majority samples. To increase the data diversity, *randomness* is added to the uncertainty sampling process. Secondly, *minority sampling* and *minority data augmentation* are adopted to oversample the minority samples.

Model Uncertainty Estimation. To perform model uncertainty estimation, this study uses the method proposed by Gal et al. [4]. The predictive distribution is then sampled based on Monte Carlo sampling by using dropout at test time to perform K times stochastic forward passes through the trained network (i.e., MC Dropout). The potential prediction of a pixel x_i can be estimated by the mean of K predictive results, and the variance of the K predictive results is taken as a measure of the model uncertainty of pixel x_i, shown in Eq. (1) and (2) respectively.

$$\widehat{y}_i = E(y_i|x_i) \approx \frac{1}{K}\sum\nolimits_{t=1}^{K} p(y_i|x_i, \widehat{\omega_t}) \tag{1}$$

$$\text{model uncertainty}, U_i = \frac{1}{K}\sum\nolimits_{t=1}^{K} \left(p(y_i|x_i, \widehat{\omega_t}) - \widehat{y}_i\right)^2 \tag{2}$$

In this study, the pixel uncertainty is calculated using Eq. (2). Furthermore, in the practical implementation, training is performed on a 4-class model which includes the background, nucleus, lymphocyte, plasma cells. Thus, the uncertainty computed on class c, $U(c)$ is measured through the summation of the uncertainties of all the pixels predicted as class c, as shown in Eq. (3), where N is the total number of pixels in the patch.

$$\text{Image uncertainty}_{\text{class}}c, \ U(c) = \sum\nolimits_{i=0}^{N} U_i * 1_c(x_i), \quad 1_c(x_i) = \begin{cases} 1, \textit{if } x_i \in c \\ 0, \textit{if } x_i \notin c \end{cases} \tag{3}$$

The $1_c(x_i)$ here the indicator about whether pixel x_i is predicted as class c. The total uncertainty U of an image patch can also be computed through the summation of U(c) of all of the classes. The is, $U = \sum_c U(c)$.

Randomness. **Randomness** is used to increase the selected data diversity and is added to the query strategy of uncertainty sampling. That is, uncertainty sampling querying R oracle samples and then randomly selects r (R > r) samples for labeling in each iteration T. As the learning process continues, the query strategy for the U-Net tends to learn more informative samples. Thus, the randomness should be decreased to focus more on the oracle uncertainty samples. Accordingly, in the present study, the value of R is updated using the following rule: if T = 0, set R equal to an initial value, R_0; else if

$T > 0$, update R in accordance with $R_0 - dr * T$, where dr is used to control the randomness decay rate.

Minority Sampling. Although the Make Balanced step queries the uncertainty samples from each class, it is still very likely that false-positive (FP) minority samples are selected, since the uncertainty samples imply the more unstable cases, which are likely to be misclassified. To compensate for this, the IEAL algorithm further uses minority sampling and minority data augmentation to oversample the minority samples. Minority sampling intentionally adds the samples predicted to contain the minority class, here the plasma cells, instead of the high uncertainty samples since these samples are more likely to include true-positive (TP) samples (i.e., the plasma cells). Thus, minority sampling prevents the training data from having too many FP samples. On the other hand, Minority data augmentation is performed to increase the number of minority samples by horizontal and vertical reflections and transposition. In addition, to take account of slide-wise stain differences and perform wider generalization, hue-based and brightness-based data augmentation are also applied to the digital histopathology dataset. These measures are helpful for learning the invariance. However, in terms of preserving the original tissue features in the digital histopathology images, scaling and affine transformation data augmentation are not recommended.

2.3 Minority Sensitive Dice Loss

In addition to the use of the IEAL algorithm, a minority-sensitive loss function is also employed to increase the contribution of the imbalanced class to the loss during U-Net training. The modified dice loss is computed for each class separately and is then averaged in order to increase the minority class contribution, i.e.,

$$Minority\ Sensitive\ Dice\ loss = \frac{1}{C}\sum_{c=0}^{C}\left(1 - DSC\left(P^{(c)}, G^{(c)}\right)\right) \tag{4}$$

where $DSC\left(P, G\right) = \frac{2\sum_i^N P_i G_i}{\sum_i^N P_i^2 + \sum_i^N G_i^2}$ is the dice coefficient. Note that P and G denote prediction and ground truth, respectively.

2.4 Training

In contrast to standard learning procedures, active learning is a continuous fine-tuning process, where new data are continually added to the training data. Previous research has shown that such a continuous fine-tuning learning process is more robust and more efficient [9]. The training process can be described as follows. Initially, the training data contain only a few labeled samples from each class and these data are used to pretrain the network. The training process then includes two training stages as new data are added. In the early stage, all of the layers are trainable for fine-tuning the new added data. However, in the later stage, associated with a greater amount of training data, the training efficiency is reduced and the performance of the model shows only a slight improvement. Thus, to improve the efficiency, the model focuses only on fine-tuning the parameters of the last two decoder units and the final layer. In other words, the training data focus only on learning the high-level features. Meanwhile, to prevent overfitting, early stopping is used when the validating loss ceases to reduce.

The present study employs a classical encoder-decoder model, namely U-Net, since it shows a good performance for medical images [18, 20–24]. To produce a probabilistic segmentation output, encoder-decoder neural network architectures are generally modified to Bayesian convolutional neural networks (CNNs), in which dropout is added to the middle five encoder/decoder units for training and testing purposes. Figure 1 presents an overview of the training process. During the early update stage, the model parameters are set as follows: Adam optimization with a learning rate of 10^{-3}, batch size 8, and number of training iterations 3–4k. Moreover, R_0 is set to 50, and K_c, K_m are both set to 7. In total, 30 images are added to the training data in each update step. In the later update step, the learning rate is maintained as 10^{-3}, but the number of training iterations is increased to 5–7k and K_n is set to 30.

Input:
$L = \{X_l, Y_l\}$, $l \in [1, n]$: initial training data; $UL = \{X_j\}$, $j \in [m + 1, n]$: unlabeled samples;
K_c, $c \in [class1, class2, class3,]$: the uncertain sample selection size in each class;
K_{mi}: the minority sample selection size;
K_n: the uncertain samples selection size in total class;
R: randomness value; R_0: initial randomness value, dr: decay rate;
Output:
W_T: updated weight at iteration T
Initialize:
W_0: initial trained weight with initial training data L
$T \leftarrow 0$: iteration number
repeat:
    $T = T + 1$
    If in the early updated stage
            $R = R_0 - dr * T$
    1.   Querying R oracle uncertain samples in each class from UL based on U(c) Eq. (3), then randomly select K_c samples from the R samples.
    2.   Random select K_{mi} minority samples from UL.
    3.   Random select 2 samples.
    4.   Move the selected samples $(3 * K_c + K_{mi} + 2)$ from U to the training set L.
        •   Trainable layer = [all layers]
        •   Update W_T via training trainable layer according to Eq. (4) with L.
    else
    1.   select K_n oracle uncertain samples from UL based on *total class uncertainty* U
    2.   Move the selected samples from UL to the training set L.
        •   Trainable layer = [layer 8-10]
        •   Update W_T via fine-tune trainable layer according to Eq. (4) with L.
until stopping criteria
return W_T

Fig. 1. Algorithm of proposed method

3   Experimental Results and Discussion

The dataset used in this study contained 764 images with a size of 1024 × 1024 pixels cropped from 43 whole slide images (WSIs) with hematoxylin-and-eosin (H&E) stains of normal liver tissue. The 764 images were split into an initial training set with 30 images, an evaluation set with 75 images, and an unlabeled set with 659 images. The ground truth (GT) of each image was pixel-accurately labeled as nucleus, lymphocyte, plasma cell, or background. (Note that the GTs were confirmed by a skilled pathologist.) During the training procedure, each image was cut into 9 patches with a size of 512 × 512 using an overlapping stride of 256 pixels. In the procedure of evaluation and uncertainty estimation, the images were cut into 4 patches with a size of 512 × 512 pixels.

The detection performance was evaluated using three criteria, namely the Precision (P), the Recall (R) and the F1 score. The aim of the proposed method was to detect the nucleus, lymphocyte and plasma cells in each patch. Hence, the study separately assessed the detection performance of each class. In particular, for each class, a detected cell centroid point was considered to be true-positive (TP) if the point was located within a 5-pixel radius of the annotated cell centroid; and was considered to be false-positive (FP) otherwise.

3.1   Performance Evaluation Results

To demonstrate the effectiveness of IEAL, the detection performance was compared with that achieved when training with all the dataset (AL_ALL) (i.e., the upper bound performance); classical active learning (AL_UNC); and randomly selected data for labeling (AL_RAND) (i.e., the lower bound performance).

The F1 scores of the various methods for detecting plasma cells and lymphocytes are shown in Figs. 2(a) and (b), respectively. As shown in Fig. 2(a), the proposed IEAL model obviously outperforms the other methods in plasma cell detection. Notably, it achieves the upper bound performance, AL_ALL, using only 50% of the number of annotated samples. The plasma cell detection problem involves an extreme class imbalance problem. Thus, the traditional active learning method (AL_UNC) cannot be effectively improved after 6 update iterations because the samples queried in each iteration may not be precisely useful for the imbalanced class. Without active learning (i.e., AL_RAND), the performance is further degraded.

For the lymphocyte detection problem (see Fig. 2(b), IEAL again achieves the upper bound performance using only 50% of the number of samples. Furthermore, IEAL achieves an F1 score of 92%, a number exceeding that obtained when using all the data for training. In other words, IEAL selects more balanced and informative data for training purposes.

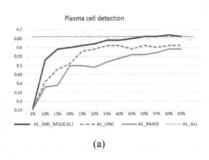
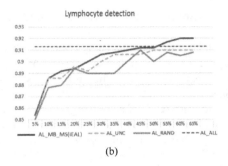

Fig. 2. F1 scores of various methods for (a) plasma cell detection and (b) lymphocyte detection in each iteration

3.2   Make Balanced vs. Minority Sampling

This section compares the detection performance when using *Make Balanced* only (denoted as AL_MB) and IEAL (using both *Make Balanced* and *Minority Sampling,* denoted as AL_MB_MS) to select minority samples. As shown in Fig. 3, AL_MB_MS, which additionally adopts Minority Sampling, achieves a better performance than AL_MB. In the early update stage, AL_MB (without Minority Sampling) has a particularly poor performance since it results in many false positives. However, Minority Sampling is beneficial in discovering a greater number of true positive samples during the early update iterations, and hence the proposed IEAL model (based on both MB and MS) achieves a better performance.

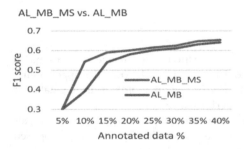

Fig. 3. Comparison of AL_MB_MS (IEAL) and AL_MB

3.3   Performance of High Skew Ratio in Minority Samples

The performance of the various schemes was investigated for different degrees of class imbalance. The corresponding results are presented in Table 1, where the skew number s indicates that the unlabeled dataset has s majority instances for each minority instance. In other words, a larger value of s implies a greater degree of imbalance. Note that in Table 1, for each skew, each query strategy is trained for the best performance and the F1 scores of the minority samples are computed. Furthermore, training is performed 5 times for each method and the mean value is

Table 1. Performance (F1 score) of different query strategies in datasets with different skews

Method	Skew		
	10	20	50
AL_RAND	0.56	0.54	0.46
AL_UNC	0.61	0.58	0.49
AL_MB_MS (IEAL)	0.66	0.63	**0.61**

reported in every case. The results presented in Table 1 show that IEAL significantly outperforms the other strategies at high skews. The performances of AL_UNC and AL_RAND at high skew values of s = 50 are particularly poor.

4 Conclusions

This paper has presented an Imbalance Effective Active Learning (IEAL) algorithm for solving the annotation loading problem and mitigating the effects of class imbalance in automatic nucleus, lymphocyte and plasma detection. Compared to traditional active learning methods, IEAL selects informative samples based on a consideration of each individual class. IEAL also incorporates a minority sampling and minority data augmentation approach to further mitigate the class imbalance problem. The experimental results have shown that the proposed IEAL algorithm significantly outperforms standard active learning schemes in extreme class imbalance problems. In particular, IEAL achieves an equal or better performance than these standard methods using less than 50% of the number of annotations.

References

1. Settles, B.: Active learning literature survey. Department of Computer Sciences, University of Wisconsin-Madison (2009)
2. Scheffer, T., Decomain, C., Wrobel, S.: Active hidden markov models for information extraction. In: Hoffmann, F., Hand, D.J., Adams, N., Fisher, D., Guimaraes, G. (eds.) International Symposium on Intelligent Data Analysis, pp. 309–318. Springer, Heidelberg (2001). https://doi.org/10.1007/3-540-44816-0_31
3. Shannon, C.E.: A mathematical theory of communication. Bell Syst. Tech. J. **27**(3), 379–423 (1948)
4. Gal, Y., Ghahramani, Z.: Dropout as a Bayesian approximation: representing model uncertainty in deep learning. In: International Conference on Machine Learning, pp. 1050–1059 (2016)
5. Ertekin, S., Huang, J., Bottou, L., Giles, L.: Learning on the border: active learning in imbalanced data classification. In: Proceedings of the Sixteenth ACM Conference on Information and Knowledge Management, pp. 127–136. ACM (2007)
6. Fu, C., Qu, W., Yang, Y.: Actively learning from mistakes in class imbalance problems. IFAC Proc. Vol. **46**(13), 341–346 (2013)
7. Zhang, X., Yang, T., Srinivasan, P.: Online asymmetric active learning with imbalanced data. In: SIGKDD (2016)
8. Zhang, Y.: Online adaptive asymmetric active learning for budgeted imbalanced data. In: SIGKDD (2018)
9. Sadafi, A., et al.: Multiclass deep active learning for detecting red blood cell subtypes in brightfield microscopy. In: Shen, D., Liu, T., et al. (eds.) MICCAI 2019. LNCS, vol. 11764, pp. 685–693. Springer, Cham (2019). https://doi.org/10.1007/978-3-030-32239-7_76
10. He, K., Girshick, R., Dollár, P.: Rethinking imagenet pre-training. arXiv preprint arXiv: 1811.08883 (2018)

11. Cao, H., Bernard, S., Heutte, L., Sabourin, R.: Improve the performance of transfer learning without fine-tuning using dissimilarity-based multi-view learning for breast cancer histology images. In: Campilho, A., Karray, F., ter Haar Romeny, B. (eds.) ICIAR 2018. LNCS, vol. 10882, pp. 779–787. Springer, Cham (2018). https://doi.org/10.1007/978-3-319-93000-8_88

12. Zhou, Z., Shin, J., Zhang, L., Gurudu, S., Gotway, M., Liang, J.: Fine-tuning convolutional neural networks for biomedical image analysis: actively and incrementally. In: Proceedings of the IEEE Conference on Computer Vision and Pattern Recognition, pp. 7340–7351 (2017)

13. Wang, K., Zhang, D., Li, Y., Zhang, R., Lin, L.: Cost-effective active learning for deep image classification. IEEE Trans. Circuits Syst. Video Technol. 27(12), 2591–2600 (2016)

14. Gorriz, M., Carlier, A., Faure, E., Giro-i-Nieto, X.: Cost-effective active learning for melanoma segmentation. arXiv preprint arXiv:1711.09168 (2017)

15. Mackowiak, R., Lenz, P., Ghori, O., Diego, F., Lange, O., Rother, C.: Cereals-cost-effective region-based active learning for semantic segmentation. arXiv preprint arXiv:1810.09726 (2018)

16. Yang, L., Zhang, Y., Chen, J., Zhang, S., Chen, D.Z.: Suggestive annotation: a deep active learning framework for biomedical image segmentation. In: Descoteaux, M., Maier-Hein, L., Franz, A., Jannin, P., Collins, D.L., Duchesne, S. (eds.) MICCAI 2017. LNCS, vol. 10435, pp. 399–407. Springer, Cham (2017). https://doi.org/10.1007/978-3-319-66179-7_46

17. Ozdemir, F., Peng, Z., Tanner, C., Fuernstahl, P., Goksel, O.: Active learning for segmentation by optimizing content information for maximal entropy. In: Stoyanov, D., et al. (eds.) DLMIA/ML-CDS-2018. LNCS, vol. 11045, pp. 183–191. Springer, Cham (2018). https://doi.org/10.1007/978-3-030-00889-5_21

18. Zhu, R.X., Seto, W.K., Lai, C.L., Yuen, M.F.: Epidemiology of hepatocellular carcinoma in the Asia-Pacific region. Gut Liver 10(3), 332 (2016)

19. Fridman, W.H., Pages, F., Sautes-Fridman, C., Galon, J.: The immune contexture in human tumours: impact on clinical outcome. Nat. Rev. Cancer 12(4), 298 (2012)

20. Ishak, K., et al.: Histological grading and staging of chronic hepatitis. J. Hepatol. 22(6), 696–699 (2012)

21. https://kknews.cc/zh-tw/health/pbk2xp.html

22. Milletari, F., Navab, N., Ahmadi, S. A. V-Net: fully convolutional neural networks for volumetric medical image segmentation. In: 2016 Fourth International Conference on 3D Vision (3DV), pp. 565–571. IEEE (2016)

23. Oktay, O., et al.: Attention U-Net: learning where to look for the pancreas. arXiv preprint arXiv:1804.03999 (2018)

24. Ronneberger, O., Fischer, P., Brox, T.: U-Net: convolutional networks for biomedical image segmentation. In: Navab, N., Hornegger, J., Wells, William M., Frangi, Alejandro F. (eds.) MICCAI 2015. LNCS, vol. 9351, pp. 234–241. Springer, Cham (2015). https://doi.org/10.1007/978-3-319-24574-4_28

Labeling of Multilingual Breast MRI Reports

Chen-Han Tsai[1]([✉]), Nahum Kiryati[2], Eli Konen[3], Miri Sklair-Levy[3], and Arnaldo Mayer[3]

[1] School of Electrical Engineering, Tel Aviv University, Tel Aviv-Yafo, Israel
maxwelltsai@yahoo.com
[2] The Manuel and Raquel Klachky Chair of Image Processing, School of Electrical Engineering, Tel-Aviv University, Tel Aviv-Yafo, Israel
[3] Diagnostic Imaging, Sheba Medical Center, Affiliated to the Sackler School of Medicine, Tel-Aviv University, Tel Aviv-Yafo, Israel

Abstract. Medical reports are an essential medium in recording a patient's condition throughout a clinical trial. They contain valuable information that can be extracted to generate a large labeled dataset needed for the development of clinical tools. However, the majority of medical reports are stored in an unregularized format, and a trained human annotator (typically a doctor) must manually assess and label each case, resulting in an expensive and time consuming procedure. In this work, we present a framework for developing a multilingual breast MRI report classifier using a custom-built language representation called LAMBR. Our proposed method overcomes practical challenges faced in clinical settings, and we demonstrate improved performance in extracting labels from medical reports when compared with conventional approaches.

Keywords: Labeling · Medical reports · Transfer learning · Breast MRI · LAMBR

1 Introduction

The introduction of the Electronic Medical Record (EMR) has improved convenience in accessing and organizing medical reports. With the increasing demand for biomedical tools based on deep learning, obtaining large volumes of labeled data is essential for training an effective model. One major category where such deep learning models excel is in the area of computer assisted diagnosis (CADx), and several works (e.g. [1, 4, 12]) have demonstrated effective utilization of weakly labeled data to achieve promising performance. Since understanding medical data requires specialized training, datasets often contain a small subset of all past exams, that are manually relabeled by doctors for the target task. Not only is this a labour-intensive process, but the resulting dataset is often too small to represent the true distribution, resulting in underperforming models.

© Springer Nature Switzerland AG 2020
J. Cardoso et al. (Eds.): iMIMIC 2020/MIL3ID 2020/LABELS 2020, LNCS 12446, pp. 233–241, 2020.
https://doi.org/10.1007/978-3-030-61166-8_25

In this work, we present a framework for developing multilingual breast MRI report classifiers by using a customized language representation called LAMBR. LAMBR is first obtained by pre-training an existing language representation on a large quantity of breast MRI reports. Fine-tuning is then applied to obtain separate classifiers that can perform tasks such as: (1) determining whether the corresponding patient in the report has been suggested to undergo biopsy or (2) predicting BI-RADS[1] score for the reported lesion (see Fig. 1). With such classifiers, one may avoid the manual labeling required from doctors, and instead, automatically extract a large number of weak labels from existing medical reports for training weakly supervised breast MRI CADx models.

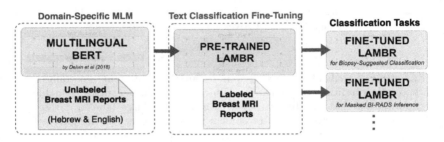

Fig. 1. An overview of training stages presented in our framework. Pre-training is performed on the multilingual BERT with unlabeled breast MRI reports to obtain a pre-trained LAMBR. The pre-trained LAMBR is then fine-tuned using a small number of labeled reports to obtain classifiers for specific downstream text classification tasks.

Prior to our work, text classification has been explored extensively by several studies such as ULMFIT [5] and SiATL [2]. ELMo [9], BERT [3], and XLNet [16] have also demonstrated adequate approaches towards text classification using the notion of a generalized language representation. However, the majority of these approaches require pre-training an encoder on a massive text corpora, and this is a time consuming and resource intensive procedure that is impractical for a clinical setting [10]. Moreover, the majority of prior works perform encoder pre-training on widely available natural language texts which differ greatly from the scarcely available medical texts.

To overcome the difference in distribution between medical texts and natural texts, BioBERT [7] introduced a pre-training objective that relied on a large collection of PubMed abstracts and PMC articles. Although BioBERT demonstrated improved performance compared with BERT, their method does not avoid the above resource intensive pre-training. Within English medical reports, ALARM [14] proposed a simple approach for labeling head MRI reports by utilizing a pre-trained Bio-BERT and this avoids the expensive pre-training often required. Yet for multilingual medical reports, such Bio-BERT does not exist,

---

[1] Breast Imaging-Reporting and Data System: a score between 0–6 indicating the level of severity of a breast lesion.

and in this work, we present a solution based on the multilingual BERT. Our novel approach introduces an inexpensive pre-training objective that yields favorable text classification performance when fine-tuned, and in our experiments, we demonstrate the robustness of the resulting classifiers even in cases where parsing errors exist (see Fig. 2). The remaining sections of our paper are organized as follows: the proposed framework is presented in Sect. 2, experimental results are reported in Sect. 3, and the conclusion is given in Sect. 4.

Example 1: A Complete Report

MRI שדיים שד ימין- קרצינומה Triple pos. להערכה של היקף. בוצעה סריקה של שני השדיים. ממצאים: הודגמה רקמת שדיים צפופה וציסטות פזורות בשני הצדדים. לאחר מתן ח"נ - מוקדי צביעה פזורים בשני השדיים, צביעת רקע ניכרת (2) BPE). שד ימין- רביע פנימי אמצעי- 1(1*1.7 ס"מ. 2)במרחק 0.5 ס"מ מדיאלית, גוש 0.8 ס"מ. 3)0.7 ס"מ 1/3) אחורי 73.45- POS), סה"כ 3 הגושים לאורך 3 ס"מ. רביע חיצוני אמצעי- צביעה בפיזור סגמנטלי 2*6 ס"מ 1/3) קדמי אמצעי ואחורי int.mammary 0.5 בלוטה). POS -39.45 ס"מ מימני 15.45) POS). שד שמאל- לא נראו בברור מוקדי צביעה חשודים. בתי שחי- לא נראו בלוטות לימפה מוגדלות. לסיכום: שד ימין- רביע פנימי- גידול רב מוקדי, רביע חיצוני- מומלץ ביופסיה בהנחיית BIRADS 6 MRI.

Label: Recommended for biopsy, BIRADS score of 6

Example 2: An Incomplete Report (missing assessments)

MRI שדיים סיבת ההפניה: סיפור משפחתי של סרטן שד. 2012 קרצינומה של שד שמאל, למפקטומיה. 4/2019 ביופסיה בהנחיית MRI משד ימין, Fibrocystic changes with foci of apocrine papillary metaplasia and usual ductal hyperplasia.. ביקורת. הודגמה רקמת שדיים פיברוגלנדזלרית בעיקרה. לאחר הזרקת חומר ניגוד מודגמת האדרת רקע קלה 1-BPE). שד ימין - ללא האדרה גושית או האדרה לא גושית. שד שמאל - ללא האדרה גושית או האדרה לא גושית. ללא הגדלת בלוטות בבתי השחי.

Label: Biopsy not required, BIRADS score of 1

Fig. 2. Examples of breast MRI reports written in Hebrew and English (read from right to left). Example 1 is complete report parsed from the EMR, and Example 2 is missing the final assessments due to incorrect parsing. The patient in Example 1 is recommended for biopsy, and patient in Example 2 is not required to perform biopsy.

2 Methods

2.1 BERT Recap

BERT is a language representation based on the Transformer-Encoder [13]. The input to the Transformer-Encoder is a sequence of tokens $\{x_i\}$ generated by WordPiece Embeddings [15] from a given series of sentences. Special tokens are inserted and position encodings are added, and the output is a sequence of bi-directional embeddings that represents each input token [3]. In order to obtain the BERT language representation, Masked Language Modeling (MLM) and Next Sentence Prediction (NSP) pre-training objectives were introduced. MLM applies random masking on the 15% of the input tokens, and BERT is trained to identify the original token of the masked token by attending to other tokens of the same sequence. The NSP objective trains BERT in understanding sentence coherence by randomly replacing the second sentence of an existing sentence pair, and BERT has to determine whether the pair are neighboring sentences.

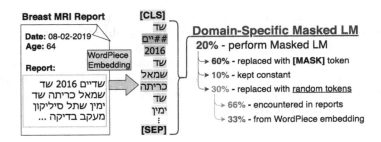

Fig. 3. An example of the tokens generated from breast MRI report using WordPiece Embeddings. During DS-MLM, a portion of the tokens are augmented and the pre-training objective is to correctly identify the original token prior to augmentation (performed only on tokens selected for augmentation).

2.2 Domain-Specific Masked Language Modeling

The Domain-Specific Masked Language Modeling (DS-MLM) we propose is a modification of the MLM pre-training objective introduced in BERT. The multilingual BERT was trained using monolingual corpora from 104 languages, and DS-MLM aims to retrain the multilingual BERT to better model the language observed in breast MRI reports written in Hebrew and English. Unlike BioBert [7], which relies on pre-training over massive biomedical corpora, we perform DS-MLM solely from the available breast MRI reports stored in the hospital's EMR.

For each medical report, tokens are generated using WordPiece Embeddings (see Fig. 3). The [CLS] and [SEP] tokens are appended to the beginning and the end of the generated tokens ([SEP] tokens are not added between sentences). Since the multilingual BERT is already trained on a general domain corpora, we select 20% of the generated tokens for MLM. Of the selected tokens, 60% are masked using the [MASK] token, 30% are replaced with existing tokens and the remaining 10% are left unchanged. In order to expose our model to more frequent tokens observed in breast MRI reports, of the 30% of tokens selected for replacement, two thirds are replaced with existing tokens encountered in breast MRI reports, and one third is replaced with tokens from the complete vocabulary (may include tokens corresponding to other languages). Dynamic masking is applied to allow more exposure to a broad range of tokens.

Since most medical reports contain sentences not adhering to a strict flow of ideas, we do not incorporate NSP into the pre-training objective of our framework. In addition, RoBERTa [8] demonstrated that the removal of NSP may even improve downstream task performance, and therefore, the pre-training objective of the LAMBR language representation is simply DS-MLM.

2.3 Text Classification Fine-Tuning

Text Classification Fine-Tuning (TCFT) is a series of techniques to fine-tune a pre-trained LAMBR for performing text classification. We propose a simple classifier head to add on top of the Transformer-Encoder, and we present a method to fine-tune the complete text classifier (Transformer-Encoder along with classifier head) using a pre-trained LAMBR (see Fig. 4).

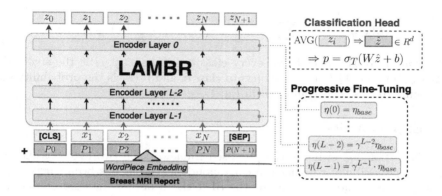

Fig. 4. An illustration of LAMBR encoding a breast MRI report. Tokens $\{x_i\}$ are generated from the report using WordPiece Embeddings, and the sum of the positional encoding and the token embedding are input to LAMBR. During TCFT, the learning rates for each layer are progressively tuned so that higher level features (layers near 0) are updated more compared to lower-level features. The classifier head takes the average of the token embeddings $\hat{z}$, applies an affine transform, and passes it into a tempered Softmax (σ_T) to generate the class probabilities.

Classifier Head. For a given token sequence $\{x_i\}$, we obtain the output embedding sequence $\{z_i\}$ from the pre-trained LAMBR. The average of the output token embeddings $\hat{z}$ is computed and is fed to an affine layer which undergoes a Tempered Softmax operation (σ_T) to obtain the outputs class probabilities p. Namely:

$$p = \sigma_T(W\hat{z} + b) \quad \leftarrow \quad \hat{z} = \frac{1}{N}\sum_{i=1}^{N} z_i \qquad (1)$$

where $\hat{z}, b \in R^d$, $W \in R^{c \times d}$, and $p \in R^K$.

Progressive Fine-Tuning. Inspired by [5,17], we propose a method for fine-tuning the complete text classifier. The learning rates are adjusted such that high-level features will be updated with a higher learning rate compared to lower level features. Specifically, for a Transformer-Encoder with L encoding layers $\{l_i\}_{i=0}^{L-1}$ (l_0 indicates the top-most layer), the layer-dependent learning rate $\eta(l)$ is formulated as:

$$\eta(l) = \eta_{base} \cdot \gamma^l \qquad (2)$$

where η_{base} is the base learning rate and γ is the decay factor valued between 0 and 1. Similarly, the classifier head is updated with learning rate η_{base}.

Fine-tuning is performed by optimizing the weighted Label Smoothing Loss [11]:

$$L(x, y(x)) = -\sum_{c=1}^{K} w(c) \cdot \left[(1 - \epsilon)y_c(x) + \frac{\epsilon}{K} \right] \cdot \log(p_c(x)) \tag{3}$$

where $w(c)$ are the weights for every class $c \in K$, $\epsilon \in [0, 1)$ is the smoothing term, $y_c(x) \in \{0, 1\}$ is 1 if x belongs to class c, and $p_c(x)$ is the probability of x belonging to class c as computed in Eq. 1.

3 Experiments

In this section, we evaluate the proposed framework on two text classification tasks: (1) classifying whether the corresponding patient has been suggested to undergo biopsy and (2) predicting the BI-RADS score for the lesion reported.

The data is a curated list of medical reports from breast MRI examinations carried out at the Sheba Medical Center, Israel. Cases that were initially diagnosed as containing potential malignant tumors have all been suggested to undergo biopsy. Breast examinations from the years 2016–2019 were involved, and a total of 10,529 medical reports were collected. Of the 10,529 breast MRI reports, 541 reports were labeled with the relevant BI-RADS score for the (single) lesion reported, and each case was labeled with whether the patient had been suggested for biopsy.

3.1 Training Setup

Pre-training. Pre-training was performed using DS-MLM as mentioned in Sect. 2.2. Of the 9,988 reports used for pre-training, 85% of the reports were randomly designated as the training set, and the remaining for validation. Cross Entropy loss was used for DS-MLM pre-training, and the multilingual BERT was trained for 70 epochs which took approximately 33 h to complete using on an NVIDIA GTX 1070 8 GB GPU.

Biopsy-Suggested Classification. The goal of this task to identify whether the patient in the report had been suggested to undergo biopsy or not. We perform fine-tuning as proposed in Sect. 2.3. Due to dataset imbalance (26.6% of the cases were suggested for biopsy), class weights were set to the inverse of the counts per class. Evaluation was performed using 5-fold cross validation, and stratified sampling was applied to ensure equal class distribution between the training and validation sets.

Training was performed using the Adam optimizer [6] with a base learning rate of 1e−4 and a batch size of 8. Decay factor γ was set to 1/4, softmax temperature T was set to 1, and the smoothing term ϵ was set to 0. The best performing model from training for 70 epochs (approx 25 min) was evaluated.

Masked BI-RADS Prediction. Masked BI-RADS Prediction is a classification task to assign the appropriate BI-RADS score given the lesion description in the report. For reports that were parsed correctly, the BI-RADS score is written in the assessments, and a simple keyword-based tagging is often enough to label the reports with the appropriate score. However, reports might also contain BI-RADS keywords that refer to previous BI-RADS scores (for the same lesion or removed lesion), which would lead to incorrect inference if the keyword based approach was used. In addition, errors encountered during parsing would occasionally miss out sections containing the BI-RADS score, rendering the keyword-based approach useless. The text classifier we propose in our framework relies on the report descriptions alone, and is thus robust against such potential obstacles.

Keyword search was performed on all the reports and any revealing BI-RADS scores (in the reports) were removed. This modified report was then fed into a pre-trained LAMBR for fine-tuning, and a 5-fold cross validation was performed. There were a total of 6 classes (no reports with BI-RADS score 5). Class weights were computed as the inverse of the class counts, and stratified sampling was performed to ensure equal class distribution between training and validation sets. Optimization was performed using the Adam optimizer with a base learning rate of $1e-4$ and batch size of 8. The decay factor γ was set to $1/3$, Softmax temperature T was set to $\sqrt{2}$, and the smoothing term ϵ was set to $1/3$. The best performing model over a training period of 70 epochs was selected for evaluation.

Table 1. Several metrics following the five-fold cross validation for Biopsy Suggested Classification and Masked BI-RADS Prediction are presented. We compare the performance between LAMBR and BERT for both classification tasks (same classification head design, but different language representations). The baseline for Biopsy Suggested Classification is a keyword matching algorithm. Notice that in both tasks, LAMBR consistently outperforms their counterparts.

Fine-tuning tasks (5-fold average)	Accuracy	ROC-AUC	Macro Avg F1	MCC
Biopsy suggested - LAMBR	0.965	0.989	0.935	0.913
Biopsy suggested - BERT	0.949	0.987	0.904	0.8733
Biopsy suggested - Baseline	0.828	–	0.718	0.6035
BI-RADS prediction - LAMBR	0.8576	–	0.7158	0.7594
BI-RADS prediction - BERT	0.795	–	0.572	0.672

3.2 Experimental Results

The experimental results for our proposed framework are listed in Table 1, and we include a comparison with BERT and a baseline algorithm.

In Biopsy Suggested Classification, the keyword matching algorithm aims to label each report in accordance with keywords that hint of a potential biopsy suggestion. Of the 541 labeled reports, 90 reports were misparsed, which contributes to a 16% drop in accuracy. In contrast, the classifier trained and fine-tuned using our proposed framework performs consistently across all five folds (see Fig. 5) despite misparsed reports. We also trained a classifier with the same classification head from Sect. 2.3 using a multi-lingual BERT, and we demonstrate a better classification performance with our approach.

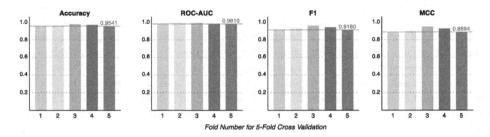

Fig. 5. Detailed visualization of the evaluation metrics for Biopsy Suggested Classification following the 5-fold cross validation. Notice the consistent performance across a 5 folds.

In the task of Masked BI-RADS Prediction, the classifier trained using our framework was able to correctly predict the BI-RADS score for most of the reports. Unlike the previous task where the BI-RADS score was available, this task requires the classifier to attend to relevant context clues in the medical report for prediction (hence, the keyword-tagging algorithm does not work). An additional comparison was made between LAMBR and the pre-trained multi-lingual BERT, and the results in Table 1 demonstrate a clear difference the two language representations partake in training a BI-RADS classifier.

4   Conclusion

In this work, we explore the task of labeling breast MRI reports written primarily in Hebrew with occasional English texts through the use of multilingual language representations. To avoid the expensive pre-training required in obtaining a generalized language representation, the Domain-Specific Masked Language Modeling objective pre-trains a multilingual BERT on existing breast MRI reports alone to obtain the LAMBR language representation. A simple classification head is integrated onto the Transformer-Encoder, and Progressive Fine-Tuning is applied to train the classifier for its specific text classification task.

In our experiments, we train two separate classifiers to perform two classification tasks based on breast MRI reports. In the first task, we trained a classifier to determine whether the patient described in the report has been suggested to undergo biopsy. When compared with past methods, our approach demonstrates better classification performance despite parsing errors in a portion of the reports. In the second task, we trained a classifier to predict the BI-RADS score based on the lesion description in the report. Despite the absence of the BI-RADS score in the report, our classifier was able to infer the correct BI-RADS score in the majority of the cases.

Future works may include labeling medical reports for additional pathologies written in different languages. Additional tasks (apart from text classification) such as named-entity recognition for medical reports and summary generation for biomedical texts may also be investigated. In this work, we focus on the task of medical text classification, and we believe our proposed framework may assist in generating large numbers of labels for weakly supervised training tasks required for breast MRI CADx development.

References

1. Bien, N., et al.: Deep-learning-assisted diagnosis for knee magnetic resonance imaging: development and retrospective validation of MRNet. Plos Med. **15**(11), e1002699 (2018). https://doi.org/10.1371/journal.pmed.1002699
2. Chronopoulou, A., Baziotis, C., Potamianos, A.: An embarrassingly simple approach for transfer learning from pretrained language models (2019)
3. Devlin, J., Chang, M.W., Lee, K., Toutanova, K.: Bert: pre-training of deep bidirectional transformers for language understanding (2018)
4. Gozes, O., et al.: Rapid AI development cycle for the coronavirus (covid-19) pandemic: initial results for automated detection & patient monitoring using deep learning CT image analysis (2020)
5. Howard, J., Ruder, S.: Universal language model fine-tuning for text classification (2018)
6. Kingma, D.P., Ba, J.: Adam: a method for stochastic optimization (2014)
7. Lee, J., et al.: BioBERT: a pre-trained biomedical language representation model for biomedical text mining. Bioinformatics **36**, 1234–1240 (2019). https://doi.org/10.1093/bioinformatics/btz682
8. Liu, Y., et al.: Roberta: a robustly optimized BERT pretraining approach (2019)
9. Peters, M.E., et al.: Deep contextualized word representations. CoRR abs/1802.05365 (2018)
10. Sharir, O., Peleg, B., Shoham, Y.: The cost of training NLP models: a concise overview. ArXiv abs/2004.08900 (2020)
11. Szegedy, C., Vanhoucke, V., Ioffe, S., Shlens, J., Wojna, Z.: Rethinking the inception architecture for computer vision (2015)
12. Tsai, C.H., Kiryati, N., Konen, E., Eshed, I., Mayer, A.: Knee injury detection using MRI with efficiently-layered network (ELNet). ArXiv abs/2005.02706 (2020)
13. Vaswani, A., et al.: Attention is all you need (2017)
14. Wood, D.A., et al.: Automated labelling using an attention model for radiology reports of MRI scans (ALARM) (2020)
15. Wu, Y., et al.: Google's neural machine translation system: bridging the gap between human and machine translation (2016)
16. Yang, Z., Dai, Z., Yang, Y., Carbonell, J., Salakhutdinov, R., Le, Q.V.: XLNet: generalized autoregressive pretraining for language understanding (2019)
17. Yosinski, J., Clune, J., Bengio, Y., Lipson, H.: How transferable are features in deep neural networks? (2014)

Predicting Scores of Medical Imaging Segmentation Methods with Meta-learning

Tom van Sonsbeek$^{(\boxtimes)}$ and Veronika Cheplygina

Eindhoven University of Technology, Eindhoven, The Netherlands
t.j.v.sonsbeek@gmail.com, v.cheplygina@tue.nl

Abstract. Deep learning has led to state-of-the-art results for many medical imaging tasks, such as segmentation of different anatomical structures. With the increased numbers of deep learning publications and openly available code, the approach to choosing a model for a new task becomes more complicated, while time and (computational) resources are limited. A possible solution to choosing a model efficiently is meta-learning, a learning method in which prior performance of a model is used to predict the performance for new tasks. We investigate meta-learning for segmentation across ten datasets of different organs and modalities. We propose four ways to represent each dataset by meta-features: one based on statistical features of the images and three are based on deep learning features. We use support vector regression and deep neural networks to learn the relationship between the meta-features and prior model performance. On three external test datasets these methods give Dice scores within 0.10 of the true performance. These results demonstrate the potential of meta-learning in medical imaging.

Keywords: Meta-learning · Segmentation · Feature extraction

1 Introduction

Deep learning algorithms have become state-of-the-art methods in numerous medical image analysis tasks [13] and have shown to outperform experts on many tasks [14]. Different models have been developed, such as various extensions of convolutional neural networks, recurrent neural networks, and generative adversarial networks, with their respective strengths. Since no model can perform the best on all problems [8], for new datasets still new models are being developed. This has led to a dramatic increase of literature: every day around 30 new papers in this field of study are published. There is therefore a need to generalize from all this experience, when selecting a good model for a new medical imaging problem. We propose to do this using meta-learning, a learning method in which prior performance of a model is used to predict the performance for new tasks [12,25,27].

© Springer Nature Switzerland AG 2020
J. Cardoso et al. (Eds.): iMIMIC 2020/MIL3ID 2020/LABELS 2020, LNCS 12446, pp. 242–253, 2020.
https://doi.org/10.1007/978-3-030-61166-8_26

Currently learning from previous experience is largely done through transfer learning. It is possible to outperform training from scratch by transferring model weights or re-using a model for a different task, as shown by Tajbakhsh et al. [24] and Shin et al. [20]. However, the quality of that newly created model can only be assessed after training and evaluation, which is costly, both in terms of time and resources. Meta-learning, which has mainly been studied in machine learning field, offers a potential solution. However, it is largely unknown in medical imaging. A Google Scholar search[1] on "medical imaging" and "deep learning" shows that only 270 papers - less than 1% - are also related to meta-learning. This is possibly due to the complexity of the data or limited differences between datasets. Another reason is data availability - although datasets and models performances are increasingly being shared online, it is only a recent development that challenges focus on multiple applications, for example [21].

We propose to use meta-learning to predict segmentation scores across ten datasets of different organs and modalities. We propose four ways to represent each dataset by meta-features: one based on statistical features of the images and three are based on deep learning features. We use support vector regression and a deep neural network to learn the relationship between the meta-features and prior model performance.

1.1   Related Work

A common application of meta-learning in computer vision is prediction ranking between methods [6,19,23]. An extension to this is predicting the result of a model. Guerra et al. predicted the outcome of multi-layer perceptron networks using regression models as a meta-learner. This was achieved using compressed representations of datasets, called meta-features. The regression model learns a relationship between the metafeature and prior performance information. Similar approaches were followed by Doan et al. [5] for predicted running time of algorithms and Soares et al. for predicting the outcome of clustering algorithms [23]. Gomes et al. [7] and Soares et al. [22] used meta-learning to predict parameter settings for support vector machines.

Meta-learning has been applied to a small number of problems in medical imaging. Campos et al. used meta-learning to predict segmentation scores of photos of wounds [3]. While this was not a typical medical dataset, it showed the possibilities of meta-learning in the medical domain. Hu et al. created a meta-learning method which initialised weights for finetuning of classification methods in medical imaging [9], thus reducing the need for data. Cheplygina et al. characterized medical image segmentation problems in meta-feature space defined by performances of classical classifiers [4], but only predicted whether datasets originate from the same source. To the best of our knowledge no attempts have yet been made to recommend models by predicting the performance of typical segmentation problems in medical imaging by using meta-learning.

---

[1] Search done in March 2020.

2   Methods

We assume we are given a collection of datasets $\{D_i\}_1^N$, for example segmentation tasks of different organs and modalities. We also assume we have a collection of models $\{U_j\}_1^M$, for example various U-Net-type architectures, and results $\{y_{ij}\}_{i=1,j=1}^{N,M}$ of these models on the datasets. The challenge is, given a previously unseen dataset D_{N+1}, to predict the model scores $\{y_{N+1,j}\}_{j=1}^{M}$.

The overall method is illustrated in Fig. 1. First a meta-feature extractor f summarizes a subset s of each dataset, such that $\mathbf{x}_i = f(D_i)$ is a q-dimensional feature vector. Using a fixed subset size ensures invariance to dataset size. Then for the j-th model, a classifier g_j is trained on the meta-feature vectors $\mathbf{x}_i$ and the meta-labels y_{ij}. Predicting the model's score for an unseen dataset is done via $\hat{y} = g_j(f(D_{N+1}))$

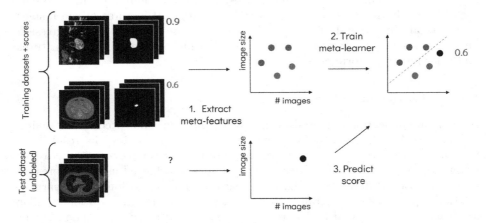

Fig. 1. Proposed meta-learning method: training is done on extracted meta-features and meta-labels (scores of segmentation algorithms). At test time, the trained meta-learner can predict scores for a previously unseen dataset. For simplicity here we illustrate a classification problem, but throughout the paper regression is used.

2.1   Meta-feature Extraction

We investigate three broad types of meta-features, described in more detail below:

- Classical meta-features, similar to meta-features used in prior work.
- Deep learning based meta-features with three different architectures: VGG16, ResNet50, MobileNetV1.
- Task-specific meta-features, which provide context about a given segmentation problem; added to both the classical and deep learning meta-features.

Classical Meta-features. We used meta-features from classical (non-imaging) applications of meta-learning. [3,16,25]. Typical examples are mean pixel value, dataset correlation and entropy. These meta-features will be referred to as classical meta-features. A selection requirement for each classical meta-feature is that it should be visually different between datasets. To check whether this criteria has been met a visual inspection of each meta feature will be done. This selection led to meta-features $\mathbf{x}_i = f_{CLAS}(D_i)$ with $\mathbf{x}_i \in \mathbb{R}^{33}$ is available in the Supplementary Material.

Deep Learning Meta-features. Deep learning is successful for feature extraction at an image level, we therefore also investigated how it can be applied to extract features at a dataset level. Initial attempts using networks pretrained on ImageNet without fine-tuning did not lead to distinctive features for datasets of the same modality. Therefore we added a fine-tuning step with a U-Net-like encoder-decoder network, where the encoder is the feature extractor and the decoder is the original U-Net [18]. This encoder-decoder network works as a binary segmentation network that fine-tunes the weights of the feature extractor (Fig. 2).

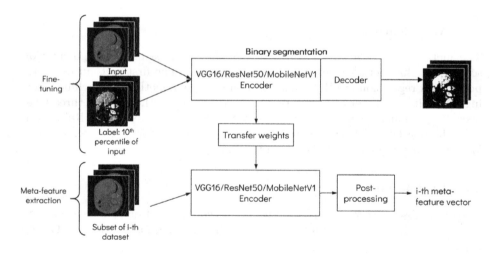

Fig. 2. Deep learning based meta-feature extraction.

An important property of the fine-tuning step is that labels from the test data cannot be used, since the meta-learning method should be able to generalize to datasets without ground-truth segmentations. Instead, we introduce an auxiliary task by thresholding the voxels intensities at the 10-th percentile to create rough segmentation masks. These masks are not accurate in terms of segmentation, but provide a good enough estimation of the image structure.

Using this strategy, we fine-tune three different models, pretrained on ImageNet: VGG16, ResNet50, and MobileNetV1. An intermediate step of these meta-features $x_i = f_{DL}(D_i))$ with $\mathbf{x}_i \in \mathbb{R}^{(z,7,7)}$, $z \in \{512, 2048, 1024\}$ consist of the output of the last layers of these models, averaged over the number of images on which the meta-feature is computed. The 7×7 feature maps are binarized yielding final meta-features $\mathbf{x}_i \in \mathbb{R}^z$. This binarization is done by thresholding using computing feature map correlation across datasets during training time, with an empirically determined threshold: $\alpha = 0.80$. To improve the meta-feature quality univariate feature selection is applied, where the optimal selection threshold is based on SVM classifier weights during training time.

Task-Specific Meta-features. Additionally, we supplement both classical and deep learning meta-features above with task-specific meta-features which capture basic properties of the datasets, such as the modality. These meta-features could be queried from the user, or detected with simple classifiers. For simplicity, here we have set the following features between 0 and 1 based on exploratory analysis of each dataset: imaging modality, whether the segmentation is location-dependent, how sphere-shaped is the segmentation, relative size of the segmentation, and presence of multiple segmentation objects.

2.2   Meta-learner

The meta-learner is a model which relates the meta-features to segmentation scores. For this meta-learner two methods are used. The first method uses support vector regression (SVR) [15], a common regression method in machine learning. Default parameter settings are used. The second method uses a three layer deep fully connected multi-layer perceptron network (DNN), with ReLu activated hidden layers of sizes 50 and 30. A dropout rate of 50% is used. The last layer is sigmoid activated to result in the final prediction.

2.3   Evaluation

The mean absolute error (MAE) is used as the scoring function. This is a common metric in similar meta-learning methods which use regression methods [7,17], see Eq. 1:

$$MAE = \frac{1}{n} \sum_{i=1}^{n} |y_i - \hat{y}_i|. \tag{1}$$

Furthermore to assess whether the meta-learner is not simply predicting the same score for every dataset (prediction towards the mean), we use the normalized mean absolute error (NMAE) [22]. A NMAE (Eq. 2) score of 1 means performance is equal to always predicting the mean performance of the training datasets. NMAE values higher than 1 mean that the meta-learners performs worse than the mean performance prediction. Values lower than 1 are desired.

Using this metric meta-learners on different problems can be compared.

$$NMAE = \frac{\sum_{i=1}^{n} |y_i - \hat{y}_i|}{\sum_{i=1}^{n} |y_i - \bar{y}_i|}. \tag{2}$$

3   Experiments

For the first part of the experiments, we use data from the Medical Segmentation Decathlon (MSD) challenge [21]. The goal of this challenge was to develop a model which could, after a fine-tuning step, segment several distinct segmentation problems. Ten datasets with varying anatomical regions and imaging modalities (CT and MR) were included. We used these datasets, and performances of challenge participants, for our meta-learning method. Additionally, we used performances of the winning participant [10] on three public datasets, as held-out test datasets: LiTS (liver CT) [2], ACDC (heart MR) [1], CHAOS (liver CT) [11].

3.1   Meta-feature Generation

A meta-feature vector is based on a subset of $s = 20$ images, sampled from the dataset. A total of 100 subsets, and thus meta-feature vectors are sampled from each dataset. We first examined the quality of the different meta-feature types using the t-stochastic nearest neighbor (t-SNE) embeddings for the MSD datasets. The results are shown in Fig. 3. The embeddings show that all meta-features are able to separate the datasets well, but the deep learning meta-features provide more well-defined separation.

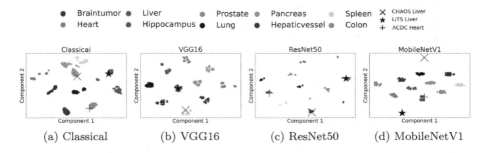

(a) Classical   (b) VGG16   (c) ResNet50   (d) MobileNetV1

Fig. 3. t-SNE embeddings of meta-features from MSD datasets and test datasets.

3.2   Cross-validation MSD

We then performed experiments with the MSD datasets to determine the performance of different meta-features and meta-learners. We used cross-validation with 7 datasets for training and 3 datasets for testing.

Table 1. Cross-validation result of SVR and DNN meta-learners on MSD data for four different meta-features. Bold = best result per column.

↓ Feature extractor	MAE		NMAE	
	SVR	DNN	SVR	DNN
Task-specific only	0.22 ± 0.13	0.26 ± 0.12	1.13 ± 0.66	1.33 ± 0.61
Statistical	0.21 ± 0.08	0.24 ± 0.10	1.07 ± 0.46	1.19 ± 0.51
VGG16	0.13 ± 0.09	0.14 ± 0.04	0.65 ± 0.46	0.70 ± 0.21
ResNet50	$\mathbf{0.12 \pm 0.07}$	$\mathbf{0.12 \pm 0.03}$	$\mathbf{0.62 \pm 0.36}$	$\mathbf{0.62 \pm 0.15}$
MobileNetV1	0.15 ± 0.09	$0.14 \pm \mathbf{0.03}$	0.76 ± 0.46	$0.70 \pm \mathbf{0.15}$

We show the performances of the different combinations in Table 1. Consistent with the t-SNE embeddings, we see that the deep learning meta-features lead to lower errors than the classical meta-features. Out of the deep learning features, ResNet50 leads to the best results. Furthermore, we see that the SVR and DNN meta-learners perform on par with each other. In general, the lower the intra-variability of segmentation scores within a dataset, the higher the predictive accuracy. Results for individual datasets and challenge participants can be found in the Supplementary Material.

To further examine the behavior of different methods, we plot the predicted Dice scores against the true Dice scores in Fig. 4. Here we can see that the overall correlation is positive, but for some datasets the predictions are better than for others. Datasets with "average" scores consistently yield low prediction errors.

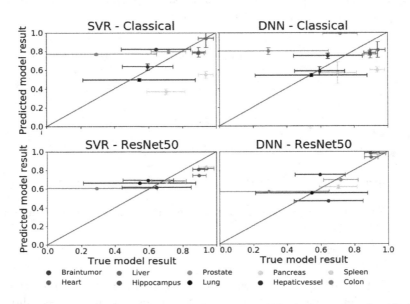

Fig. 4. Examples of results of cross-validation on MSD datasets for classifical and ResNet50 meta-features, and SVR and DNN meta-learners.

Table 2. MAE scores of SVR and DNN meta-learners on test datasets for different types of metafeatures.

↓ Feature extractor	Liver (LiTS)		Heart (ACDC)		Liver (CHAOS)		Mean MAE		Mean NMAE	
	SVR	DNN	SVR	DNN	SVR	DNN	SVR	DNN	SVR	DNN
Task-specific only	0.10	0.46	0.17	0.42	0.02	0.44	0.10	0.44	0.61	2.72
Statistical	**0.09**	0.14	0.16	0.16	**0.01**	0.04	**0.09**	0.11	**0.54**	0.71
VGG16	0.20	0.13	0.22	0.05	0.15	**0.01**	0.19	**0.06**	1.19	**0.40**
ResNet50	0.23	0.50	0.23	**0.02**	0.15	0.06	0.20	0.19	1.27	1.21
MobileNetV1	0.14	**0.01**	**0.14**	0.13	0.07	0.07	0.14	0.07	0.88	0.44

3.3   Held-Out Test Data

We then do a similar experiment as before, but instead of cross-validation on 10 datasets, we train the meta-learners on the MSD data, and test them on three held-out datasets.

The MAE results are shown in Table 2. Prediction results can be found in the Supplementary Material. Comparing the meta-features, we see that the classical meta-features are best for two out of three datasets when SVR is used, and the three deep learning features are best once when DNN is used. Averaging the results across the datasets, the DNN meta-learner has the lowest error.

4   Discussion

We investigated whether meta-learners can predict the performance of segmentation algorithms, based on various meta-feature representations of datasets. We found that the predicted Dice scores are within a 0.10 of the true results, which is a promising result. While such a method would not help between distinguishing among the top few methods for a particular segmentation problem, it could eliminate some alternatives that are not suitable.

The proposed study still has some limitations. One issue is that the datasets are quite sparse, with a low number and large differences between datasets. We would recommend including more datasets which share either task and/or modality with the existing datasets.

Furthermore, our method assumes all segmentation methods under consideration have been tested on all the available datasets. This scenario is still limited to challenges, although a platform where different datasets and models are shared, such as OpenML [26], could be a possibility in the future. Furthermore, meta-learners which can be trained with missing data, could also be investigated.

5   Conclusion

Prediction of performance using these meta-features yields promising results. The error margins of the methods are still too large for decision-making based on the outcome of this meta-learning method, but is is clearly shown that prior performance of methods in combination with dataset characteristics is a predictor of performance and can lead to a more efficient way of development.

Supplementary Materials

A   Classical Metafeatures

List of 33 classical features used to compose the classical metafeatures.

Table A.1. Classical Metafeatures used in the Support Vector Regression method. M = mean, STD = standard deviation, CVAR = coefficient of variation.

Classical metafeatures
Number of instances
Voxel value M
Voxel value STD
Voxel value CVAR
Skew M
STD
Skew CVAR
Kurtosis M
Kurtosis STD
Kurtosis CVAR
Entropy M
Entropy STD
Entropy CVAR
Median M
Median STD
Mutual information M
Mutual information STD
Mutual information CVAR
Mutual information maximum value
Correlation M
Correlation STD
Correlation CVAR
Sparsity M
Sparsity STD
Sparsity CVAR
Slice size M
Slice size STD
Slice size CVAR
Number of slices M
Number of slices STD
Number of slices CVAR
Equivalent number of features
Noise signal ratio

B Full Results MSD Cross-validation with SVR and DNN meta-learner

Full results of SVR and DNN meta-learners. Consists of: MAE scores per MSD dataset and MAE scores per MSD challenge participant.

Table B.1. Total MAE results of cross-validation on MSD datasets per MSD dataset using different types of meta-features and SVR and DNN meta-learner

MSD datasets ↓	Mean absolute error ↓							
	Classical		VGG16		ResNet50		MobileNetV1	
Meta-learner →	SVR	DNN	SVR	DNN	SVR	DNN	SVR	DNN
1 Braintumor	0.17 ± 0.05	0.06 ± 0.06	0.07 ± 0.01	0.09 ± 0.01	0.08 ± 0.01	0.22 ± 0.06	0.14 ± 0.01	0.22 ± 0.06
2 Heart	0.07 ± 0.01	0.18 ± 0.05	0.06 ± 0.05	0.09 ± 0.01	0.14 ± 0.06	0.05 ± 0.01	0.15 ± 0.07	0.11 ± 0.02
3 Liver	0.29 ± 0.15	0.55 ± 0.25	0.08 ± 0.17	0.03 ± 0.00	0.11 ± 0.10	0.04 ± 0.00	0.28 ± 0.15	0.24 ± 0.06
4 Hippocampus	0.07 ± 0.01	0.09 ± 0.01	0.04 ± 0.07	0.07 ± 0.01	0.08 ± 0.07	0.10 ± 0.01	0.06 ± 0.07	0.07 ± 0.01
5 Prostate	0.14 ± 0.03	0.27 ± 0.09	0.09 ± 0.00	0.21 ± 0.05	0.05 ± 0.00	0.07 ± 0.01	0.04 ± 0.00	0.05 ± 0.01
6 Lung	0.09 ± 0.01	0.09 ± 0.01	0.21 ± 0.11	0.23 ± 0.06	0.17 ± 0.07	0.10 ± 0.02	0.09 ± 0.09	0.09 ± 0.02
7 Pancreas	0.19 ± 0.03	0.07 ± 0.01	0.10 ± 0.01	0.09 ± 0.01	0.06 ± 0.01	0.09 ± 0.01	0.05 ± 0.01	0.16 ± 0.04
8 Hepatic vessel	0.12 ± 0.01	0.10 ± 0.05	0.26 ± 0.03	0.21 ± 0.06	0.13 ± 0.04	0.21 ± 0.05	0.13 ± 0.06	0.09 ± 0.02
9 Spleen	0.46 ± 0.18	0.40 ± 0.18	0.08 ± 0.14	0.03 ± 0.00	0.10 ± 0.11	0.03 ± 0.00	0.21 ± 0.10	0.12 ± 0.02
10 Colon	0.54 ± 0.23	0.56 ± 0.32	0.31 ± 0.28	0.34 ± 0.13	0.31 ± 0.25	0.33 ± 0.15	0.36 ± 0.30	0.25 ± 0.08
Total	0.21 ± 0.08	0.24 ± 0.09	0.13 ± 0.09	0.14 ± 0.04	0.12 ± 0.07	0.12 ± 0.03	0.15 ± 0.09	0.14 ± 0.03

Table B.2. Total MAE results of cross-validation on MSD datasets per MSD participant using different types of meta-features and SVR and DNN meta-learners

Participants ↓	Mean absolute error ↓							
	Classical		VGG16		ResNet50		MobileNetV1	
Meta-learner →	SVR	DNN	SVR	DNN	SVR	DNN	SVR	DNN
Participant 1	0.25 ± 0.08	0.29 ± 0.29	0.08 ± 0.11	0.11 ± 0.02	0.18 ± 0.10	0.20 ± 0.08	0.19 ± 0.12	0.21 ± 0.06
Participant 2	0.18 ± 0.06	0.19 ± 0.19	0.11 ± 0.05	0.11 ± 0.02	0.10 ± 0.05	0.12 ± 0.03	0.13 ± 0.05	0.13 ± 0.02
Participant 3	0.20 ± 0.05	0.21 ± 0.07	0.12 ± 0.06	0.10 ± 0.02	0.11 ± 0.05	0.11 ± 0.02	0.13 ± 0.06	0.13 ± 0.03
Participant 4	0.27 ± 0.09	0.24 ± 0.10	0.11 ± 0.14	0.14 ± 0.03	0.14 ± 0.12	0.17 ± 0.04	0.18 ± 0.14	0.13 ± 0.03
Participant 5	0.19 ± 0.70	0.21 ± 0.05	0.12 ± 0.06	0.12 ± 0.03	0.11 ± 0.05	0.09 ± 0.02	0.12 ± 0.06	0.11 ± 0.02
Participant 6	0.24 ± 0.10	0.25 ± 0.11	0.13 ± 0.09	0.15 ± 0.03	0.14 ± 0.08	0.11 ± 0.02	0.19 ± 0.10	0.18 ± 0.04
Participant 7	0.21 ± 0.07	0.29 ± 0.09	0.16 ± 0.12	0.20 ± 0.05	0.15 ± 0.09	0.12 ± 0.03	0.15 ± 0.11	0.13 ± 0.02
Participant 8	0.15 ± 0.03	0.16 ± 0.02	0.07 ± 0.03	0.07 ± 0.01	0.09 ± 0.03	0.07 ± 0.01	0.12 ± 0.03	0.11 ± 0.02
Participant 9	0.24 ± 0.10	0.23 ± 0.08	0.14 ± 0.08	0.15 ± 0.03	0.12 ± 0.07	0.08 ± 0.01	0.22 ± 0.08	0.15 ± 0.04
Participant 10	0.21 ± 0.08	0.27 ± 0.10	0.17 ± 0.10	0.19 ± 0.06	0.14 ± 0.08	0.16 ± 0.06	0.13 ± 0.10	0.12 ± 0.02
Participant 11	0.24 ± 0.09	0.26 ± 0.14	0.17 ± 0.09	0.18 ± 0.07	0.14 ± 0.08	0.16 ± 0.06	0.15 ± 0.1	0.18 ± 0.05
Participant 12	0.26 ± 0.10	0.26 ± 0.15	0.17 ± 0.11	0.20 ± 0.06	0.16 ± 0.09	0.16 ± 0.04	0.17 ± 0.11	0.15 ± 0.04
Participant 13	0.18 ± 0.06	0.23 ± 0.60	0.11 ± 0.06	0.11 ± 0.02	0.01 ± 0.05	0.12 ± 0.03	0.14 ± 0.06	0.15 ± 0.03
Participant 14	0.15 ± 0.04	0.18 ± 0.03	0.08 ± 0.04	0.09 ± 0.01	0.09 ± 0.03	0.08 ± 0.01	0.10 ± 0.04	0.10 ± 0.02
Participant 15	0.17 ± 0.04	0.19 ± 0.04	0.09 ± 0.05	0.08 ± 0.01	0.10 ± 0.04	0.10 ± 0.01	0.12 ± 0.05	0.10 ± 0.02
Participant 16	0.22 ± 0.09	0.26 ± 0.12	0.15 ± 0.08	0.14 ± 0.04	0.12 ± 0.07	0.12 ± 0.03	0.14 ± 0.09	0.13 ± 0.03
Participant 17	0.20 ± 0.08	0.24 ± 0.09	0.13 ± 0.08	0.13 ± 0.04	0.12 ± 0.07	0.11 ± 0.03	0.14 ± 0.08	0.16 ± 0.04
Participant 18	0.28 ± 0.12	0.31 ± 0.15	0.21 ± 0.15	0.25 ± 0.07	0.15 ± 0.11	0.22 ± 0.07	0.19 ± 0.13	0.22 ± 0.06
Participant 19	0.22 ± 0.08	0.25 ± 0.09	0.16 ± 0.09	0.15 ± 0.05	0.12 ± 0.08	0.11 ± 0.02	0.14 ± 0.09	0.13 ± 0.03
Total	0.21 ± 0.08	0.24 ± 0.10	0.13 ± 0.09	0.14 ± 0.04	0.13 ± 0.07	0.13 ± 0.03	0.15 ± 0.09	0.14 ± 0.03

C Full Prediction Results Independent Test Datasets

Prediction results of SVR and DNN meta-learners on external test datasets (LiTS, ACDC and CHAOS) using different types of meta-features.

Table C.1. Prediction result of SVR and DNN meta-learners on test datasets for different types of meta-features.

True model result → ↓Meta-feature extractor	Liver (LiTS)		Heart (ACDC)		Liver (CHAOS)	
	0.96		0.96		0.89	
	SVR	DNN	SVR	DNN	SVR	DNN
Classical	0.87	0.82	0.80	0.8	0.90	0.85
VGG16	0.76	0.83	0.74	0.91	0.74	0.9
ResNet50	0.73	0.46	0.73	0.94	0.74	0.95
MobileNetV1	0.82	0.97	0.75	0.83	0.82	0.96

References

1. Bernard, O., et al.: Deep learning techniques for automatic MRI cardiac multi-structures segmentation and diagnosis: is the problem solved? IEEE Trans. Med. Imaging **37**(11), 2514–2525 (2018)
2. Bilic, P., et al.: The liver tumor segmentation benchmark (lits). arXiv preprint arXiv:1901.04056 (2019)
3. Campos, G.F., Barbon, S., Mantovani, R.G.: A meta-learning approach for recommendation of image segmentation algorithms. In: 2016 29th SIBGRAPI Conference on Graphics, Patterns and Images (SIBGRAPI), pp. 370–377. IEEE (2016)
4. Cheplygina, V., Moeskops, P., Veta, M., Dashtbozorg, B., Pluim, J.P.W.: Exploring the similarity of medical imaging classification problems. In: Cardoso, M.J., et al. (eds.) LABELS/CVII/STENT -2017. LNCS, vol. 10552, pp. 59–66. Springer, Cham (2017). https://doi.org/10.1007/978-3-319-67534-3_7
5. Doan, T., Kalita, J.: Predicting run time of classification algorithms using meta-learning. Int. J. Mach. Learn. Cybern. **8**(6), 1929–1943 (2016). https://doi.org/10.1007/s13042-016-0571-6
6. Finn, C., Yu, T., Zhang, T., Abbeel, P., Levine, S.: One-shot visual imitation learning via meta-learning. arXiv preprint arXiv:1709.04905 (2017)
7. Gomes, T.A., Prudêncio, R.B., Soares, C., Rossi, A.L., Carvalho, A.: Combining meta-learning and search techniques to select parameters for support vector machines. Neurocomputing **75**(1), 3–13 (2012)
8. Ho, Y., Pepyne, D.: Simple explanation of the no-free-lunch theorem and its implications. J. Optim. Theory Appl. **115**(3), 549–570 (2002). https://doi.org/10.1023/A:1021251113462
9. Hu, S., Tomczak, J., Welling, M.: Meta-learning for medical image classification (2018)
10. Isensee, F., Petersen, J., Kohl, S.A.A., Jäger, P.F., Maier-Hein, K.H.: nnU-Net: breaking the spell on successful medical image segmentation. CoRR abs/1904.08128 (2019)

11. Kavur, A.E., et al.: CHAOS challenge-combined (CT-MR) healthy abdominal organ segmentation. arXiv preprint arXiv:2001.06535 (2020)
12. Lemke, C., Budka, M., Gabrys, B.: Metalearning: a survey of trends and technologies. Artif. Intell. Rev. **44**(1), 117–130 (2015)
13. Litjens, G., et al.: A survey on deep learning in medical image analysis. Med. Image Anal. **42**, 60–88 (2017)
14. Liu, X., et al.: A comparison of deep learning performance against health-care professionals in detecting diseases from medical imaging: a systematic review and meta-analysis. Lancet Digit. Health **1**, e271–e297 (2019)
15. Pedregosa, F., et al.: Scikit-learn: machine learning in Python. J. Mach. Learn. Res. **12**, 2825–2830 (2011)
16. Peng, Y., Flach, P.A., Soares, C., Brazdil, P.: Improved dataset characterisation for meta-learning. In: Lange, S., Satoh, K., Smith, C.H. (eds.) DS 2002. LNCS, vol. 2534, pp. 141–152. Springer, Heidelberg (2002). https://doi.org/10.1007/3-540-36182-0_14
17. Prudêncio, R.B., Ludermir, T.B.: Meta-learning approaches to selecting time series models. Neurocomputing **61**, 121–137 (2004)
18. Ronneberger, O., Fischer, P., Brox, T.: U-Net: convolutional networks for biomedical image segmentation. CoRR abs/1505.04597 (2015)
19. Rossi, A.L.D., de Leon Ferreira, A.C.P., Soares, C., De Souza, B.F., et al.: MetaStream: a meta-learning based method for periodic algorithm selection in time-changing data. Neurocomputing **127**, 52–64 (2014)
20. Shin, H.C., et al.: Deep convolutional neural networks for computer-aided detection: CNN architectures, dataset characteristics and transfer learning. IEEE Trans. Med. Imaging **35**(5), 1285–1298 (2016)
21. Simpson, A.L., et al.: A large annotated medical image dataset for the development and evaluation of segmentation algorithms. arXiv preprint arXiv:1902.09063 (2019)
22. Soares, C., Brazdil, P.B., Kuba, P.: A meta-learning method to select the kernel width in support vector regression. Mach. Learn. **54**(3), 195–209 (2004). https://doi.org/10.1023/B:MACH.0000015879.28004.9b
23. Soares, R.G.F., Ludermir, T.B., De Carvalho, F.A.T.: An analysis of meta-learning techniques for ranking clustering algorithms applied to artificial data. In: Alippi, C., Polycarpou, M., Panayiotou, C., Ellinas, G. (eds.) ICANN 2009. LNCS, vol. 5768, pp. 131–140. Springer, Heidelberg (2009). https://doi.org/10.1007/978-3-642-04274-4_14
24. Tajbakhsh, N., et al.: Convolutional neural networks for medical image analysis: full training or fine tuning? IEEE Trans. Med. Imaging **35**(5), 1299–1312 (2016)
25. Vanschoren, J.: Meta-learning: a survey. arXiv preprint arXiv:1810.03548 (2018)
26. Vanschoren, J., van Rijn, J.N., Bischl, B., Torgo, L.: OpenML: networked science in machine learning. SIGKDD Explor. **15**(2), 49–60 (2013)
27. Vilalta, R., Drissi, Y.: A perspective view and survey of meta-learning. Artif. Intell. Rev. **18**(2), 77–95 (2002). https://doi.org/10.1023/A:1019956318069

Labelling Imaging Datasets on the Basis of Neuroradiology Reports: A Validation Study

David A. Wood[1]([✉]), Sina Kafiabadi[2], Aisha Al Busaidi[2], Emily Guilhem[2], Jeremy Lynch[2], Matthew Townend[3], Antanas Montvila[2], Juveria Siddiqui[2], Naveen Gadapa[2], Matthew Benger[2], Gareth Barker[4], Sebastian Ourselin[1], James H. Cole[4,5], and Thomas C. Booth[1,2]

[1] School of Biomedical Engineering, King's College London, London, UK
david.wood@kcl.ac.uk
[2] King's College Hospital, London, UK
[3] Wrightington, Wigan and Leigh NHSFT, Wigan, UK
[4] Institute of Psychiatry, Psychology and Neuroscience, King's College London, London, UK
[5] Centre for Medical Image Computing, Dementia Research, University College London, London, UK

Abstract. Natural language processing (NLP) shows promise as a means to automate the labelling of hospital-scale neuroradiology magnetic resonance imaging (MRI) datasets for computer vision applications. To date, however, there has been no thorough investigation into the validity of this approach, including determining the accuracy of report labels compared to image labels as well as examining the performance of non-specialist labellers. In this work, we draw on the experience of a team of neuroradiologists who labelled over 5000 MRI neuroradiology reports as part of a project to build a dedicated deep learning-based neuroradiology report classifier. We show that, in our experience, assigning binary labels (i.e. normal vs abnormal) to images from reports alone is highly accurate. In contrast to the binary labels, however, the accuracy of more granular labelling is dependent on the category, and we highlight reasons for this discrepancy. We also show that downstream model performance is reduced when labelling of training reports is performed by a non-specialist. To allow other researchers to accelerate their research, we make our refined abnormality definitions and labelling rules available, as well as our easy-to-use radiology report labelling tool which helps streamline this process.

MR Imaging abnormality Deep learning Identification study (MIDI) Consortium.

Electronic supplementary material The online version of this chapter (https://doi.org/10.1007/978-3-030-61166-8_27) contains supplementary material, which is available to authorized users.

J. Cardoso et al. (Eds.): iMIMIC 2020/MIL3ID 2020/LABELS 2020, LNCS 12446, pp. 254–265, 2020.
https://doi.org/10.1007/978-3-030-61166-8_27

Keywords: Natural language processing · Deep learning · Labelling

1   Introduction

Deep learning-based computer vision systems hold promise for a variety of applications in neuroradiology. However, a rate-limiting step to clinical adoption is the labelling of large datasets for model training, a laborious task requiring considerable domain knowledge and experience. Following recent breakthroughs in natural language processing (NLP), it is becoming feasible to automate this task by training text classification models to derive labels from radiology reports and to assign these labels to the corresponding images [7,12–14]. To date, however, there has been no investigation into the general validity of this approach, including determining the accuracy of report labels compared to image labels as well as assessing the performance of non-specialist labellers.

In this work we draw on the experience of a team of neuroradiologists who labelled over 5000 magnetic resonance imaging (MRI) neuroradiology reports as part of a project to build a dedicated deep learning-based neuroradiology report classifier. In particular, we examine several aspects of this process which have hitherto been neglected, namely (i) the degree to which radiology reports faithfully reflect image findings (ii) whether the labelling of reports for model training can be reliably outsourced to clinicians who are not specialists (here we examined whether the performance of a neurologist or radiology trainee (UK registrar grade; US resident equivalent) is similar to that of a neuroradiologist) (iii) the difficulty of creating an exhaustive and consistent set of labelling rules, and (iv) the extent to which abnormalities labelled on the basis of examination-level reports are detectable on MRI sequences likely to be available to a computer vision model.

Overall, our findings support the validity of deriving image labels from neuroradiology reports, but with several important caveats. We find that, contrary to basic assumptions often made for this methodology, radiological reports are often less accurate than image findings. Indeed, certain categories of neuroradiological abnormality are inaccurately reported. We conclude that, in our experience assigning binary labels (i.e. normal vs abnormal) to images from reports alone is very accurate. The accuracy of more granular labelling, however, is dependent on the category, and we highlight reasons for this discrepancy.

We also find that several aspects of model training are more challenging than is suggested by a review of the literature. For example, designing a complete set of clinically relevant abnormalities for report labelling, and the rules by which these were applied, took our team of four neuroradiologists more than six months to complete with multiple iterations, and involved the preliminary inspection of over 1,000 radiology reports. To allow other researchers to bypass this step and accelerate their research, we make our refined abnormality definitions and labelling rules available. We also make our radiology report labelling tool available which helps streamline this manual annotation process. Importantly, we found that even when enabled with the labelling tool and set of abnormalities and rules, report annotation for model training must be performed by experienced neuroradiologists, because a considerable reduction in model performance was seen when labelling was performed by a neurologist or a radiology trainee

2   Related Work

NLP models have previously been employed to assign image labels in the context of training computer vision models for neuroradiology applications using radiology reports from both computed tomography (CT) [9,12,14] and MRI [13] examinations. In all cases, classification performance was reported for the primary objective of labelling reports. However, there was no comparison of either the predicted or annotated labels with the images. The closest published work to our paper is therefore a conference abstract highlighting discrepancies between the findings detailed in chest radiograph reports and the corresponding images when labelling a limited set of abnormalities [8]. To the best of our knowledge no such investigation has been performed in the context of neuroradiology, nor have the challenges of creating an NLP labelling tool for neuroradiology applications been described.

Previous work has investigated the accuracy of using crowdsourcing to label images in the context of general [5] as well as medical [4] computer vision tasks. However, we know of no work in the context of neuroradiology which investigates the level of expertise required for accurate manual annotation of reports. Although it might seem obvious that experienced neuroradiologists are required for this task, previous works have instead employed post-graduate radiology and neurosurgery residents [14] or attending physicians [9,12], without providing any insight into the possible reduction in labelling accuracy that such delegation may invite.

Automated brain abnormality detection using either T_2-weighted or diffusion-weighted images (DWI) and employing supervised [10,11] and unsupervised [2] deep learning models has previously been reported. However, in each case only a limited set of abnormalities were available during training and testing, and there was no investigation into the range of abnormalities likely to be detected by the computer vision system using only these sequences. In fact, to the best of our knowledge no investigation has determined what fraction of abnormalities are visible to expert neuroradiologists inspecting only a limited number of sequences. Resolving this point could help narrow the architecture search space for future deep learning-based abnormality detection systems.

3   Data and Methods

The UK's National Health Research Authority and Research Ethics Committee approved this study. 126,556 radiology reports produced by expert neuroradiologists (UK consultant grade; US attending equivalent), consisting of all adult (>18 years old) MRI head examinations performed at Kings College Hospital NHS Foundation Trust, London, UK (KCH) between 2008 and 2019, were included in this study. The reports were extracted from the Computerised Radiology Information System (CRIS) (Healthcare Software Systems, Mansfield, UK) and all data was de-identified. Over the course of more than twelve months, 5000 reports were annotated by a team of neuroradiologists to generate reference standard report labels to train the neuroradiology report classifier described in [13] (ALARM classifier). Briefly, each unstructured report was typically composed of 5–10 sentences of image interpretation, and sometimes included information from the scan protocol, comments regarding the patient's clinical history, and recommended actions for the referring doctor. In the current paper, we refer to these reference standard labels generated on the basis of manual inspection of radiology reports as "silver reference standard labels". Prior to manual labelling, a complete set of clinically relevant categories of neuroradiological abnormality, as well as the rules by which reports were labelled, were generated following six months of iterative experiments involving the inspection of over 1000 radiology reports. The complete set of abnormalities, grouped by category, are presented in the supplemental material.

Three thousand reports were independently labelled by two neuroradiologists for the presence or absence of any of these abnormalities. We refer to this as the 'coarse dataset' (i.e. normal vs. abnormal). Agreement between these two labellers was 94.9%, with a consensus classification decision made with a third neuroradiologist where there was disagreement. Separately, 2000 reports were labelled by a team of three neuroradiologists for the presence or absence of each of 12 more specialised categories of abnormality (mass e.g. tumour; acute stroke; white matter inflammation; vascular abnormality e.g. aneurysm; damage e.g. previous brain injury; Fazekas small vessel disease score [6]; supratentorial atrophy; infratentorial atrophy; foreign body; haemorrhage; hydrocephalus; extra-cranial abnormality). We refer to this as the "granular dataset". There was unanimous agreement between these three labellers across each category for 95.3% of reports, with a consensus classification decision made with all three neuroradiologists where there was disagreement.

We manually inspected 500 images (comprising, on average, 6 MRI sequences) to generate reference standard image labels. We refer to labels generated in this way as "gold reference standard labels". 250 images were labelled for the presence or absence of any abnormality, systematically following the same criteria as that used to generate the coarse report dataset. Similarly, 250 images were examined and given 12 binary labels corresponding to the presence or absence of each of the more granular abnormality categories.

Our team designed a complete set of clinically relevant categories capable of accurately capturing the full range of pathologies which present on brain MRI scans. The aim here was to try and emulate the behaviour of a radiologist in the real world, guided by the need for clinical intervention for an abnormal finding. To help other researchers bypass this step, and to encourage standardization across research groups of abnormality definitions, we make our abnormality categories, as well as all clinical rules, available in the supplemental material. Our manual labelling campaign was considerably aided by our development of a dedicated labelling app. This tool allows easy visualisation and labelling of reports through a graphical user interface (GUI), and includes functionality for flagging difficult cases for group consensus/review. Two apps were developed - one for binary labelling (Fig. 1), and one for more granular labelling (Fig. 2) - and we make both available to other researchers at https://github.com/MIDIconsortium/RadReports.

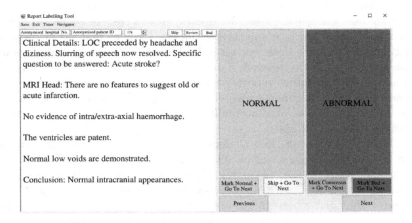

Fig. 1. Binary report labelling tool for the MR Imaging abnormality deep learning identification (MIDI) study. The example report should be marked as normal.

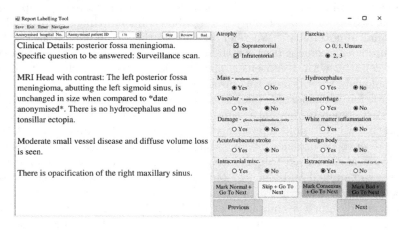

Fig. 2. Granular report labelling tool for the MIDI study. The correct labels for this example report have been selected.

4 Results

4.1 Impact of Annotator Expertise

To assess the level of expertise required to perform manual annotation of reports for training a text classification model, two experiments were performed.

First, we compared the coarse labels (i.e. normal vs. abnormal) generated by a hospital doctor with ten years experience as a stroke physician and neurologist, who was trained by our team of neuroradiologists over a six month period, with neuroradiologist-generated labels. The rationale for determining the performance was twofold. Neurologists and stroke physicians frequently interpret reports held on the Electronic Patient Record during patient consultations, therefore it is expected that they would be able to differentiate, and therefore label, normal or abnormal reports accurately. Moreover, given that there are less neuroradiologists than neurologists or stroke physicians, with a ratio of 1:4 in the UK, it is likely to be easier to recruit such physicians to perform such labelling tasks.

We found a reduction in performance of neurologist labelling when compared to the labels created by an expert neuroradiologist (Table 1). Based on classification and evaluation methodology in [13], the state-of-the-art ALARM classifier was trained using these neurologist-derived labels and, for comparison, labels generated by a blinded neuroradiologist (Fig. 3). The corresponding reduction in classification performance on a hold-out test set of silver reference-standard labels (i.e. reports with consensus) at an arbitrarily fixed sensitivity of 90% (Table 2) demonstrates the impact of what we have shown to be a sub-optimal labelling strategy. In summary, there is optimal performance when the classifier is trained with reports labelled by an experienced neuroradiologist.

Table 1. Labelling performance of a stroke physician and neurologist.

Accuracy (%)	Sensitivity (%)	Specificity (%)
92.7	77.2	98.9

Table 2. Accuracy, specificity, and F1 score of a neuroradiology report classifier trained using data labelled by either a neurologist or neuroradiologist operating at a fixed sensitivity of 90%. Best performance in bold.

Annotator	Accuracy (%)	Specificity (%)	F1 (%)
Neurologist	89.8	89.5	75.8
Neuroradiologist	**96.4**	**97.7**	**90.3**

As a second experiment, a 3rd year radiology trainee who was also trained by our team over a six month period to label neuroradiology reports, generated labels for our 'granular dataset'. There was a reduction in radiology trainee

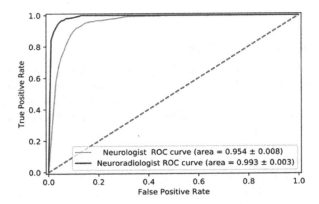

Fig. 3. ROC curve for a neuroradiology report classifier trained on labels generated by a neurologist (cyan) and a neuroradiologist (blue). The area under the curve (AUC) is shown. (Color figure online)

performance, averaged across all 12 binary labels, when compared to the silver reference standard labels created by our team of expert neuroradiologists (Table 3). The sensitivity of these labels is clearly too low to be used for model training.

Table 3. Labelling performance of a radiology trainee on the 'granular dataset', averaged across all 12 binary labels.

Sensitivity (%)	Specificity (%)	F1 (%)
64.4	98.3	70.8

It is worth highlighting that reliability (inter-rater agreement) and accuracy (performance) should not be conflated for labelling tasks. We demonstrate this in a further experiment where the same neurologist previously described also generated labels for our 'granular dataset'. The Fleiss κ score for the radiology trainee and the neurologist averaged over all 12 binary categories was 0.64, which is above the threshold previously employed to establish neuroradiology label reliability [14]. Substantial inter-rater agreement (commonly taken as $\kappa > 0.6$), therefore, does not necessarily equate to label accuracy as this experiment has shown.

4.2   Report Validation

To determine the validity of assigning image labels on the basis of radiology reports, the granular labels derived from reports (silver reference standard) were compared to those derived by inspecting the corresponding images (gold reference standard) for 500 cases (Table 4). Although the false positive rate of report labelling is very low for the 12 granular categories of interest, it is clear that the sensitivity of radiology report labelling is category dependent and can be low. On further analysis, we found that insensitive labelling for any given category typically reflects the absence of any reference in the report to that particular category rather than a discrepancy in interpretation. The categories with low sensitivity include hydrocephalus, haemorrhage, extra-cranial abnormalities, and infratentorial atrophy. The reasons for this are discussed below.

Table 4. Accuracy of silver reference standard report labels for granular categories when compared to the corresponding gold standard image labels. Categories with sensitivity >80% in bold.

Category	Sensitivity (%)	Specificity (%)	F1 (%)
Fazekas	**90.5**	95.6	93.2
Mass	**97.9**	93.6	95.9
Vascular	**83.3**	88.4	86.5
Damage	**82.4**	92.7	87.8
Acute stroke	**94.4**	99.5	94.4
Haemorrhage	69.2	99.6	78.3
Hydrocephalus	70.0	99.6	77.8
White matter inflammation	**95.6**	100	97.7
Foreign body	**100.0**	99.6	96.6
Extracranial abnormality	60.0	94.7	54.5
Supratentorial atrophy	**100**	94.6	76.9
Infratentorial atrophy	77.7	94.3	54.5
Macro-average	85.1	96.0	82.8

Importantly, silver standard binary labels indicating the presence or absence of any abnormality in a report (i.e. normal vs. abnormal) were accurate when compared to the image (gold reference standard label) (Table 5).

Table 5. Accuracy of silver reference standard report labels for binary categories (i.e. normal vs abnormal) relative to the corresponding gold standard image labels.

Category	Sensitivity (%)	Specificity (%)	F1 (%)
Normal vs. abnormal	98.7	96.6	98.5

4.3   MRI Sequences and Abnormality Visibility

In another experiment we examined the utility of assigning examination-level labels derived from radiology reports to different MRI sequences. In general, neuroradiology reports detail findings from multi-modality (i.e. multiple MRI sequences) imaging examinations, with individual sequences providing complementary information to discriminate specific tissues, anatomies and pathologies. For example, the signal characteristics of blood changes over time, the rate of which is sequence dependent. Therefore analysis of images from multiple sequences allows the chronicity of a haemorrhage to be deduced. Assigning the same label to all images in a multi-modality examination can confound computer vision classification if a model isn't optimised to take as its input the individual sequence from which a particular examination-level label was derived. Therefore, we wished to determine whether a minimal number of sequences would be sufficient for use with report-derived labels. At our institution, axial T_2-weighted and DWI images are typically obtained for routine image review, with over 78% of patients receiving both images during an examination. We sought to determine what fraction of abnormalities are visible to a neuroradiologist inspecting only the T_2-weighted and DWI images. Binary labels (i.e. normal vs. abnormal) for 250 examinations were generated by inspecting only these sequences, and compared to labels derived from all available sequences for the same examinations. The agreement between these two labels was 97.8%, showing that these two sequences would be sufficient for use with report-derived labels for most abnormality detection tasks. Examples of the wide range of abnormalities identified on the basis of T_2-weighted and DWI imaging appear in the supplemental material, along with reports describing abnormalities which weren't visible on either of these two sequences.

5   Discussion

In this work we have examined several assumptions which are fundamental to the process of deriving image labels from radiology reports. Overall, our findings support the validity of deriving image labels from neuroradiology reports. In particular, assigning binary labels (i.e. normal vs abnormal) to images from reports alone is highly accurate and therefore acceptable. Until now this has been assumed but has not been thoroughly investigated. The accuracy of more granular labelling, however, is dependent on the category. For example, labelling of acute stroke, mass, neuro-degeneration, and vascular disorders, is shown to be accurate.

The low labelling accuracy seen in some granular labelling categories is a result of low sensitivity. Low sensitivity typically reflects the absence of any reference in the report to that particular category rather than a discrepancy in interpretation. A qualitative analysis by our team of neuroradiologists has determined several reasons for low sensitivity in some categories.

First, in the presence of more clinically important findings, neuroradiologists often omit descriptions of less critical abnormalities which may not necessarily change the overall conclusion or instigate a change in the patient's management. For example, on follow-up imaging of previously resected tumours, we have found that the pertinent finding as to whether there is any progressive or recurrent tumour is invariably commented on. In contrast, the presence of white matter changes secondary to previous radiotherapy appears less important within this clinical context. If unchanged from the previous imaging, a statement to the effect of "otherwise stable intracranial appearances" is typical in these cases.

A second source of low sensitivity is the observation that radiology reports are often tailored to specific clinical contexts and the referrer. A report aimed at a neurologist referrer who is specifically enquiring about a neurodegenerative process in a patient with new onset dementia, for example, may make comments about subtle parenchymal atrophy. In contrast, parenchymal volumes may not be scrutinised as closely in the context of someone who has presented with a vascular abnormality, such as an aneurysm, and a report is aimed at a vascular neurosurgeon. Both sources of low sensitivity mentioned above often reflect a "satisfaction of search error" where the radiologist has failed to appreciate the full gamut of abnormalities. After identifying one or two abnormalities the task may appear complete and there is less desire to continue to interrogate the image [1]. It is also noteworthy that abnormalities which are identified by the neuroradiologist by chance may be judiciously omitted from the report on a case by case basis when such "incidentalomas" are thought to be of little consequence. Because of these sources of low sensitivity, labelling categories of abnormality from radiology reports remains challenging for haemorrhage (note that acute haemorrhage is typically detected by CT; MRI reports were often insensitive to those haemorrhages associated with non-critical findings such as micro-haemorrhages), hydrocephalus, extracranial abnormalities and infratentorial atrophy.

In addition to examining the accuracy of radiology reports compared to image findings, we have also demonstrated that most abnormalities typical of a real-world triage environment are picked up using only T_2-weighted and DWI sequences. This observation may help narrow the architecture search-space for future deep learning-based brain abnormality detection systems, and allow a more accurate comparison of model performance across research groups. However, there are certain abnormalities which may not be visible on these sequences. For example, the presence of microhaemorrhages or blood breakdown products (hemosiderin), are sometimes only visible on gradient echo (T_2^*-weighted) or susceptibility weighted imaging (SWI) [3]. Furthermore, foci of pathological enhancement on post contrast T_1-weighted imaging can indicate underlying disease which may not be apparent on other sequences. Therefore, whilst we have shown that using T_2-weighted and DWI sequences alone allows almost all abnormalities to be identified visually, and that plausibly this will translate to efficient computer vison training tasks, it is important to be aware that there are potential limitations.

We briefly discuss several logistical aspects of the report labelling process which were not covered by our more quantitative investigations. Our team designed a complete set of clinically relevant categories capable of accurately capturing the full range of pathologies which present on brain MRI scans. The aim here was to try and emulate the behaviour of a radiologist in the real world, guided by the need for clinical intervention for an abnormal finding. This process, however, was more onerous than is often presented in the literature, requiring the inspection of over 1000 radiology reports by our team of experienced neuroradiologists over the course of more than six months before an exhaustive and consistent set of abnormality categories, as well as the rules by which reports were to be labelled, could be finalised. The rules and definitions constantly evolved during the course of the practice labelling experiments. To allow other researchers to bypass this step and accelerate their research, we make our refined abnormality definitions and labelling rules available as well as our dedicated labelling easy-to-use app.

6   Conclusion

We conclude that in our experience, assigning binary labels (i.e. normal vs abnormal) to images from reports alone is highly accurate. Importantly, we found that even when enabled with the labelling tool and set of abnormalities and rules, annotation of reports for model training must be performed by experienced neuroradiologists, because a considerable reduction in model performance was seen when labelling was performed by a neurologist or a radiology trainee. In contrast to the binary labels, the accuracy of more granular labelling is dependent on the category.

References

1. Berbaum, K., Franken, E., Caldwell, R., Schartz, K., Madsen, M.: Satisfaction of Search in Radiology, 2nd edn., pp. 121–166. Cambridge University Press, Cambridge (2018). https://doi.org/10.1017/9781108163781.010
2. Chen, X., Konukoglu, E.: Unsupervised detection of lesions in brain MRI using constrained adversarial auto-encoders (2018). arXiv:1806.04972
3. Chiewvit, P., Piyapittayanan, S., Poungvarin, N.: Cerebral venous thrombosis: diagnosis dilemma. Neurol. Int. **3**, e13 (2011). https://doi.org/10.4081/ni.2011.e13
4. Cocos, A., Masino, A., Qian, T., Pavlick, E., Callison-Burch, C.: Effectively crowdsourcing radiology report annotations. In: Proceedings of the Sixth International Workshop on Health Text Mining and Information Analysis, pp. 109–114. Association for Computational Linguistics, Lisbon, Portugal, September 2015. https://doi.org/10.18653/v1/W15-2614, https://www.aclweb.org/anthology/W15-2614
5. Crump, M.J.C., McDonnell, J.V., Gureckis, T.M.: Evaluating Amazon's Mechanical Turk as a tool for experimental behavioral research. PLOS One **8**(3), 1–18 (03 2013). https://doi.org/10.1371/journal.pone.0057410
6. Fazekas, F., Chawluk, J., Alavi, A., Hurtig, H., Zimmerman, R.: MR signal abnormalities at 1.5 T in Alzheimer's dementia and normal aging. AJR Am. J. Roentgenol. **149**, 351–356 (1987). https://doi.org/10.2214/ajr.149.2.351
7. Garg, R., Oh, E., Naidech, A., Kording, K., Prabhakaran, S.: Automating ischemic stroke subtype classification using machine learning and natural language processing. J. Stroke Cerebrovasc. Dis. **28**(7), 2045–2051 (2019). https://doi.org/10.1016/j.jstrokecerebrovasdis.2019.02.004, http://www.sciencedirect.com/science/article/pii/S1052305719300485

8. Olatunji, T., Yao, L., Covington, B., Rhodes, A., Upton, A.: Caveats in generating medical imaging labels from radiology reports. CoRR (2019). http://arxiv.org/abs/1905.02283, arXiv:1905.02283
9. Ong, C.J., et al.: Machine learning and natural language processing methods to identify ischemic stroke, acuity and location from radiology reports. PLOS ONE **15**(6), 1–16 (2020). https://doi.org/10.1371/journal.pone.0234908
10. Rauschecker, A.M., et al.: Artificial intelligence system approaching neuroradiologist-level differential diagnosis accuracy at brain MRI. Radiology **295**(3), 626–637 (2020). https://doi.org/10.1148/radiol.2020190283, pMID: 32255417
11. Rezaei, M., Yang, H., Meinel, C.: Brain abnormality detection by deep convolutional neural network (2017). arXiv:1708.05206
12. Shin, B., Chokshi, F.H., Lee, T., Choi, J.D.: Classification of radiology reports using neural attention models. In: 2017 International Joint Conference on Neural Networks (IJCNN), pp. 4363–4370. IEEE (2017)
13. Wood, D.A., et al.: Automated labelling using an attention model for radiology reports of MRI scans (ALARM) (2020). arXiv:2002.06588
14. Zech, J., et al.: Natural language-based machine learning models for the annotation of clinical radiology reports. Radiology **287**, 171093 (2018). https://doi.org/10.1148/radiol.2018171093

Semi-supervised Learning for Instrument Detection with a Class Imbalanced Dataset

Jihun Yoon[1], Jiwon Lee[1], SungHyun Park[1], Woo Jin Hyung[1,2], and Min-Kook Choi[1(✉)]

[1] hutom, Seoul, Republic of Korea
mkchoi@hutom.io
[2] Department of Surgery, Yonsei University College of Medicine, Seoul, Republic of Korea

Abstract. The automated recognition of surgical instruments in surgical videos is an essential factor for the evaluation and analysis of surgery. The analysis of surgical instrument localization information can help in analyses related to surgical evaluation and decision making during surgery. To solve the problem of the localization of surgical instruments, we used an object detector with bounding box labels to train the localization of the surgical tools shown in a surgical video. In this study, we propose a semi-supervised learning-based training method to solve the class imbalance between surgical instruments, which makes it challenging to train the detectors of the surgical instruments. First, we labeled gastrectomy videos for gastric cancer performed in 24 cases of robotic surgery to detect the initial bounding box of the surgical instruments. Next, a trained instrument detector was used to discern the unlabeled videos, and new labels were added to the tools causing class imbalance based on the previously acquired statistics of the labeled videos. We also performed object tracking-based label generation in the spatio-temporal domain to obtain accurate label information from the unlabeled videos in an automated manner. We were able to generate dense labels for the surgical instruments lacking labels through bidirectional object tracking using a single object tracker; thus, we achieved improved instrument detection in a fully or semi-automated manner.

Keywords: Surgical instrument detection · Semi-supervised learning · Class imbalanced problem

1 Introduction

Recently, the improvements in the performance of visual recognition technology has led to the widespread use of computer-assisted surgery (CAS). CAS was mainly used to deliver specific information, such as the location of lesions required for a surgical procedure. However, in recent times, the concept of CAS

© Springer Nature Switzerland AG 2020
J. Cardoso et al. (Eds.): iMIMIC 2020/MIL3ID 2020/LABELS 2020, LNCS 12446, pp. 266–276, 2020.
https://doi.org/10.1007/978-3-030-61166-8_28

has been expanded for postoperative feedback through surgical procedures or behavior analysis. Currently, the localization information of the surgical instruments in the surgical video is essential for understanding the surgical process or for providing useful information to the surgeon during the surgical process [1]. In particular, as the demand for robotic surgery that can reduce the burden on the surgeon increases, the need for information technologies, such as the navigation or the analysis of surgical conditions during surgery using recognition information, that occurs in the surgical process also increases.

With the increase in the need for the automated localization of surgical instruments, datasets, including instance segmentation labels of instruments for laparoscopic cholecystectomy [2,3] and robotic surgery using the DaVinci Xi in abdominal porcine procedures, have been published for the instance segmentation challenge [4]. However, in the case of both datasets, sufficient label information for accurate localization was not provided. In the case of robotic surgery, only the partial process of the procedure using pigs was included, thereby limiting the appearance information of the recognizable device. Additionally, a dataset, including video and kinematic log information for analyzing the motion of robotic surgical instruments, has been released. However, it does not contain localization information in actual surgery [5].

We propose a semi-supervised learning-based instrument detection methodology to efficiently solve instrument localization problems in robotic surgery while overcoming the drawbacks of the existing dataset. The localization information of the surgical tools in the robotic surgery video is a database with an extreme class imbalance problem, which makes it difficult to train some instruments. Because some instruments are used only for a specific purpose in the surgical process, even if a large number of surgical videos are secured, some instruments appear very rarely depending on the surgical situation or the surgeon's preference. The proposed methodology is designed to operate in a fully or semi-automated manner and addresses class imbalance issues while minimizing human intervention and maximizing the amount of training data that is lacking.

To effectively apply the proposed semi-supervised learning technique to surgical instrument localization, we used an algorithm based on the detection and tracking strategy [6,7] among the semi-supervised learning techniques for object detection [6–8]. As demonstrated by [6], the goal was to perform robust tracking to collect reliable positive examples for utilizing the detection and tracking strategy. In this case, [6] aims to obtain reliable positive examples during training for a robust classifier; however, the recently proposed CNN (Convolutional Neural Networks)-based detector [9–12] may not be suitable because it intentionally manipulates training inputs to obtain difficult positive examples during training. In the case of a CNN-based detector, positive examples given as input through data augmentation are transformed to improve the generalization performance. For automated label generation for CNN-based detectors, we created training labels for surgical instruments with class imbalance problems using an object tracker [21] with a balanced performance in terms of speed and accuracy.

[7] proposed a learning methodology that can efficiently train with sparse annotations distributed in a video on the premise that there are initial ground truth bounding boxes for some objects. The case presented by [7] is similar to the proposed algorithm in that it uses a CNN-based tracking model [21]. However, we generated labels in an automated manner from unlabeled videos using a trained detector from the initial database. At this time, according to the statistics of the initial database, the automated annotation was performed for a specific surgical instrument causing class imbalance, and dense labels for a specific frame on the surgical video were generated. Figure 1 shows a schematic flow chart of our proposed detection and tracking based semi-supervised learning methodology.

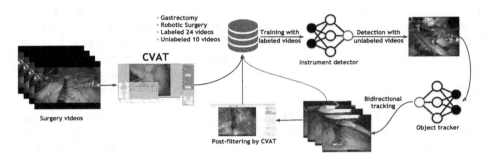

Fig. 1. Schematic representation of the proposed semi-supervised learning-based surgical instrument detector. The proposed technique is trained by adding labels in an automated manner from a trained detector with initial data by CVAT [20]. If the proposed technique follows the red arrow indicated in the flow chart, the training is done in a fully automated way, and if it follows the blue arrow, human intervention is required. Even when human intervention is required, using the proposed algorithm can significantly reduce human intervention.

We used the MMDetection library [15] with PyTorch [14] to verify the performance of instrument detection and performed training using state-of-the-art (SOTA) models. For performance evaluation, we used Faster R-CNN (Region-based Convolutional Neural Networks) and Cascade R-CNN, which are two-stage object detectors based on CNN. At the same time, for a fair comparison, we used the anchorless detector, FCOS (Fully Convolutional One-Stage detector) [18], as a representative one-stage detector among other one-stage detectors [16–18]. Finally, we performed ensembles for the trained detectors to complement each model's inference outputs. We applied joint NMS-based ensembles [19] for effective ensembles and were able to achieve improved instrument detection performance. The technical contribution of the proposed semi-supervised learning-based surgical instrument detection methodology is as follows:

– The proposed surgical instrument detection methodology solves the class imbalance problem by generating labels in an automated manner. A detection and tracking scenario is used in semi-supervised learning to obtain a label in an automated manner, and a specific scenario performs a learning process in a fully automated manner.

– The proposed technique utilizes a CNN-based SOTA object detector for effective instrument detection. We evaluated the performance of each object detector in semi-supervised learning scenarios.

– Ensemble-based test results are provided using trained detectors based on supervised/semi-supervised learning scenarios to obtain the final performance. We were able to achieve the best performance among the proposed techniques by utilizing the ensemble technique.

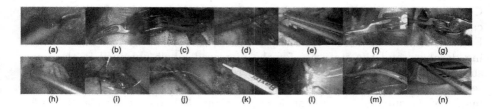

Fig. 2. List of instruments to be detected in gastrectomy for gastric cancer and captured images of the corresponding instruments. There are 14 types of instruments collected in the instrument detection database, consisting of robotic surgical instruments, laparoscopic instruments for surgical assistance, and consumables. The types of instruments are from the top left (a) Harmonic Ace, (b) Maryland Bipolar Forceps, (c) Cadiere Forceps, (d) Curved Atraumatic Grasper, (e) Stapler, (f) Medium-Large Clip Applier, (g) Small Clip Applier, (h) Suction-Irrigation, (i) Needle, (j) Needle Holder, (k) Baxter, (l) Specimen Bag, (m) Drain Tube, and (n) Covidien Ultrasonic. Each instrument is labeled with a minimum pixel size bounding box that contains the instrument.

2   Data Collection

We used the original videos of gastrectomy for 24 cases of gastric cancer performed by a skilled specialist to construct an initial training database for robotic surgical instrument detection. Of the videos included in the initial training data, 14 videos were recorded using the da Vinci Si system, and the remaining ten videos were recorded using the da Vinci Xi system. The gastrectomy videos included in the initial database were recorded between a minimum of 1 h 30 min and 4 h depending on the type of patient and the surgical procedure. For the surgical instrument labeling, 1f/s sampling was applied to all frames where the instrument appeared to obtain a label for 1 frame per second on average, and all annotations were performed by annotators skilled in using the labeling tool. The Computer Vision Annotation Tool (CVAT) [20] was used as the labeling tool, and labeling was conducted by receiving annotation input from multiple annotators and confirming it from at least one inspector. Each surgical instrument was labeled with a minimum pixel size bounding box containing a visually identifiable instrument.

3   Statistics of Instrument Detection Database

Figure 2 shows the list of surgical instruments included in the instrument detection database. There are 14 instruments used in gastrectomy for gastric cancer: Harmonic Ace, Maryland Bipolar Forceps, Cadiere Forceps, Curved Atraumatic Grasper, Stapler, Medium-Large Clip Applier, Small Clip Applier, Suction-Irrigation, Needle, Needle Holder, Baxter, Specimen Bag, Drain Tube, and Covidien Ultrasonic. The surgical instruments include not only robotic surgical instruments but also assistive laparoscopic instruments and consumables. Figure 3 shows the label statistics for each instrument in the initial training database and the statistics for the increased labels obtained in the semi-supervised learning process. We constructed evaluation data with a validation set of three videos that were not included in the training to verify the performance of the surgical instrument detector. Figure 4 shows the data statistics for the validation set.

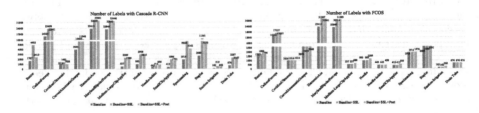

Fig. 3. Changes in label statistics according to the proposed training scenario. This figure shows the initial distribution of surgical instruments obtained from 24 videos and the distribution of surgical instruments for each model after bidirectional tracking-based labeling. *SSL* refers to data that generate labels after bidirectional tracking, and *Post* refers to post-correction with human intervention.

4   Semi-supervised Learning for Instrument Detection

To verify the scenarios using the semi-supervised learning shown in Fig. 1, we conducted CNN-based object detector training using the MMDetection library [15]. The model trained to check the baseline performance includes a two-stage model, Faster R-CNN [9] and Cascade R-CNN [12] as well as an one-stage model, FCOS [18]. We used SiamMask as a tracker to apply semi-supervised learning based on detection and tracking. Using the initial bounding box obtained through

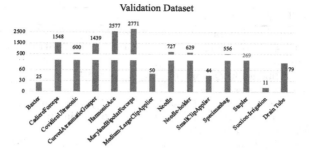

Fig. 4. Instrument label distribution included in a validation set of 3 videos. To verify the performance of the trained instrument detector, 3 videos were divided into a validation set in 24 videos. This figure shows the distribution of the number of labels for each instrument included in the verification video set.

the detector as an input, we performed bidirectional tracking through SiamMask [13] to acquire additional labels. Algorithm 1 shows an automated detection and tracking algorithm for obtaining additional labels for surgical instruments with a small number of quantities. Algorithm 1 is used in the same way for backward tracking to complete bidirectional tracking. Figure 5 shows an example of label information obtained using an automated method through the proposed detection and tracking based algorithm. Through the proposed method, the additional labeling for 10 unlabeled videos was performed to obtain additional labels for class imbalanced surgical instruments.

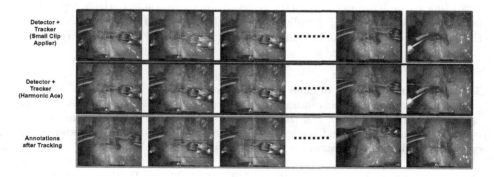

Fig. 5. Example of creating dense annotations based on bidirectional tracking. The initial detection result for the Small Clip Applier was received as input, and labels were generated through bidirectional tracking. In order to generate dense labels, bidirectional tracking of the detected Harmonic Ace was generated in an automated manner. The blue box shows the case where the tracking is done, and the red box shows the situation where the tracking ends.

Algorithm 1: Forward tracking

Input: Action clip (A), pretrained tracking model (T), a set of bbox for initial input (BB), threshold of tracking score (ρ_1), threshold of IoU between two pair of tracked bbox (ρ_2)

Output: A set of bboxes in Q from T

Initialize an empty queue Q

while *bbox $b_{c,o,i}$ with class c and detector o at i-th frame from BB* **do**

    Get a list of frames FF in forward from i-th frame in A;

    Initialize T with $b_{c,o,i}$ from A;

    Initialize a variable pre_s with a size of $b_{c,o,i}$ to store a size of object from T at previous frame;

    while *each frame in FF* **do**

        Get a bbox $b_{c,t,k}$ with t index from T at k-th frame;

        $crnt_s :=$ a size of $b_{c,t,k}$;

        if $IoU(prev_s, crnt_s) \geq \rho_1$ *and $b_{c,o,k}$ not exists* **then**

            $prev_s := crnt_s$;

            Add $b_{c,t,k}$ to Q;

        else

            break;

        end

    end

end

---

5   Experimental Results

Training Details. The training of all detectors and trackers was performed according to the training parameter settings shown in Table 1. HRNet [21] was used as the backbone CNN applied for the training of the object detector, and the structure of the feature pyramid network [22] was used for effective training.

Table 1. Training details of each detector and tracker. *lr* represents the learning rate. The learning schedule is indicated as (scheduler, drop rate) [drop epoch1:drop epoch2:max epoch]. *Det-thr* and *Track-thr* are the threshold values for the output reliability of the detector and tracker, respectively, and *Track-IoU* is the termination condition specified in Algorithm 1.

Model	Backbone	Backbone param.	Optimizer (lr)	LR scheduler	Det-thr	Track-thr	Track-IoU
Cascade R-CNN	HRNet	V2p-W32	SGD (0.02)	(step, 0.1) [16:19:20]	0.9	0.75	0.25
Faster R-CNN	HRNet	V2p-W32-GN-head	SGD (0.02)	(step, 0.1) [8:11:12]	0.7	0.75	0.25
FCOS	HRNet	V2p-W32	SGD (0.01)	(step, 0.1) [8:11:12]	0.7	0.75	0.25

Data Statistics. Figure 3 shows the change in the amount of training labels after applying the proposed technique, and Fig. 4 shows the number of labels for each tool included in the verification set. For all the performance evaluations,

Table 2. mAP changes according to the proposed training scenarios. mAP is evaluated under the IoU of [0.5 : 0.05 : 0.95]. The highest-performance detector is indicated in bold font.

Model	Backbone	Dataset	mAP
Faster R-CNN	HRNet	Model only	46.5
–	HRNet	Model+SSL	48.7
Cascade R-CNN	HRNet	Model only	51.0
–	HRNet	Model+SSL	49.1
–	HRNet	Model+SSL+Post	**52.3**
FCOS	HRNet	Model only	35.5
–	HRNet	Model+SSL	38.1
–	HRNet	Model+SSL+Post	49.1

we calculated the mAP for all classes under the [0.5 : 0.05 : 0.95] IoU value by referring to the evaluation method of the COCO dataset [23].

Semi-supervised Instrument Detection. Table 2 shows the performance change according to the semi-supervised learning scenarios. The detectors used in the proposed learning technique are Cascade R-CNN, Faster R-CNN, and FCOS with a HRNet backbone, and they are composed of two types of training scenarios, as shown in Fig. 1. The first is a model that adds labels without human intervention until learning by performing detection and tracking in a fully automated manner, which is named model+SSL. The other method is a label applied with human post-correction on the additionally obtained label, which is named model+SSL+post. When the proposed method was applied, the faster R-CNN and FCOS were able to obtain improved results in overall performance without post-correction. In the case of the Cascade R-CNN detector, when the learning technique was applied in a fully automated manner, the overall performance was degraded. However, when post-correction was applied, a significant improvement in performance was obtained.

Human Intervention. Table 3 shows the level of human intervention required during the annotation process at the frame level. By applying the proposed learning methodology, we reduced the level of intervention by more than half compared to the case of labeling 10 new videos based on 1f/s. In the case of fully automated semi-supervised learning, an improvement in performance was achieved for most class imbalanced surgical tools and in some networks, without any human intervention.

Ensemble and Visualization. Table 4 shows the ensemble performance in the inference process for models obtained from the proposed learning methodology. Figure 6 shows the visualization of the final output of the model for each learning scenario.

Table 3. Difference between human interventions according to the proposed training scenario and detector. The intervention ratio was calculated as a percentage of the total number of frames collected at 1f/s from the unlabeled videos.

Model	Total number of frames with human intervention	Intervention ratio (%)
Manual labeling (1f/s)	82,412	–
Cascade R-CNN	41,498	50.35
Faster R-CNN	44,441	53.93
FCOS	18,080	**21.94**

Table 4. Performance evaluation result of model ensemble according to each learning scenario. As a result of the ensemble of all models, the highest performance was achieved, and the two-stage model recorded higher performance than the one-stage model.

Model	Backbone	Dataset	mAP
Faster R-CNN ensemble	HRNet	Model only/Model+SSL	52.3
Cascade R-CNN ensemble	HRNet	Model only/Model+SSL/Model+SSL+Post	54.0
FCOS ensemble	HRNet	Model only/Model+SSL/Model+SSL+Post	43.8
All ensemble	HRNet	Model only/Model+SSL/Model+SSL+Post	**54.2**

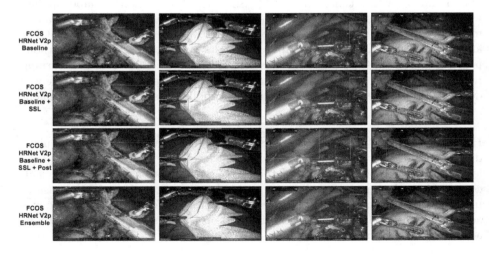

Fig. 6. Visualization of detectors trained with the proposed learning methodology. This figure shows the visualized results for the output of model only, model+SSL, model+SSL+post, and ensemble in order from the first row. The threshold for the output reliability for visualization was set to 0.5.

Per Instrument Analysis. Table 5 shows the change in performance for each instrument obtained according to the semi-supervised learning scenario. Bold surgical instruments are tools that cause severe class imbalance problems in the initial database. If the proposed learning technique is used, performance improvement can be achieved for most class imbalanced surgical instruments. At the same time, Table 3 shows the correlation between the increased label amount and the change in performance for each instrument.

Table 5. Changes of AP for each surgical instrument in the training scenario using the FCOS detector. The instrument index is shown in Fig. 2. M stands for Model only, MS stands for Model+SSL, and MSP stands for Model+SSL+Post. The values in parentheses indicate the performance difference from Model only. In most cases, the ensembles of all the models demonstrated improved performance.

Model	inst. (a)	inst. (b)	inst. (c)	inst. (d)	inst. (e)	inst. (f)	inst. (g)
FCOS_M	71.6	71.0	56.6	20.1	60.8	33.4	45.9
FCOS_MS	75.5 (+3.9)	69.4 (−1.6)	53.3 (−3.3)	17.7 (−2.4)	57.9 (−2.9)	43.8 (+10.4)	36.1 (−9.8)
FCOS_MSP	78.1 (+6.5)	74.6 (+3.6)	66.5 (+9.9)	27.1 (+7.0)	72.4 (+11.6)	35.0 (+1.6)	45.2 (−0.7)
FCOS_ensemble	77.1 (+5.5)	74.1 (+3.1)	63.0 (+6.4)	23.8 (+3.7)	68.1 (+7.3)	49.4 (+16.0)	50.5 (+4.6)
All_ensemble	**79.8** (+8.2)	**78.1** (+7.1)	**70.0** (+13.4)	**33.0** (+12.9)	**76.4** (+15.6)	**75.0** (+41.6)	**79.1** (+33.2)
Model	inst. (h)	inst. (i)	inst. (j)	inst. (k)	inst. (l)	inst. (m)	inst. (n)
FCOS_M	4.7	2.5	0.2	60.3	35.8	1.0	33.3
FCOS_MS	28.2 (+23.5)	2.1 (−0.4)	0.7 (+0.5)	72.8 (+12.5)	28.8 (−7.0)	0.7 (−0.3)	46.3 (+13.0)
FCOS_MSP	31.9 (+27.2)	2.9 (+0.4)	0.0 (−0.2)	72.3 (+12.0)	46.6 (+10.8)	11.1 (+10.1)	48.6 (+15.3)
FCOS_ensemble	30.9 (+26.2)	2.9 (+0.4)	0.6 (+0.4)	74.6 (+14.3)	42.3 (+6.5)	6.2 (+5.2)	**49.9** (+16.6)
All_ensemble	**42.9** (+38.2)	**12.2** (+9.7)	**9.6** (+9.4)	**83.0** (+22.7)	**56.9** (+21.1)	**30.8** (+29.8)	32.1 (−1.2)

6 Conclusion

In this study, we proposed a semi-supervised learning-based training methodology to solve the problem of surgical instrument localization in robotic surgery. The proposed methodology was confirmed to successfully alleviate the severe class imbalance problem caused by the nature of the surgical videos. However, there was a limitation in that an automated labeling process could not completely solve the class imbalance problem. It is possible to consider other parameterization methods, such as approaching the problem by defining it as pixel-level segmentation and approaching surgical instrument localization as an object detection problem to ensure improvements in future research. In the current experiment scenarios, the results applying to only one cycle among the scenarios shown in Fig. 1 are reported. However, it is necessary to analyze the performance and problems of the proposed algorithm in the iterative scenarios.

References

1. Jin, A., et al.: Tool detection and operative skill assessment in surgical videos using region-based convolutional neural networks. In: Proceedings of WACV (2018)
2. Twinanda, A.P., Shehata, S., Mutter, D., Marescaux, J., de Mathelin, M., Padoy, N.: EndoNet a deep architecture for recognition tasks on laparoscopic videos. IEEE Trans. Med. Imaging **36**(1), 86–97 (2017)
3. Allan, M., et al.: 2017 robotic instrument segmentation challenge. arXiv: 1902.06426 (2019)
4. Ahmidi, N., et al.: A dataset and benchmarks for segmentation and recognition of gestures in robotic surgery. Trans. Biomed. Eng. **64**(9), 2025–2041 (2017)
5. Gao, Y., et al.: The JHU-ISI gesture and skill assessment working set (JIGSAWS): a surgical activity dataset for human motion modeling. In: Proceedings of MICCAIW (2014)
6. Misra, I., Shrivastava, A., Hebert, M.: Watch and learn: semi-supervised learning for object detectors from video. In: Proceedings of CVPR (2015)
7. Yoon, J., Hong, S., Jeong, S., Choi, M.-K.: Semi-supervised object detection with sparsely annotated dataset. arXiv:2006.11692 (2020)
8. Choi, M.-K., et al.: Co-occurrence matrix analysis-based semi-supervised training for object detection. In: Proceedings of ICIP (2018)
9. Ren, S., He, K., Girshick, R., Sun, J.: Faster R-CNN towards real-time object detection with region proposal networks. In: Proceedings of NIPS (2015)
10. Dai, J., Li, Y., He, K., Sun, J.: R-FCN: object detection via region-based fully convolutional networks. In: Proceedings of NIPS (2016)
11. Dai, J., Qi, H., Xiong, Y., Li, Y., Zhang, G., Hu, H., Wei, Y.: Deformable convolutional networks. In: Proceedings of ICCV (2017)
12. Cai, Z., Vasconcelos, N.: Cascade R-CNN: delving into high quality object detection. In: Proceedings of CVPR (2018)
13. Wang, Q., Zhang, L., Bertinetto, L., Hu, W., Torr, P.H.S.: Fast online object tracking and segmentation: a unifying approach. In: Proceedings of CVPR (2019)
14. Chen, K., et al.: MMDetection: open MMLab detection toolbox and benchmark. arXiv:1906.07155 (2019)
15. Paszke, A., et al.: PyTorch: an imperative style, high-performance deep learning library. In: Proceedings of NeurIPS (2019)
16. Liu, W., et al.: SSD: single shot multibox detector. In: Leibe, B., Matas, J., Sebe, N., Welling, M. (eds.) ECCV 2016. LNCS, vol. 9905, pp. 21–37. Springer, Cham (2016). https://doi.org/10.1007/978-3-319-46448-0_2
17. Lin, T.-Y., Goyal, P., Girshick, R., He, K., Dollar, P.: Focal loss for dense object detection. In: Proceedings of ICCV (2017)
18. Tian, Z., Shen, C., Chen, H., He, T.: FCOS: fully convolutional one-stage object detection. In: Proceedings of ICCV (2019)
19. Jung, H., Choi, M.-K., Jung, J., Lee, J.-H., Kwon, S., Jung, W.Y.: ResNet-based vehicle classification and localization in traffic surveillance systems. In: Proceedings of CVPRW (2017)
20. Computer Vision Annotation Tool (CVAT). https://github.com/opencv/cvat
21. Wang, J., et al.: Deep high-resolution representation learning for visual recognition. arXiv:1908.07919 (2019)
22. Lin, T.-Y., Dollár, P., Girshick, R., He, K., Hariharan, B., Belongie, S.: Feature pyramid networks for object detection. In: Proceedings of CVPR (2017)
23. Lin, T.-Y., et al.: Microsoft COCO: common objects in context. In: Fleet, D., Pajdla, T., Schiele, B., Tuytelaars, T. (eds.) ECCV 2014. LNCS, vol. 8693, pp. 740–755. Springer, Cham (2014). https://doi.org/10.1007/978-3-319-10602-1_48

Paying Per-Label Attention
for Multi-label Extraction
from Radiology Reports

Patrick Schrempf[1,2]([✉]), Hannah Watson[1], Shadia Mikhael[1], Maciej Pajak[1],
Matúš Falis[1], Aneta Lisowska[1], Keith W. Muir[3], David Harris-Birtill[2],
and Alison Q. O'Neil[1,4]

[1] Canon Medical Research Europe, Edinburgh, UK
patrick.schrempf@eu.medical.canon
[2] University of St Andrews, St Andrews, UK
[3] Institute of Neuroscience & Psychology, University of Glasgow, Glasgow, UK
[4] University of Edinburgh, Edinburgh, UK

Abstract. Training medical image analysis models requires large
amounts of expertly annotated data which is time-consuming and expensive to obtain. Images are often accompanied by free-text radiology
reports which are a rich source of information. In this paper, we tackle
the automated extraction of structured labels from head CT reports for
imaging of suspected stroke patients, using deep learning. Firstly, we
propose a set of 31 labels which correspond to radiographic findings (e.g.
hyperdensity) and clinical impressions (e.g. haemorrhage) related to neurological abnormalities. Secondly, inspired by previous work, we extend
existing state-of-the-art neural network models with a label-dependent
attention mechanism. Using this mechanism and simple synthetic data
augmentation, we are able to robustly extract many labels with a single model, classified according to the radiologist's reporting (positive,
uncertain, negative). This approach can be used in further research to
effectively extract many labels from medical text.

Keywords: NLP · Radiology report labelling · BERT

1 Introduction

Training medical imaging models requires large amounts of expertly annotated
data which is time-consuming and expensive to obtain. Fortunately, medical
images are often accompanied by free-text reports written by radiologists sum-
marising their main *findings* (what the radiologist sees in the image e.g. "hyperdensity") and *impressions* (what the radiologist diagnoses based on the findings

Electronic supplementary material The online version of this chapter (https://
doi.org/10.1007/978-3-030-61166-8_29) contains supplementary material, which is
available to authorized users.

e.g. "haemorrhage"). This information can be converted to structured labels which are used to train image analysis algorithms to detect the findings and to predict the impressions. Image-level labels have previously been provided to train image analysis algorithms e.g. as part of the RSNA haemorrhage detection challenge [17] and the CheXpert challenge for automated chest X-Ray interpretation [9]. The task of reading the radiology report and assigning labels is not trivial and requires a certain degree of medical knowledge on the part of the human annotator. An alternative is to automatically extract labels, and in this paper we study the task of automatically labelling head computed tomography (CT) radiology reports.

Automatic extraction has traditionally been accomplished using expert medical knowledge to engineer a feature extraction and classification pipeline [24]; this was the approach taken by Irvin et al. to label the CheXpert dataset of Chest X-Rays [9] and by Gorinski et al. in the EdIE-R method for labelling head CT reports [8]. Such pipelines separate the individual tasks such as named entity recognition and negation detection.

An alternative approach is to design an end-to-end machine learning model that will learn to extract the final labels directly from the text. Simple approaches have been demonstrated using word embeddings or bag of words feature representations followed by logistic regression [25] or decision trees [22]. More complex recurrent neural networks (RNNs) have been shown to be effective for document classification by many authors [3,23] and Drozdov et al. [7] show that a bidirectional long short term memory (Bi-LSTM) network with a single attention mechanism also works well for a binary task. However, with recent developments of transformer natural language processing (NLP) models such as Bidirectional Encoder Representations from Transformers (BERT) [6], it is easier than ever before to use existing pre-trained models that have learnt underlying language patterns and fine-tune them on small domain-specific datasets. This was the approach taken by Wood et al. in the Automated Labelling using an Attention model for Radiology reports of MRI scans (ALARM) model for labelling head magnetic resonance imaging (MRI) reports [21]. Specifically, they use BioBERT [1] as the base model, which has been pretrained on PubMed abstracts rather than Wikipedia, to obtain contextualised embeddings for each input token and then apply a further attention mechanism to this embedding. Wood et al. perform a binary classification of normal versus abnormal radiology report, which is determined by a number of criteria during data annotation. BERT has also been used for multi-label classification of radiology reports by Smit et al. [19]. They show that BERT can outperform the previous state of the art for labelling 13 different labels on the CheXpert open source dataset [9].

Mullenbach et al. proposed per-label attention in a similar document classification task (for clinical coding) in their Convolutional Attention for Multi-Label classification (CAML) model [15]. In this paper, inspired by [15], we extend existing state-of-the-art models with a label-dependent attention mechanism. Our contributions are to:

- Propose a set of radiographic findings and clinical impressions for labelling of head CT scans for suspected stroke patients.
- Show that a multi-headed model with per-label attention improves the accuracy compared to a simple multi-label softmax output.
- Show that simple synthetic data significantly improves task performance, especially for classification of rarer labels.

2   Data

Below we describe the three datasets used in this work.

NHS GGC Dataset: Our target dataset contains 230 radiology reports supplied by the NHS Greater Glasgow and Clyde (GGC) Safe Haven. We have the required ethical approval[1] to use this data. A synthetic example report with similar format to the NHS GGC reports can be seen in Fig. 1.

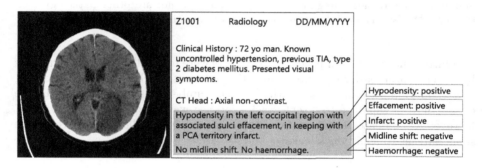

Fig. 1. Example radiology report. The image (left) shows a slice from an example CT scan (Case courtesy of Dr David Cuete, Radiopaedia.org, rID: 30225); there is a visible darker patch indicating an infarct. The synthetic radiology report (middle) has a similar format to the NHS GGC data. We manually filter relevant sentences, highlighted with blue background. The boxes (right) indicate which labels are annotated. (Color figure online)

A list of 31 radiographic findings and clinical impressions found in stroke radiology reports was collated by a clinical researcher; this is the set of labels that we aim to classify. Figure 2 shows a complete list of these labels. Each sentence is labelled for each finding or impression as "positive", "uncertain", "negative" or "not mentioned" - the same certainty classes as used by Smit et al. [19]. The most common labels such as "haemorrhage", "infarct" and "hyperdensity" have between 200–400 mentions (100–200 negative, 0–50 uncertain, 100–200 positive) while the rarest labels such as "abscess" or "cyst" only occur once in the dataset.

---

[1] iCAIRD project number: 104690; University of St Andrews: CS14871.

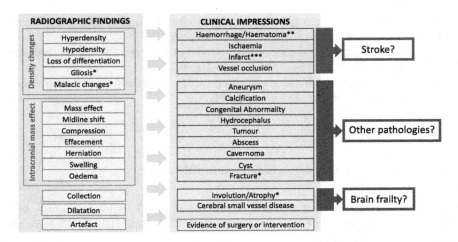

Fig. 2. Label schematic: 13 radiographic findings, 14 clinical impressions and 4 crossover labels (finding→impression links not shown). *These labels fit both the finding and impression categories. **Haematoma can indicate other pathology (e.g. trauma). ***Established infarcts indicate brain frailty [10].

During the annotation process, the reports were manually split into sentences by the clinical researcher, resulting in 1,353 sentences which we split into training and validation datasets (due to the limited number of annotated reports, we do not have a separate test set). Each sentence was annotated independently, however we allocate sentences from the same original radiology report to the same dataset to avoid data leakage.

Synthetic Dataset: We augment our training dataset by synthesising 5 sentences for each label as follows:

- "There is [label]." → positive
- "There is [label] in the brain." → positive
- "[Label] is evident in the brain." → positive
- "There may be [label]." → uncertain
- "There is no [label]." → negative

For the labels "haemorrhage/haematoma/contusion", "evidence of surgery/ intervention", "vessel occlusion (embolus/thrombus)", and "involution/ atrophy", we synthesise sentences for each variant. There are 180 synthetic sentences total.

MIMIC-III Dataset: To pre-train the word embedding, we use clinical notes from the MIMIC dataset [11]; in total 2,083,180 documents from 46,146 patients. The datasets are summarised in Table 1.

Table 1. Summary statistics for the datasets used in this work.

Dataset	#patients	#reports	#sentences
NHS GGC – Training	138	138	838
NHS GGC – Validation	92	92	515
Synthetic data	–	–	180
MIMIC-III	46,146	2,083,180	99,718,301

3 Methods

Below we describe the methods which are compared in this paper (implemented in Python). We denote our set of labels as L and our set of certainty classes as C, such that the number of labels $n_L = |L|$ and the number of certainty classes $n_C = |C|$. For the NHS GGC dataset, $n_L = 31$ and $n_C = 4$. For all methods, data is pre-processed by extracting sentences and words using the NLTK library [13], removing punctuation, and converting to lower case. Hyperparameter search was performed through manual tuning on the validation set, based on the micro-averaged F1 metric.

3.1 Simple Machine Learning Approaches

BoW + RF: The Bag of Words + Random Forest (BoW + RF) model uses a bag of words representation as its input. We train one model per label since this gives the most accurate results, resulting in 31 random forest classifiers. Random forest classifiers are quick to train and apply so multiple models are still practical in a real use case. We use the sci-kit learn library [16] implementation with 100 estimators, a maximum depth of 10, and 200 maximum features.

Word2Vec: The Word2Vec [14] baseline uses a pre-trained word embedding of size e. The embedding is pre-trained on the MIMIC dataset described in Sect. 2 for 30 epochs using the gensim [18] library; the vocabulary size is 107,497 words. The word vectors for the input sentence are averaged and passed through a fully connected single layer neural network mapping to an output layer of size $n_L \times n_C$. This network is trained with a constant learning rate of 0.001, batch size of 16 and an embedding size of 200. This and all following models are trained for a maximum of 200 epochs with early stopping patience of 25 epochs on F1 micro.

3.2 Deep Learning: Per-Label Attention Mechanism

When training neural networks, we find that accuracy can be reduced where there are many classes. Here we describe the per-label attention mechanism [2] as seen in Fig. 3, an adaptation of the multi-label attention mechanism in the CAML model [15]. We can apply this to the output of any given neural network subarchitecture. We define the output of the subnetwork as $r \in \mathbb{R}^{n_{tok} \times h}$ where n_{tok} is the number of tokens and h is the hidden representation size. The parameters we learn are the weights $W_0 \in \mathbb{R}^{h \times h}$ and bias $b_0 \in \mathbb{R}^h$. Furthermore, for each label l we learn an independent $v_l \in \mathbb{R}^h$ to calculate an attention vector $\alpha_l \in \mathbb{R}^{n_{tok}}$.

$$u = \tanh(W_0 r + b_0)$$

$$\alpha_l = \text{softmax}(v_l^T u)$$

$$s_l = \sum \alpha_l r$$

The attended output $s_l \in \mathbb{R}^h$ is then passed through n_L parallel classification layers reducing dimensionality from h to n_C.

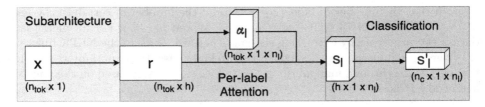

Fig. 3. Simplified model diagram: the subarchitecture is a CNN, Bi-GRU or BERT variant and maps from input x to a hidden representation r; per-label attention maps to a separate representation s_l for each label before classification.

3.3   Deep Learning: Neural Network Models

We pre-process the data before input to the neural network architectures. Each input sentence is limited to n_{tok} tokens and padded with zeros to reach this length if the input is shorter. We choose $n_{tok} = 50$ as this is larger than the maximum number of words in any of the sentences in the NHS GGC dataset. The neural network models all finish with n_L softmax classifier outputs, each with n_C classes.

Models are trained using a weighted categorical cross entropy loss and Adam optimiser [12]. We weight across the labels but not across classes, as this did not give any improvements. Given a parameter β, the number of sentences n and the number of "not mentioned" occurrences of a label o_l, we calculate the weights for each label using the training data as follows:

$$w_{l,\text{"not mentioned"}} = \left(\frac{n}{o_l}\right)^\beta \qquad w_{l,\text{"mentioned"}} = \left(\frac{n}{n - o_l}\right)^\beta$$

CAML: The CAML model follows the implementation by Mullenbach et al. [15] and uses an embedding that is initialised to the same pre-trained weights as for the Word2Vec baseline. The embedded input passes through a convolutional layer of graduated filter sizes applied in parallel (see below), followed by max-pooling operations across each graduated set of filters, to produce our intermediate representation r. This is then passed through the per-label attention mechanism introduced by the CAML model. For the convolutional layer, we chose 512 CNN filter maps with kernel sizes of 2 and 4. The model was trained with a learning rate of 0.0005 and a batch size of 16.

Bi-GRU: The embedding is initialised to the same pre-trained weights as used for the Word2Vec baseline. The embedded sentence x passes through a bidirectional GRU (Bi-GRU) network [5] with hidden size of $h/2$. The outputs from both directions are concatenated to produce a representation r for each input sentence. For **Bi-GRU + single attention**, this representation is passed through a single attention mechanism. For **Bi-GRU + per-label attention**, this representation is passed through the per-label attention mechanism. The model was trained with a learning rate of 0.0005, batch size of 16 and hidden size $h = 1024$.

BERT and BioBERT: The BERT model is a standard pre-trained BERT model, "bert-base-uncased" - weights are available for download online[2] - we use the huggingface [20] implementation. We take the output representation for the CLS token of size 768×1 at position 0 and follow with the n_L softmax outputs. The model was trained with a learning rate of 0.0001 and batch size of 32. For **BioBERT**, we use a Bio-/ClinicalBERT model pretrained on both PubMed abstracts and the MIMIC-III dataset[3] with the huggingface BERT implementation. We use the same training parameters as for BERT (above).

ALARM: Our implementation of the ALARM [21] model uses the BioBERT model (and training parameters) described above. Following the implementation details of Wood et al., instead of using a single output vector of size 768×1, we extract the entire learnt representation of size $768 \times n_{tok}$. For the **ALARM + softmax** model, we pass this through a single attention vector and then through three fully connected layers to map from 768 to 512 to 256 to the $n_L \times n_C$ outputs. For **ALARM + per-label-attention**, we employ n_L per-label attention mechanisms instead of a single shared attention mechanism before passing through three fully connected layers *per label*.

4 Results

Tables 2 and 3 show the results. We report the micro-averaged F1 score as our main metric, calculated across all labels. We also report the macro-averaged F1

---

[2] https://github.com/google-research/bert.
[3] https://github.com/th0mi/clinicalBERT.

score; this is F1 score averaged across all labels with equal weighting for each label. We note that although we used micro F1 as our early stopping criterion, we do not observe an obvious difference in the scores if F1 macro is used for early stopping. We exclude the "not mentioned" certainty from our metrics, similar to the approach used by Smit et al. [19] - we denote $C' = C\backslash\{\text{"not mentioned"}\}$, so $n_{c'} = n_C - 1$. When we report our F1 metrics for a single certainty class we report the usual F1 metric, whereas when we report metrics for all classes and labels we report an average per certainty class.

For all experiments, we use a machine with NVIDIA GeForce GTX 1080 Ti GPU (11 GB of VRAM), Intel Xeon CPU E5 v3 (6 physical cores, maximum clock frequency of 3.401 GHz) and 32 GB of RAM. Training run times range from 14 s for the Random Forest model to 376 s for the Bi-GRU + per-label attention model and 1448 s for the ALARM + per-label-attention model. For details of all run times, see Table 1 in the supplementary material.

Table 2. Micro-averaged F1 results as mean$_{\text{standard deviation}}$ of 5 runs with different random seeds. "All" combines the classes "negative", "uncertain" and "positive". Bold indicates the best model for each metric.

Model	All	Negative	Uncertain	Positive
BoW + RF	$0.871_{0.003}$	$0.936_{0.003}$	$0.119_{0.021}$	$0.889_{0.003}$
Word2Vec	$0.808_{0.005}$	$0.900_{0.007}$	$0.328_{0.023}$	$0.812_{0.008}$
CAML [15]	$0.838_{0.005}$	$0.866_{0.011}$	$0.135_{0.050}$	$0.873_{0.001}$
Bi-GRU	$0.868_{0.009}$	$0.936_{0.011}$	$0.488_{0.017}$	$0.872_{0.009}$
Bi-GRU + single attention	$0.863_{0.009}$	$0.924_{0.017}$	$0.424_{0.032}$	$0.873_{0.006}$
Bi-GRU + per-label attention	$0.921_{0.003}$	$\mathbf{0.970_{0.006}}$	$0.573_{0.011}$	$0.932_{0.004}$
BERT	$0.907_{0.003}$	$0.953_{0.004}$	$0.585_{0.035}$	$0.916_{0.002}$
BioBERT	$0.915_{0.005}$	$0.959_{0.003}$	$0.627_{0.040}$	$0.922_{0.007}$
ALARM + softmax	$0.899_{0.008}$	$0.948_{0.002}$	$0.570_{0.028}$	$0.909_{0.010}$
ALARM + per-label attention	$\mathbf{0.928_{0.008}}$	$0.965_{0.004}$	$\mathbf{0.689_{0.039}}$	$\mathbf{0.936_{0.008}}$

Per-Label Attention: The micro- and macro-averaged F1 scores (Tables 2 and 3) show that for both BioBERT and the Bi-GRU models, adding *per-label* attention to the models improves performance consistently over the models with a single attention mechanism (p-values of < 0.05). We also show the breakdown in accuracies across certainty classes (negative, uncertain and positive) in our results tables. It can be seen that the per-label attention provides large gains in accuracy across all classes. The macro F1 metric amplifies this because all labels are weighted equally, giving an idea of how the model performs for the rarer labels, several of which have fewer than 10 training samples each.

Table 3. Macro-averaged F1 results as mean$_{\text{standard deviation}}$ of 5 runs with different random seeds. "All" combines the classes "negative", "uncertain" and "positive". Bold indicates the best model for each metric.

Model	All	Negative	Uncertain	Positive
BoW + RF	$0.477_{0.013}$	$0.667_{0.019}$	$0.052_{0.025}$	$0.711_{0.001}$
Word2Vec	$0.455_{0.011}$	$0.581_{0.034}$	$0.164_{0.048}$	$0.619_{0.029}$
CAML [15]	$0.394_{0.013}$	$0.435_{0.017}$	$0.086_{0.050}$	$0.661_{0.025}$
Bi-GRU	$0.631_{0.025}$	$0.718_{0.042}$	$0.404_{0.051}$	$0.718_{0.011}$
Bi-GRU + single attention	$0.522_{0.039}$	$0.666_{0.065}$	$0.223_{0.051}$	$0.677_{0.018}$
Bi-GRU + per-label attention	$0.708_{0.014}$	$0.796_{0.027}$	$0.524_{0.023}$	$0.803_{0.016}$
BERT	$0.673_{0.015}$	$0.773_{0.004}$	$0.457_{0.050}$	$0.790_{0.025}$
BioBERT	$0.673_{0.041}$	$0.730_{0.038}$	$0.529_{0.094}$	$0.761_{0.017}$
ALARM + softmax	$0.652_{0.025}$	$0.767_{0.009}$	$0.441_{0.071}$	$0.749_{0.007}$
ALARM + per-label attention	$\mathbf{0.766}_{0.028}$	$\mathbf{0.818}_{0.029}$	$\mathbf{0.661}_{0.061}$	$\mathbf{0.818}_{0.021}$

a) the lateral ventricles are minimally asymmetric as previously which may be congenital or secondary to the left basal ganglia haemorrhage .

b) the lateral ventricles are minimally asymmetric as previously which may be congenital or secondary to the left basal ganglia haemorrhage .

c) the lateral ventricles are minimally asymmetric as previously which may be congenital or secondary to the left basal ganglia haemorrhage .

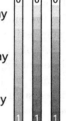

Fig. 4. Visualisation of attention for (a) per-label attention vectors, (b) a single attention vector and (c) per-label attention from a model trained without synthetic data. Model (a) detects congenital (yellow) and haemorrhage (green) separately. Model (b) detects both keywords in the single attention vector (blue). Model (c) does not detect the "congenital" keyword. (Color figure online)

Figure 4 compares the attention learnt by a single attention model to per-label attention models. We see that the single attention vector (Fig. 4b) attends to the correct words - "congenital" and "haemorrhage" - however the model incorrectly predicts both labels as "not mentioned". In comparison, the model with per-label attention (Fig. 4a) recognises the same keywords separately within the respective label attention mechanisms, and correctly predicts both labels as "positive". This makes sense because the single attention mechanism does not have separate follow-on s_l representations and therefore features for all labels are entangled in one representation. Finally, the model trained without synthetic data (Fig. 4c) does not recognise the "congenital" keyword and does not make the correct prediction for this label.

Synthetic Data and Importance of Pre-training: To investigate the effect of the synthetic training data, we train models on only the synthetic data, only NHS GGC data, and both combined. The results for macro F1 in Fig. 5 clearly show an improvement when the synthetic data is used alongside the original data - this is consistent across both of our best models (p-values of < 0.05). For numerical results see Tables 2 and 3 in the supplementary material.

We also investigated the effect of the embedding pre-training. A model with randomly initialised embeddings (maintaining the same vocabulary and embedding size) performs 0.028 worse for the micro-averaged F1 compared to a model using a pre-trained embedding (p-value of < 0.05).

Error Analysis: When investigating the prediction errors of our best model, we identify that approximately 30% of errors are due to missed labels, 10% are due to falsely predicted labels, and the remaining 60% are due to confusion between certainty classes (negative, uncertain, positive). Many of the missed labels are caused by previously unseen synonyms or subtypes, for instance "arteriovenous malformation" is an instance of "congenital abnormality" which is a diverse class. There are also many ways of expressing certainty which are subtly different; for instance positive might be expressed as "probable", "likely", "indicates", "suggestive of", "is consistent with" whereas uncertainty might be expressed as "possible", "may represent", "could indicate", "is suspicious of" and other subtly different expressions. Errors might be mitigated with the use of a larger training dataset and richer data synthesis, potentially by exploiting medical knowledge bases such as UMLS [4] to augment the synthetic dataset with a rich synonym set.

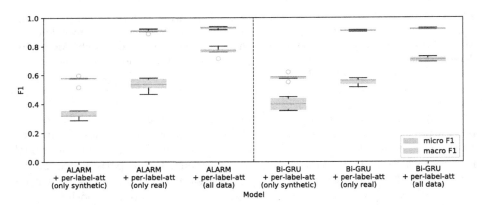

Fig. 5. Graph showing effect of synthetic data on micro-averaged F1 (blue) and macro-averaged F1 (orange). Synthetic data gives consistent improvement. (Color figure online)

5   Conclusions and Future Work

We have introduced a set of radiographic findings and clinical impressions that are relevant for stroke and can be extracted from head CT radiology reports. For deep learning approaches, we have shown that per-label attention and a simple synthetic dataset each improve accuracy for our multi-label classification task, yielding a recipe for scalable learning of many labels. In future work, we intend to annotate a larger dataset as well as leveraging knowledge bases to create a richer synthetic dataset. Furthermore, the labels generated by our models should be used to train an image analysis algorithm on the associated head CT scans.

Acknowledgements. This work is part of the Industrial Centre for AI Research in digital Diagnostics (iCAIRD) which is funded by Innovate UK on behalf of UK Research and Innovation (UKRI) [project number: 104690]. We would like to thank the Glasgow Safe Haven for assistance in creating and providing this dataset. Thanks also to The Data Lab for support and funding.

References

1. Alsentzer, E., et al.: Publicly available clinical BERT embeddings. In: Proceedings of the 2nd Clinical Natural Language Processing Workshop, pp. 72–78. Association for Computational Linguistics, Minneapolis, Minnesota, USA, Jun 2019. https://doi.org/10.18653/v1/W19-1909
2. Bahdanau, D., Cho, K., Bengio, Y.: Neural machine translation by jointly learning to align and translate. In: Bengio, Y., LeCun, Y. (eds.) 3rd International Conference on Learning Representations, ICLR, San Diego, CA, USA, 7–9 May 2015, Conference Track Proceedings (2015)
3. Banerjee, S., Akkaya, C., Perez-Sorrosal, F., Tsioutsiouliklis, K.: Hierarchical transfer learning for multi-label text classification. In: Proceedings of the 57th Annual Meeting of the Association for Computational Linguistics, pp. 6295–6300 (2019)
4. Bodenreider, O.: The unified medical language system (UMLS): integrating biomedical terminology. Nucleic Acids Res. **32**(90001), 267D–270 (2004). https://doi.org/10.1093/nar/gkh061
5. Cho, K., van Merriënboer, B., Bahdanau, D., Bengio, Y.: On the properties of neural machine translation: Encoder-decoder approaches. In: Proceedings of SSST-8, Eighth Workshop on Syntax, Semantics and Structure in Statistical Translation, pp. 103–111. Association for Computational Linguistics, Doha, Qatar, October 2014. https://doi.org/10.3115/v1/W14-4012
6. Devlin, J., Chang, M.W., Lee, K., Toutanova, K.: BERT: pre-training of deep bidirectional transformers for language understanding. In: Proceedings of the 2019 Conference of the North American Chapter of the Association for Computational Linguistics: Human Language Technologies, Volume 1 (Long and Short Papers), pp. 4171–4186. Association for Computational Linguistics, Minneapolis, Minnesota, June 2019. https://doi.org/10.18653/v1/N19-1423
7. Drozdov, I., et al.: Supervised and unsupervised language modelling in chest x-ray radiological reports. Plos One **15**(3), e0229963 (2020)
8. Gorinski, P.J., et al.: Named entity recognition for electronic health records: a comparison of rule-based and machine learning approaches. arXiv preprint arXiv:1903.03985 (2019)

9. Irvin, J., et al.: CheXpert: a large chest radiograph dataset with uncertainty labels and expert comparison. In: Proceedings of the AAAI Conference on Artificial Intelligence, vol. 33, pp. 590–597 (2019)

10. IST-3 collaborative group: Association between brain imaging signs, early and late outcomes, and response to intravenous alteplase after acute ischaemic stroke in the third International Stroke Trial (IST-3): secondary analysis of a randomised controlled trial. Lancet Neurol. **14**, pp. 485–496 (2015). https://doi.org/10.1016/S1474-4422(15)00012-5

11. Johnson, A.E., et al.: MIMIC-III, a freely accessible critical care database. Sci. Data **3**, 160035 (2016)

12. Kingma, D.P., Ba, J.: Adam: a method for stochastic optimization. In: Bengio, Y., LeCun, Y. (eds.) 3rd International Conference on Learning Representations, ICLR, San Diego, CA, USA, 7–9 May 2015, Conference Track Proceedings (2015)

13. Loper, E., Bird, S.: NLTK: the natural language toolkit. In: Proceedings of the ACL Workshop on Effective Tools and Methodologies for Teaching Natural Language Processing and Computational Linguistics. Association for Computational Linguistics, Philadelphia (2002)

14. Mikolov, T., Sutskever, I., Chen, K., Corrado, G.S., Dean, J.: Distributed representations of words and phrases and their compositionality. In: Advances in Neural Information Processing Systems, pp. 3111–3119 (2013)

15. Mullenbach, J., Wiegreffe, S., Duke, J., Sun, J., Eisenstein, J.: Explainable prediction of medical codes from clinical text. In: Proceedings of the 2018 Conference of the North American Chapter of the Association for Computational Linguistics: Human Language Technologies, Volume 1 (Long Papers), pp. 1101–1111. Association for Computational Linguistics, New Orleans, Louisiana, Jun 2018. https://doi.org/10.18653/v1/N18-1100

16. Pedregosa, F., et al.: Scikit-learn: machine learning in Python. J. Mach. Learn. Res. **12**, 2825–2830 (2011)

17. Radiological Society of North America: RSNA Intracranial Hemorrhage Detection (Kaggle challenge). https://www.kaggle.com/c/rsna-intracranial-hemorrhage-detection/overview

18. Řehůřek, R., Sojka, P.: Software framework for topic modelling with large corpora. In: Proceedings of the LREC 2010 Workshop on New Challenges for NLP Frameworks, ELRA, Valletta, Malta, pp. 45–50, May 2010

19. Smit, A., Jain, S., Rajpurkar, P., Pareek, A., Ng, A.Y., Lungren, M.P.: CheXbert: combining automatic labelers and expert annotations for accurate radiology report labeling using BERT. arXiv preprint arXiv:2004.09167 (2020)

20. Wolf, T., et al.: HuggingFace's transformers: state-of-the-art natural language processing. ArXiv abs/1910.03771 (2019)

21. Wood, D., et al.: Automated labelling using an attention model for radiology reports of MRI scans (ALARM). In: Medical Imaging with Deep Learning (2020). https://openreview.net/forum?id=UFnWZTbM5t

22. Yadav, K., Sarioglu, E., Choi, H., Cartwright IV, W.B., Hinds, P.S., Chamberlain, J.M.: Automated outcome classification of computed tomography imaging reports for pediatric traumatic brain injury. Acad. Emerg. Med. **23**(2), 171–178 (2016). https://doi.org/10.1111/acem.12859

23. Yang, Z., Yang, D., Dyer, C., He, X., Smola, A., Hovy, E.: Hierarchical attention networks for document classification. In: Proceedings of the 2016 Conference of the North American Chapter of the Association for Computational Linguistics: Human Language Technologies, pp. 1480–1489 (2016)

24. Yetisgen-Yildiz, M., Gunn, M.L., Xia, F., Payne, T.H.: A text processing pipeline to extract recommendations from radiology reports. J. Biomed. Inf. **46**(2), 354–362 (2013)
25. Zech, J., et al.: Natural language-based machine learning models for the annotation of clinical radiology reports. Radiology **287**(2), 570–580 (2018)

Correction to: Interpretable and Annotation-Efficient Learning for Medical Image Computing

Jaime Cardoso⬤, Hien Van Nguyen⬤, Nicholas Heller⬤,
Pedro Henriques Abreu⬤, Ivana Isgum⬤, Wilson Silva⬤,
Ricardo Cruz⬤, Jose Pereira Amorim⬤, Vishal Patel,
Badri Roysam, Kevin Zhou, Steve Jiang, Ngan Le, Khoa Luu,
Raphael Sznitman⬤, Veronika Cheplygina, Diana Mateus⬤,
Emanuele Trucco⬤, and Samaneh Abbasi⬤

Correction to:
J. Cardoso et al. (Eds.): *Interpretable and Annotation-Efficient Learning for Medical Image Computing*, **LNCS 12446,**
https://doi.org/10.1007/978-3-030-61166-8

The original version of this book was revised. The following corrections were implemented:

The acronym was corrected to "MIL3ID" throughout the book.

The equation on page 5 of Chapter 14 was modified to improve its accuracy and readability.

The updated version of the book can be found at
https://doi.org/10.1007/978-3-030-61166-8_14
https://doi.org/10.1007/978-3-030-61166-8

Author Index

Printed in the United States
By Bookmasters